1,001 Chemistry Practice Problems

FOR

DUMMIES

A Wiley Brand

by Heather Hattori and Richard H. Langley, PhD

FOR

DUMMIES

A Wiley Brand

1,001 Chemistry Practice Problems For Dummies®

Published by: **John Wiley & Sons, Inc.,** 111 River Street, Hoboken, NJ 07030-5774, www.wiley.com

Copyright © 2014 by John Wiley & Sons, Inc., Hoboken, New Jersey

Media and software compilation copyright © 2014 by John Wiley & Sons, Inc. All rights reserved.

Published simultaneously in Canada

For general information on our other products and services, please contact our Customer Care Department within the U.S. at 877-762-2974, outside the U.S. at 317-572-3993, or fax 317-572-4002. For technical support, please visit www.wiley.com/techsupport.

Wiley publishes in a variety of print and electronic formats and by print-on-demand. Some material included with standard print versions of this book may not be included in e-books or in print-on-demand. If this book refers to media such as a CD or DVD that is not included in the version you purchased, you may download this material at http://booksupport.wiley.com. For more information about Wiley products, visit www.wiley.com.

Library of Congress Control Number: 2013949063

ISBN 978-1-118-54932-2 (pbk); ISBN 978-1-118-54933-9 (ebk); ISBN 978-1-118-54935-3 (ebk); ISBN 978-1-118-54940-7 (ebk)

Manufactured in the United States of America

10 9 8 7 6 5 4 3

Contents at a Glance

Table of Contents

Introduction

●●●

Whether you're taking your first chemistry course, you're taking your last chemistry course, or you just need a little practice before taking a test that contains chemistry questions (like a nursing, pre-med, or teacher certification test), doing problems is a fine way to prepare.

The 1,001 practice questions in this book cover topics that you might encounter in a high school chemistry course, an introductory college chemistry course, the first semester of a general chemistry course for science majors, or a general science test for entry into a pre-professional program. The types of questions here are similar to the ones you may see on homework assignments, quizzes, practice tests, or actual tests.

You can start at Question 1 or Question 121 or skip around. You may find that your instructor (or textbook) covers topics in a different order from this book. That's okay; just go to the table of contents, find the topic you need, and start there.

Completing 1,001 chemistry practice questions is no small undertaking, but the time you spend practicing valuable science, math, and chemistry skills can improve your scores and help you "know what you know."

What You'll Find

The 1,001 chemistry practice questions in this book are divided among 15 chapters, each one representing a few major subject areas in chemistry. Within each chapter, questions are grouped by topic and arranged from easy to hard, allowing you to answer beginner questions as well as multi-step and more difficult questions. Some questions are accompanied by an image or diagram that you need in order to answer the question correctly.

After you answer the questions for one chapter or subcategory — or even after you answer just one question — you can flip to the last chapter of the book and check your answers. There, you find thorough answer explanations for each problem, often covering processes, formulas, and definitions. In many cases, studying an answer explanation can help you better understand a difficult subject, so spend as much time as you need reviewing the explanations.

Beyond the Book

This book gives you plenty of chemistry questions to work on. But maybe you want to track your progress as you tackle the questions, or maybe you're having trouble with different types of questions and wish they were all presented in one place. You're in luck. Your book purchase comes with a free one-year subscription to all 1,001 practice questions online. You get on-the-go access any way you want it — from your computer, smartphone, or tablet. Track your progress and view personalized reports that show what you need to study the most. Study what, where, when, and how you want.

What you'll find online

The online practice that comes free with this book offers the same 1,001 questions and answers that are available here. The beauty of the online problems is that you can customize your online practice to focus on the topic areas that give you the most trouble. So if you need help naming binary compounds or solving combined gas law problems, just select those question types online and start practicing. Or if you're short on time but want to get a mixed bag of a limited number of problems, you can specify the number of problems you want to practice. Whether you practice a few hundred problems in one sitting or a couple dozen, and whether you focus on a few types of problems or practice every type, the online program keeps track of the questions you get right and wrong so you can monitor your progress and spend time studying exactly what you need.

You can access this online tool using an access code, as described in the next section. Keep in mind that you can create only one login with your access code. After the access code is used, it's no longer valid and is nontransferable, so you can't share your access code with other users after you establish your login credentials.

This product also comes with an online Cheat Sheet that helps you increase your odds of performing well in chemistry. Check out the free Cheat Sheet at www.dummies.com/cheatsheet/1001Chemistry. (No access code required. You can access this info before you even register.)

How to register

To gain access to additional tests and practice online, all you have to do is register. Just follow these simple steps:

1. **Find your PIN access code:**

   • **Print-book users:** If you purchased a print copy of this book, turn to the inside front cover of the book to find your access code.

   • **E-book users:** If you purchased this book as an e-book, you can get your access code by registering your e-book at www.dummies.com/go/getaccess. Go to this website, find your book and click it, and answer the security questions to verify your purchase. You'll receive an email with your access code.

2. **Go to** Dummies.com **and click** Activate Now.

3. **Find your product (*1,001 Chemistry Practice Problems For Dummies*) and then follow the on-screen prompts to activate your PIN.**

Now you're ready to go! You can come back to the program as often as you want — simply log on with the username and password you created during your initial login. No need to enter the access code a second time.

For Technical Support, please visit `http://wiley.custhelp.com` or call Wiley at 1-800-762-2974 (U.S.), +1-317-572-3994 (international).

Where to Go for Additional Help

It's easy to get overwhelmed when trying to study a subject as integrated as chemistry. But don't despair. This book is designed to break everything into less complex categories so you can concentrate on one topic at a time. Practicing in smaller areas within each topic helps you identify your strong points and your weak points.

After you use this book and identify the areas you feel need extra effort, you can start studying on your own and then come back here to answer the questions again to measure your improvement. For example, if your knowledge of molarity is a little hazy (or nonexistent), try reviewing the molar calculations in Chapter 10. Check your answers and jot down notes or questions you may have. Then research, say, how solution concentration is expressed using molarity or which biology applications might use molarity. You can look for resources at your local library or online, or you can ask a friend, coworker, or professor to coach you if he or she seems to spend a lot of time in the lab. You can also check out the *For Dummies* series for books about many of the topics covered in chemistry. Head to www.dummies.com to see the many books and articles that can help you in your studies.

1,001 Chemistry Practice Questions For Dummies gives you just that — 1,001 practice questions and answers in order for you to practice your chemistry skills. If you need more in-depth study and direction for your chemistry courses, you may want to try out the following *For Dummies* products:

- ✔ ***Chemistry Essentials For Dummies:*** This book is a quick-reference resource that outlines key topics found in a first-year high school chemistry course or a first-semester college chemistry course.

- ✔ ***Chemistry For Dummies:*** This book provides content parallel to the 1,001 chemistry practice problems found in this book.

- ✔ ***Chemistry II For Dummies:*** This book provides content similar to what you may encounter in a second-year high school chemistry course or a second-semester college chemistry course.

- ✔ ***AP Chemistry For Dummies:*** This book prepares the Advanced Placement chemistry student to take the College Board's AP Chemistry exam. It also includes tools for organizing and planning your study time.

- ✔ ***Chemistry Workbook For Dummies:*** This book includes basic instruction, chemistry problems worked step-by-step, shortcuts, and more practice problems.

Several *For Dummies* chemistry titles are also available for download to your electronic device(s).

Part I
The Questions

1,001 Questions

Visit www.dummies.com for free access to great *For Dummies* content online.

In this part . . .

One thousand one chemistry problems — that's a lot of chemistry practice. Hundreds of our former students have persevered through what we're sure they felt was even more than that. Here are the general types of questions you'll be dealing with:

- The basics of chemistry (Chapters 1–5)
- Chemical bonding (Chapters 6–8)
- Chemical reactions (Chapters 9–11)
- Acids, bases, liquids, and gases (Chapters 12–14)
- Graphing (Chapter 15)

Chapter 1

Units and Unit Conversions

∙ ∙

Many aspects of chemistry are quantitative, and you use units to measure these quantities. In many cases, you have to convert from one unit to another. Most people in the United States initially learn the English system of units, but most chemists use the *Système international d'unités* (the SI system), derived from the older metric system. Unit conversions relate these two systems. Dimensional analysis provides a systematic means to not only perform these conversions but also to work many of the other problems in this book. Dimensional analysis lets the units solve the problem for you.

The Problems You'll Work On

In this chapter, you work with units and unit conversions in the following ways:

- ✔ Choosing appropriate units
- ✔ Interpreting metric prefixes
- ✔ Converting metric and English units
- ✔ Solving problems with dimensional analysis

What to Watch Out For

Don't let common mistakes trip you up; remember the following when working on units and unit conversions:

- ✔ Always include your units when setting up equations and answering questions.
- ✔ Set up your problem so that units cancel to leave the desired units.
- ✔ Make sure your answer looks reasonable and the final units match what they describe. For example, s^{-1} represents 1/seconds, which is frequency, not time.
- ✔ When rounding your answer for significant figures, remember that many, but not all, conversions are exact numbers and therefore don't affect the number of significant figures in the answer.

Understanding Metric Prefixes and Units

1–10 Answer the questions on metric prefixes and metric units used in the laboratory.

1. What is a common metric unit of mass used in the laboratory?

2. What is a common metric unit of length used for measuring small objects in the laboratory?

3. What is a common metric unit of volume used in the laboratory?

4. What is a common metric unit of pressure used in the laboratory?

5. What is a common metric unit of energy?

6. What is the metric prefix that represents 1,000?

7. What is the metric prefix that represents $\frac{1}{1,000}$?

8. What is the metric prefix that represents $\frac{1}{100}$?

9. What is the metric prefix that represents 10^{-9}?

10. What is the metric prefix that represents 10^6?

Choosing Appropriate Units

11–20 Choose appropriate metric or English units for measuring everyday objects.

11. Which metric unit is most appropriate for expressing the mass of an adult human?

12. Which metric unit is most appropriate for recording the volume of a child's wooden block?

13. Which metric unit would a scientist use to measure the temperature on a warm autumn day?

14. Which metric unit is most often used for small doses of solid medications?

15. Which SI base unit is named after a person?

16. Which English unit is most similar in volume to a liter?

17. Which English unit is most similar in length to a meter?

18. How many fluid ounces are in a cup?

19. An Olympic swimmer competes in the 100-meter freestyle. What is the comparable English unit?

20. If a wooden board's width is 6 in., what is an appropriate metric unit to express this width?

Doing Metric Conversions

21–32 Complete the conversion between metric units.

21. How many milligrams are in 1 dg?

22. How many deciliters are in 1 L?

23. How many kilometers are in 1 m?

24. How many centimeters are in 1 m?

25. How many grams are in 1 hg?

26. How many milliliters are in 2.5 daL?

27. How many centigrams are in 49 kg?

28. How many gigawatts are in 370,000 W?

29. How many micrograms are in 0.126 Mg?

30. How many kilometers are in 80. pm?

31. How many cubic meters are in 2 L?

32. How many milliliters are in 0.64 m³?

Converting between Systems of Measurement

33–59 Convert between metric and English units.

33. How many miles are in 35 km?

34. How many inches are in 0.20 cm?

35. How many yards are in 202 m?

36. How many pounds are in 58 kg?

37. How many quarts are in 7.54 L?

38. How many centimeters are in 0.087 in.?

39. How many kilometers are in 463 mi.?

40. How many grams are in 91 lb.?

41. How many liters are in 525 gal.?

42. How many atmospheres are in 44 psi?

43. How many cups are in 2.00 L?

44. How many pounds are in 164 hg?

45. How many gallons are in 587 mL?

46. How many centimeters are in 6.02 mi.?

47. How many decigrams are in 225 lb.?

48. How many milliliters are in 6.8 qt.?

49. How many centimeters are in 15.3 ft.?

50. How many liters are in 99 pt.?

51. How many kilograms are in 1.00 short ton? (1 short ton = 2,000 lb.)

52. How many centimeters are in 6.04 yd.?

53. How many cups are in 15 cc?

54. How many millimeters are in 1,760 yd.?

55. How many pints are in 250 hL?

56. How many grams are in 0.35 slugs? (1 slug = 32.2 lb.)

57. How many kilometers are in 9,999 in.?

58. How many ounces are in 0.734 kg?

59. How many microliters are in 55 oz.?

Using Dimensional Analysis

60–75 Solve the word problem using a setup similar to those used in unit conversions.

60. How many dozen eggs are in 17,981 eggs?

61. How many years are in 6,250 days?
(1 yr. = 365.25 days)

62. How many weeks are in 2.5 centuries?
(1 yr. = 52 weeks)

63. If the average penny has a mass of 3.16 g, what is the dollar value of 1.00 short ton of pennies? (1 short ton = 2,000 lb.)

64. If an athlete runs the 100-yard dash in 10.0 s, how long will it take for the athlete to run 400 m?

65. You're planning a party and need enough soda for 60 guests. How many liters will you need, assuming each guest drinks 10. fl. oz. of soda?

66. You plan to serve sub sandwiches at a party. How many 6.0-foot subs will you need to feed the 60 guests if each person eats a 25.4-cm length of sandwich?

67. A textbook measures 230. mm long, 274 mm wide, and 60.0 mm thick. What is the volume in cubic centimeters?

68. A textbook measures 230. mm long, 274 mm wide, and 60.0 mm thick. What is the surface area of the front cover in square meters?

69. A hallway measures 10.0 ft. by 5.0 ft. How many square tiles, measuring 10.0 in. on each side, are necessary to cover the floor?

70. If a car is going 20. mph through a school zone, how many centimeters per minute is it traveling?

71. A solid sphere made of pure gold has a volume of 2.0 L. What is the mass of the sphere, in pounds, if 1.00 cm³ of gold has a mass of 19.3 g?

72. How many minutes does it take a horse to run 12 furlongs at 35.3 mph? (1 furlong = 40 rods, and 1 rod = 5.5 yd.)

73. If a pitcher throws a 96-mph fastball, how many seconds will it take to travel the 60.5 ft. from the pitcher's mound to home plate?

74. Pure gold can be made into extremely thin sheets called gold leaf. Suppose that 25 kg of gold is made into gold leaf having a surface area of 1,810 m^2. How thick is the gold leaf in millimeters? The density of gold is 19.3 g/cm^3.

75. Radio waves travel at 300,000,000 m/s. If you asked a question of someone who was on the moon, 239,000 mi. from the Earth, what is the minimum time that you would have to wait for a reply?

Chapter 2

Scientific Notation and Significant Figures

● ●

Scientific notation allows you to write very large and very small numbers, which are common in chemistry, in a simplified manner. Many chemical experiments involve very precise measurements. The significant figures are an indication of the precision of these measurements. In calculations involving more than one measurement, you need to maintain the precision inherent in the significant figures.

The Problems You'll Work On

In this chapter, you work with scientific notation and significant figures in the following ways:

- ✔ Expressing numbers in standard and scientific notation
- ✔ Doing calculations with numbers in scientific notation
- ✔ Determining significant figures
- ✔ Combining math operations with significant figures

What to Watch Out For

Remember the following when working on scientific notation and significant figures:

- ✔ All nonzero digits and zeroes between nonzero digits are significant. Zeroes to the left in the number (leading zeroes) are never significant. Zeroes to the right are significant only if they aren't just indicating the power of ten.
- ✔ Don't confuse the addition/subtraction rule with the multiplication/division rule. Be extra careful when solving mixed-operation problems.
- ✔ Most calculators convert to and from scientific notation, but double-check the answer. Calculators are complete idiots concerning the rules for significant figures.

Putting Numbers in Scientific Notation

76–80 Express the given number in scientific notation.

76. 876

77. 4,000,001

78. 0.000510

79. 900×10^4

80. 10

Taking Numbers out of Scientific Notation

81–85 Convert the given number to nonscientific notation (regular decimal form).

81. 2.00×10^2

82. 9×10^{-2}

83. 4.7952×10^3

84. 1.64×10^{-5}

85. 0.83×10^{-1}

Calculating with Numbers in Scientific Notation

86–105 Complete the calculations and record your answer in scientific notation. (If you use a calculator, choose a mode that doesn't put the numbers in scientific notation for you.)

86. $(1.26 \times 10^3) + (4.71 \times 10^3) =$

87. $(3.9 \times 10^{-1}) + (2.1 \times 10^{-1}) =$

88. $(8.9 \times 10^2) - (3.3 \times 10^1) =$

89. $(7.4 \times 10^{-1}) - (5.2 \times 10^{1}) =$

90. $(8.240 \times 10^{2}) + (3.791 \times 10^{2}) =$

91. $(1.00 \times 10^{7}) - (5.2 \times 10^{5}) =$

92. $(5.42 \times 10^{-3}) + (6.19 \times 10^{-4}) =$

93. $(8.20 \times 10^{6}) - (7.31 \times 10^{4}) + (2.846 \times 10^{5}) =$

94. $(1.0 \times 10^{-7}) \times (4.5 \times 10^{5}) =$

95. $(1.0 \times 10^{-3}) \div (1.0 \times 10^{-4}) =$

96. $(3.15 \times 10^{12}) \times (2.0 \times 10^{3}) =$

97. $(4.7 \times 10^{-2}) \div (9.6 \times 10^{-7}) =$

98. $(8.40 \times 10^{15}) \times (2.00 \times 10^{-5}) =$

99. $(1.0 \times 10^{8}) \div (3.2 \times 10^{2}) =$

100. $(9.76 \times 10^{-9}) \times (3.55 \times 10^{-3}) \div (1.8 \times 10^{-5}) =$

101. $(2.48 \times 10^{3}) \times (4.756 \times 10^{-4}) \times (9.1 \times 10^{-2}) =$

102. $(1.8 \times 10^{-4}) + (6.27 \times 10^{-2}) \times (2.9 \times 10^{-3}) =$

103. $(9.189 \times 10^{-19}) \div (0.6021 \times 10^{-13}) + (4.5 \times 10^{-11}) =$

104. $(4.115 \times 10^{2}) + (1.1 \times 10^{1}) \div (3.68 \times 10^{-6}) \div (8.2 \times 10^{4}) =$

105. $\dfrac{\left(4.6 \times 10^{2}\right) + \left(6.97 \times 10^{9}\right) \times \left(3 \times 10^{-7}\right)}{\left(5.18 \times 10^{4}\right) - \left(2.00 \times 10^{3}\right)} =$

Recognizing Significant Figures

106–115 Indicate how many significant figures (significant digits) are in the given number.

106. 343

107. 0.4592

108. 705,204

109. 0.0075

110. 248,000

111. 9,400,300

112. 1.0070

113. 3,000,000.0

114. 0.0040800

115. 0.870

Writing Answers with the Right Number of Sig Figs

116–135 Complete the calculation and express your answer using the correct number of significant figures.

116. 5,379 + 100 =

117. 12.4 + 0.59 =

118. 61.035 – 33.48 =

119. 71 + 24.87 + 0.0003 =

120. $0.387 - 467 =$

121. $0.005689 + 0.0410 =$

122. $60.0080 - 128.35429 + 7.941 =$

123. $130 + 4,600 + 395.2 =$

124. $0.0074 \div 0.000035 =$

125. $75 \times 349 =$

126. $7.98 \times 5.21 =$

127. $5.00 \div 0.0025 =$

128. $7.0 \text{ cm} \times 7 \text{ cm} =$

129. $6.48 \div 194.21 =$

130. $0.000000029 \times 0.00000745 =$

131. $\dfrac{0.0034 \times 518.27}{9.00} =$

132. $2,300.00 \times 0.854 + 110 =$

133. $\dfrac{10.78 \text{ g}}{25.0 \text{ mL} - 23.8 \text{ mL}} =$

134. $\dfrac{8.1 + 2.32 + 0.741}{2.54} =$

135. $\dfrac{250 + 12}{2.0} \times \dfrac{1.0}{3.57 - 1.2} =$

Chapter 3

Matter and Energy

● ●

Chemists deal with matter. Matter occurs in many forms with certain observable properties. One easily observable property of matter is density, which is the mass of a sample of matter divided by its volume. It's possible to alter matter either physically or chemically. All alterations involve energy.

The Problems You'll Work On

In this chapter, you work with matter and energy in the following ways:

✔ Describing phases of matter

✔ Classifying matter as substances and mixtures

✔ Understanding properties of matter

✔ Determining density

✔ Calculating energy and temperature

What to Watch Out For

Remember the following when working on matter and energy:

✔ Know the properties of solids, liquids, and gases on microscopic and macroscopic levels.

✔ Remember that a density must have a mass unit divided by a volume unit.

Phases of Matter and Phase Changes

136–143 Check your understanding of phases of matter and phase changes.

136. Which phase of matter doesn't have a definite shape or a definite volume under normal conditions?

137. Which phase of matter has a definite shape and a definite volume?

138. Which phase of matter has a definite volume but takes the shape of the container that it's in?

139. When matter changes from a liquid to a solid, which phase change is it going through?

140. When matter changes from a gas to a liquid, which phase change is it going through?

141. When matter changes from a liquid to a gas, which phase change is it going through?

142. When matter changes from a solid to a gas without becoming a liquid in between, which phase change is it going through?

143. When matter changes from a gas to a solid without becoming a liquid in between, which phase change is it going through?

Classifying Substances and Mixtures

144–152 Classify each type of matter as a pure substance or mixture. Then classify each pure substance as an element or compound and each mixture as homogeneous or heterogeneous.

144. Gold

145. Table sugar

146. Fresh air

147. Oxygen

148. Vegetable soup

149. Fruit salad

150. Calcium

151. Concrete

152. Smog

Properties of Matter

153–165 Check your understanding of the properties of matter.

153. Which type of property of matter doesn't depend on the amount of the substance that's present?

154. Which type of change involves a change in the form of a substance?

155. Which type of change involves a change in the identity of a substance?

156. Which type of property of matter depends on exactly how much of the substance is present?

157. Density is a(n) _____ (chemical/ extensive physical/intensive physical) property.

158. Length is a(n) _____ (chemical/ extensive physical/intensive physical) property.

159. Color is a(n) _____ (chemical/ extensive physical/intensive physical) property.

160. Flammability is a(n) _____ (chemical/extensive physical/intensive physical) property.

161. Mass is a(n) _____ (chemical/ extensive physical/intensive physical) property.

162. Odor is a(n) _____ (*chemical/ extensive physical/intensive physical*) property.

163. Ductility is a(n) _____ (*chemical/ extensive physical/intensive physical*) property.

164. Electrical conductivity is a(n) _____ (*chemical/extensive physical/intensive physical*) property.

165. Solubility is a(n) _____ (*chemical/extensive physical/intensive physical*) property.

Calculating Density

166–174 Perform the density calculations. Be sure to round your answers to the correct number of significant figures. (See Chapter 2 for significant figure problems.)

166. In grams per cubic centimeter, what is the density of a substance with a mass of 57.5 g and a volume of 5.0 cm³?

167. A 25.0-mL sample of a liquid has a mass of 22.1 g. What is the liquid's density in grams per milliliter?

168. The mass of 2.00 m³ of a gas is 3,960 g. What is the density of this gas in kilograms per cubic meter?

169. What is the mass, in grams, of 0.200 L of a saltwater solution with a density of 1.2 g/mL?

170. Aluminum is a metal that has a density of 2.7 g/cm³. How many grams are in a solid cube of aluminum that measures 3.00 cm per side?

171. If a block has a length of 5.0 cm, a width of 3.0 cm, a height of 2.0 cm, and a mass of 120. g, what is the block's density?

172. What is the mass, in kilograms, of 1.5 L of solid gold? Solid gold has a density of 19.3 g/cm³.

173. If a sample of gasoline has a mass of 77.0 g and a density of 0.71 g/mL, what is the volume of the gasoline in milliliters?

174. What is the length of one side of a metallic cube that has a density of 10.5 g/cm³ and a mass of 672 g?

Working with Energy

175–190 Assess your understanding of energy and related calculations.

175. Which unit represents the amount of energy necessary to raise the temperature of 1 g of water by 1°C?

176. Which SI unit is used to express the heat content of a mole of a chemical?

177. Which unit is used to express the energy found in food?

178. A bucket on a ladder has _____ (*kinetic/potential*) energy, which is the energy of _____ (*motion/position*).

179. A rolling ball has _____ (*kinetic/potential*) energy, which is the energy of _____ (*motion/position*).

180. Fuels contain _____ (*moving/stored*) energy, which is _____ (*kinetic/potential*) energy.

181. How many kilocalories are in 25,970 J?

182. How many joules are in 3.1×10^8 kilocalories?

183. Average kinetic energy can be measured in what units?

184. If water is heated to 80.0°C, how many kelvins is that?

185. The melting point of sodium chloride is 1,074 K. What temperature is this in degrees Celsius?

186. Liquid oxygen boils at –183°C. How many degrees Fahrenheit is this?

187. Normal human body temperature is considered to be around 98.6°F. How many degrees Celsius is this?

188. Adding dry ice to acetone brings the temperature down to –78°C. What is this temperature in degrees Fahrenheit?

189. One summer day, the temperature was recorded as 113°F. How many kelvins is this?

190. Room temperature is about 300. K. How many degrees Fahrenheit is this?

Chapter 4

The Atom and Nuclear Chemistry

. . . .

An atom consists of a nucleus surrounded by one or more electrons. Although the number of protons identifies the element, the electrons are the key to the chemistry. The arrangement of electrons in an atom influences the atom's ability to gain, lose, or share electrons and therefore form compounds. Quantum numbers describe the arrangement of the electrons. Unstable atoms undergo nuclear decay to transform to stable atoms. Atoms may be broken apart by fission or joined by fusion.

The Problems You'll Work On

In this chapter, you work with atoms and nuclear chemistry in the following ways:

- ✔ Counting subatomic particles
- ✔ Interpreting isotope notation
- ✔ Writing electron configurations
- ✔ Calculating average atomic mass and percent abundance
- ✔ Understanding nuclear decay and balancing nuclear equations

Note: For access to the periodic table, see the Appendix.

What to Watch Out For

Remember the following when working on atoms and nuclear chemistry:

- ✔ Note that unlike the atomic mass, the mass number is usually not found on the periodic table.
- ✔ Remember the maximum number of electrons possible in each subshell, and follow Hund's rule and the Aufbau principle when filling orbitals. Know the rules for assigning the four quantum numbers.
- ✔ Know the common nuclear decay modes.
- ✔ Remember that balancing nuclear equations depends on both the mass numbers and the atomic numbers.
- ✔ Know how to calculate the half-life and how to use it.

Isotopes and Subatomic Particles

191–219 Answer the question on isotopes and subatomic particles.

191. How many protons are in an atom of sodium?

192. How many electrons are in an atom of bromine?

193. How many electrons are in an atom of nickel?

194. How many protons are in an atom of radon?

195. How many neutrons are in an atom of isotope potassium-40?

196. How is the atomic number related to the number of protons in an atom?

197. How is the mass number of an atom related to the number of neutrons?

198. How many protons, electrons, and neutrons are in an atom of isotope copper-63?

199. An atom has a mass number of 14 and 6 electrons. How many protons and neutrons does it have?

200. An atom has 40 electrons and 51 neutrons. What is its mass number, and how many protons does it have?

201. What does the top number in isotope notation represent?

202. What does the bottom number in isotope notation represent?

203. How many protons and neutrons are in $^{18}_{9}F$?

204. How many protons and neutrons are in $^{25}_{11}$Na?

205. What is the isotope notation for an atom of carbon-12?

206. What is the isotope notation for an atom of chlorine-37?

207. What is the name of $^{35}_{18}$Ar?

208. How does an ion differ from an atom of the same element with regard to the numbers of subatomic particles?

209. When two atoms or ions have the same number of electrons, they're said to be _____.

210. When an ion has a positive charge, how do the numbers of subatomic particles differ?

211. When an ion has a negative charge, how do the numbers of subatomic particles differ?

212. How many protons and electrons does $^{40}_{20}$Ca^{2+} have?

213. How many protons and electrons does $^{131}_{53}$I^{1-} have?

214. How many protons and electrons does $^{30}_{13}$Al^{3+} have?

215. How many protons and electrons does ^{33}P^{3-} have?

216. What is the isotope notation for an ion of silver-109 with a charge of positive 1?

217. What is the isotope notation for an ion of sulfur-34 with a charge of negative 2?

218. How many protons, neutrons, and electrons are in $^{52}Cr^{6+}$?

219. How many protons, neutrons, and electrons are in $^{62}Ni^{3+}$?

Electrons and Quantum Mechanics

220–234 Answer the questions on electrons and quantum mechanics.

220. What is the name of the premise that one electron fills each orbital in a subshell until all orbitals contain one electron and then electrons are added to fill in the second available spot in the subshell?

221. What rule or principle describes the order in which electrons fill orbitals?

222. What is the maximum number of electrons that can be in the f orbitals?

223. What is the maximum number of electrons that can be in the p orbitals?

224. What is the electron configuration of carbon?

225. What is the electron configuration of magnesium?

226. What is the electron configuration of argon?

227. What is the electron configuration of bromine?

228. What is the electron configuration of zirconium?

229. What is the expected electron configuration of plutonium?

230. Which quantum number describes the spin of the electron?

231. Which quantum number describes the average distance between the nucleus and the orbital?

232. Which quantum number describes how the various orbitals are oriented in space?

233. Which quantum number describes the shape of the orbital?

234. What are the possible values for the spin quantum number?

Average Atomic Mass

235–242 Answer the questions on average atomic mass.

235. The decimal numbers in the blocks of the periodic table represent the _____.

236. What is the average atomic mass of lithium that is 7.59% lithium-6 (mass of 6.0151 amu) and 92.41% lithium-7 (mass of 7.0160 amu)?

237. What is the average atomic mass of chlorine that is 75.78% chlorine-35 (mass of 34.96885 amu) and 24.22% chlorine-37 (mass of 36.9659 amu)?

238. What is the average atomic mass of magnesium, given the information in the following table?

Isotope	Percent Abundance	Atomic Mass (amu)
$^{24}_{12}$Mg	78.99	23.985
$^{25}_{12}$Mg	10.00	24.986
$^{26}_{12}$Mg	11.01	25.983

239. What is the average atomic mass of potassium, given the information in the following table?

Isotope	Percent Abundance	Atomic Mass (amu)
$^{39}_{19}$K	93.258	38.9637
$^{40}_{19}$K	0.01170	39.9640
$^{41}_{19}$K	6.7302	40.9618

240. What is the average atomic mass of iron, given the information in the following table?

Isotope	Percent Abundance	Atomic Mass (amu)
$^{54}_{26}$Fe	5.845	53.9396
$^{56}_{26}$Fe	91.754	55.9349
$^{57}_{26}$Fe	2.119	56.9354
$^{58}_{26}$Fe	0.282	57.9333

241. What is the average atomic mass of krypton, given the information in the following table?

Isotope	Percent Abundance	Atomic Mass (amu)
$^{78}_{36}$Kr	0.350	77.9204
$^{80}_{36}$Kr	2.28	79.9164
$^{82}_{36}$Kr	11.58	81.9135
$^{83}_{36}$Kr	11.49	82.9141
$^{84}_{36}$Kr	57.00	83.9115
$^{86}_{36}$Kr	17.30	85.9106

242. If the average atomic mass of boron is 10.81 amu, what is the percent abundance of boron-11 (mass of 11.009306 amu) if the only other isotope is boron-10 (mass of 10.012937 amu)?

Nuclear Reactions and Nuclear Decay

243–252 Answer the question on aspects of nuclear reactions and nuclear decay.

243. What is the primary nuclear process that occurs in the sun?

244. In which nuclear process does a nucleus split into two or more smaller elements and possibly some extra neutrons?

245. What is it called when a helium nucleus is ejected from the nucleus of an atom during a nuclear reaction?

246. When a $^{0}_{0}\gamma$ ray is a byproduct of a nuclear reaction, what nuclear process has occurred?

247. What is the isotope (nuclear) notation for a particle that is produced from beta decay?

248. A $^{0}_{+1}$e particle is a product of which nuclear reaction?

249. $^7_4\text{Be} + ^0_{-1}\text{e} \rightarrow ^7_3\text{Li}$ is an example of what type of nuclear reaction?

250. $^{238}_{92}\text{U} \rightarrow ^{234}_{90}\text{Th} + ^4_2\text{He}$ is an example of what type of nuclear reaction?

251. $^{60}_{27}\text{Co} \rightarrow ^{60}_{27}\text{Co} + ^0_0\gamma$ is an example of what type of nuclear reaction?

252. $^{32}_{15}\text{P} \rightarrow ^{32}_{16}\text{S} + ^0_{-1}\text{e}$ is an example of what type of nuclear reaction?

Completing Nuclear Reactions

253–260 Determine the missing part of the equation.

253. _____ $\rightarrow ^{220}_{86}\text{Rn} + ^4_2\text{He}$

254. $^{247}_{96}\text{Cm} \rightarrow$ _____ $+ ^4_2\text{He}$

255. $^{241}_{94}\text{Pu} \rightarrow$ _____ $+ ^{241}_{95}\text{Am}$

256. $^{192}_{77}\text{Ir} \rightarrow ^{192}_{77}\text{Ir} +$ _____

257. _____ $+ ^0_{-1}\text{e}- \rightarrow ^{93}_{41}\text{Nb}$

258. $^2_1\text{H} +$ _____ $\rightarrow ^4_2\text{He} + ^1_0\text{n}$

259. $^4_2\text{He} + ^{14}_7\text{N} \rightarrow$ _____ $+ ^1_1\text{H}$

260. $^{235}_{92}\text{U} + ^1_0\text{n} \rightarrow ^{134}_{54}\text{Xe} + ^{100}_{38}\text{Sr} +$ _____

Half-Lives

261–270 Answer the question on half-lives.

261. After five half-lives, how many grams of a 400.-g radioactive sample remain undecayed?

262. After three half-lives, how many grams of a 50.0-g radioactive sample have decayed?

263. If 5.15 g of a radioactive sample remains undecayed after six half-lives, how many grams were in the original sample?

264. A radioactive sample starts with 1.500×10^{20} undecayed atoms. When measured again at a later date, the sample has 9.375×10^{18} undecayed atoms. How many half-lives have passed?

265. What fraction of a sample remains undecayed after 39 hours if the half-life of the sample is 13 hours?

266. Iodine-131 has a half-life of 8.02 days. If the original sample contained 25.0 g of iodine-131, how many grams have decayed after 56.14 days?

267. Strontium-90 has a half-life of 28.9 years. How many grams were in the original sample if 11 g remain undecayed after 115.6 years?

268. In 5 minutes, a radioactive sample decays from 2.56×10^{10} atoms to 8.00×10^{8} atoms. How long is the isotope's half-life?

269. What fraction of a sample has decayed after 320 days if the half-life of the sample is 160 days?

270. If a radioactive sample decays from 2.5 kg to 0.61 g and the isotope has a half-life of 9.35 hours, how much time has passed?

Chapter 5

Periodicity and the Periodic Table

● ●

*T*he periodic table is much more than a simple listing of the symbols of the elements with additional information about each element. The position of an element on the periodic table indicates many of the element's properties. In addition, the position gives information on how the properties of an element relate to those of its neighbors. However, watch out for exceptions to the general trends.

The Problems You'll Work On

In this chapter, you work with periodicity and the periodic table in the following ways:

- ✔ Recognizing element symbols
- ✔ Understanding the structure of the periodic table
- ✔ Identifying periodic trends

Note: For reference, you can find the periodic table in the Appendix.

What to Watch Out For

Don't let common mistakes trip you up; remember the following when working on periodicity and the periodic table:

- ✔ The position of an element on the periodic table gives important information.
- ✔ Don't confuse periods (rows) with groups (columns).
- ✔ Learn the basic periodic trends. Note that hydrogen is an exception to nearly all trends and that the top member of each group on the periodic table shows a slight variation to most trends.

Element Symbols and Names

271–290 Check your knowledge of the symbols and names of elements on the periodic table.

271. What is the symbol for the element carbon?

272. What is the symbol for the element chlorine?

273. What is the symbol for the element aluminum?

274. What is the symbol for the element cadmium?

275. What is the symbol for the element copper?

276. What is the symbol for the element arsenic?

277. What is the symbol for the element sodium?

278. What is the symbol for the element potassium?

279. What is the symbol for the element iron?

280. What is the symbol for the element silver?

281. What is the name of the element that has the symbol N?

282. What is the name of the element that has the symbol S?

283. What is the name of the element that has the symbol Br?

284. What is the name of the element that has the symbol P?

285. What is the name of the element that has the symbol Mn?

286. What is the name of the element that has the symbol At?

287. What is the name of the element that has the symbol Ra?

288. What is the name of the element that has the symbol Hg?

289. What is the name of the element that has the symbol Sn?

290. What is the name of the element that has the symbol Pa?

Structure of the Periodic Table

291–310 Answer the questions on the structure of the periodic table.

291. A period goes in which direction on the periodic table?

292. Where are the metalloids located in the periodic table?

293. Where on the periodic table are the alkaline earth metals located?

294. Where on the periodic table are the transition metals located?

295. Where on the periodic table are the noble gases located?

296. Where on the periodic table are the inner transition metals located?

297. Where on the periodic table are the alkali metals located?

298. Which chemist is most often recognized for basing the arrangement of the periodic table on atomic mass and other physical properties of the elements?

299. Which scientist placed the elements in order of increasing atomic number on the periodic table?

300. Where on the periodic table are most of the elements that exist as gases at room temperature?

301. To which family does the element potassium belong?

302. To which family does the element silver belong?

303. To which family does the element selenium belong?

304. To which family does the element tin belong?

305. To which family does the element iodine belong?

306. To which family does the element calcium belong?

307. To which family does the element aluminum belong?

308. To which family does the heaviest naturally occurring element belong?

309. Which family of elements contains the only metal that's liquid at room temperature?

310. Which family of elements contains the only nonmetal that's liquid at room temperature?

Periodic Trends

311–330 Examine your knowledge of periodic trends.

311. Members of which family have three valence electrons?

312. Members of which family have five valence electrons?

313. Members of which family have two valence electrons?

314. The atomic masses of the elements generally _____ (*increase/decrease/remain the same*) going from left to right in a period and _____ (*increase/decrease/remain the same*) going down a family.

315. The atomic radii of the elements _____ (*increase/decrease/remain the same*) going from left to right in a period and _____ (*increase/decrease/remain the same*) going down a family.

316. What term describes the amount of energy needed to remove an electron from a gaseous atom?

317. What term describes the energy change that results from adding an electron to a gaseous atom or ion?

318. The ionic radius of an anion is _____ (*larger than/smaller than/the same size as*) the atomic radius of the neutral atom because the ion has _____ (*more/fewer/an equal number of*) electrons compared to the atom.

319. The atomic radii of the elements _____ (*increase/decrease/remain the same*) going from left to right in a period because the effective nuclear charge _____ (*increases/decreases*).

320. Rank the following elements from smallest to largest atomic radius: Ba, Be, Ca.

321. Rank the following elements from smallest to largest atomic radius: Cl, P, S.

322. Rank the following elements from lowest to highest ionization energy: B, C, Li.

323. Rank the following elements from lowest to highest ionization energy: Br, Cl, I.

324. Rank the following elements from lowest to highest electron affinity: F, O, N.

325. Rank the following elements from lowest to highest electron affinity: S, Se, Te.

326. Rank the following elements from most to least metallic character: Cl, Si, Sn.

327. Rank the following elements from smallest to largest atomic radius: K, Mg, Na.

328. Rank the following ions from largest to smallest ionic radius: F^-, O^{2-}, S^{2-}.

329. Rank the following elements from highest to lowest ionization energy: Cs, F, Li.

330. Rank the following elements from lowest to highest ionization energy: Ba, Bi, N.

Chapter 6

Ionic Bonding

*I*onic compounds usually contain a metal (from the left side of the periodic table) and a nonmetal (from the right side of the periodic table, other than the noble gases). Some ionic compounds contain polyatomic ions, such as the sulfate ion SO_4^{2-}. Ternary compounds technically contain three elements; however, more may be present. The names of ionic compounds consist of two or more words. The last word in the name of a binary compound has an *-ide* suffix. The last word in the name of a ternary compound usually has an *-ite* or an *-ate* suffix. Pure acids are generally not ionic, but they produce ions in solution.

The Problems You'll Work On

In this chapter, you work with ionic bonding in the following ways:

- Naming binary compounds
- Naming ternary compounds
- Writing formulas for ionic compounds

Note: See the Appendix if you need to check the periodic table.

What to Watch Out For

Remember the following when working on ionic bonding problems:

- The names of inorganic compounds normally consist of two words.
- All simple binary compounds have an *-ide* suffix on the second word in the name.
- The *-ide* suffix is not common in compounds containing three or more elements. Most compounds containing three or more elements have either an *-ite* or *-ate* suffix.
- Don't use multiplying prefixes, such as *di-* and *tri-*, for ionic compounds.
- If you use the "crisscross" rule, don't forget to reduce.
- Acids follow their own rules for naming.

Naming Binary Compounds

331–347 Name the binary ionic compound given the formula.

331. What is the name of NaCl?

332. What is the name of CaO?

333. What is the name of $AlBr_3$?

334. What is the name of K_2S?

335. What is the name of Al_2O_3?

336. What is the name of Li_3N?

337. What is the name of MgI_2?

338. What is the name of SrSe?

339. What is the name of BaF_2?

340. What is the name of NaH?

341. What is the name of Zn_3P_2?

342. What is the name of $CuBr_2$?

343. What is the name of $AuCl_3$?

344. What is the name of CoS?

345. What is the name of MnF_3?

346. What is the name of Hg_2I_2?

347. What is the name of SnO_2?

Naming Compounds with Polyatomic Ions

348–364 Name the compounds containing polyatomic ions.

348. What is the name of $NaClO$?

349. What is the name of KOH?

350. What is the name of $SrSO_3$?

351. What is the name of $CaCO_3$?

352. What is the name of $AlPO_4$?

353. What is the name of $NaClO_3$?

354. What is the name of $GaPO_3$?

355. What is the name of NH_4Cl?

356. What is the name of $ZnSO_4$?

357. What is the name of $(NH_4)_2C_2O_4$?

358. What is the name of $KMnO_4$?

359. What is the name of $Be(NO_2)_2$?

360. What is the name of $Cu(CN)_2$?

361. What is the name of $AgBrO$?

362. What is the name of $(NH_4)_3PO_4$?

363. What is the name of $Ni_2(SO_4)_3$?

364. What is the name of $Pb(C_2H_3O_2)_2$?

Writing Formulas of Binary Compounds

365–390 Give the chemical formula for the binary ionic compound.

365. What is the chemical formula for cesium chloride?

366. What is the chemical formula for indium(III) fluoride?

367. What is the chemical formula for magnesium oxide?

368. What is the chemical formula for barium bromide?

369. What is the chemical formula for potassium iodide?

370. What is the chemical formula for aluminum chloride?

371. What is the chemical formula for chromium(III) fluoride?

372. What is the chemical formula for iron(II) sulfide?

373. What is the chemical formula for copper(I) nitride?

374. What is the chemical formula for lead(II) oxide?

375. What is the chemical formula for nickel(I) selenide?

376. What is the chemical formula for silver oxide?

377. What is the chemical formula for strontium bromide?

378. What is the chemical formula for magnesium nitride?

379. What is the chemical formula for lithium hydride?

380. What is the chemical formula for zinc chloride?

381. What is the chemical formula for chromium(II) sulfide?

382. What is the chemical formula for manganese(II) selenide?

383. What is the chemical formula for tin(IV) fluoride?

384. What is the chemical formula for copper(I) iodide?

385. What is the chemical formula for nickel(III) phosphide?

386. What is the chemical formula for aluminum selenide?

387. What is the chemical formula for tin(IV) oxide?

388. What is the chemical formula for calcium phosphide?

389. What is the chemical formula for iron(III) oxide?

390. What is the chemical formula for manganese(IV) sulfide?

Writing Formulas of Compounds with Polyatomic Ions

391–420 Give the formula for the compound containing a polyatomic ion.

391. What is the chemical formula for rubidium hypochlorite?

392. What is the chemical formula for beryllium carbonate?

393. What is the chemical formula for aluminum phosphite?

394. What is the chemical formula for sodium hydrogen carbonate?

395. What is the chemical formula for sodium hydroxide?

396. What is the chemical formula for potassium hydrogen sulfate?

397. What is the chemical formula for lithium perchlorate?

398. What is the chemical formula for barium oxalate?

399. What is the chemical formula for chromium(III) arsenate?

400. What is the chemical formula for silver nitrate?

401. What is the chemical formula for lead(II) sulfite?

402. What is the chemical formula for thallium(I) bromite?

403. What is the chemical formula for gold(III) phosphate?

404. What is the chemical formula for iron(II) sulfate?

405. What is the chemical formula for calcium thiosulfate?

406. What is the chemical formula for sodium peroxide?

407. What is the chemical formula for ammonium nitrite?

408. What is the chemical formula for beryllium chlorite?

409. What is the chemical formula for sodium cyanide?

410. What is the chemical formula for magnesium permanganate?

411. What is the chemical formula for ammonium dichromate?

412. What is the chemical formula for cobalt(III) periodate?

413. What is the chemical formula for lead(IV) oxalate?

414. What is the chemical formula for calcium hydrogen phosphate?

415. What is the chemical formula for iron(III) acetate?

416. What is the chemical formula for potassium thiocyanate?

417. What is the chemical formula for copper(II) hydrogen sulfite?

418. What is the chemical formula for mercury(II) peroxide?

419. What is the chemical formula for gold(I) cyanate?

420. What is the chemical formula for aluminum dihydrogen phosphate?

Chapter 7

Covalent Bonding

. . . .

In intro-level chemistry courses, most covalent compounds contain only nonmetals (elements to the right on the periodic table plus hydrogen). As with ionic compounds, the last word in the name of a covalent compound has an *-ide* suffix. Unlike ionic compounds, multiplying prefixes are typically present. For example, the compound S_2Cl_2 is disulfur dichloride, which uses the multiplying prefix *di-* twice. Organic compounds are another major group of covalent compounds, but they're beyond the scope of this text.

The Problems You'll Work On

In this chapter, you work with covalent bonding in the following ways:

- ✔ Naming compounds containing covalent bonds
- ✔ Writing formulas for covalent compounds

Note: You can go to the Appendix if you need to check the periodic table.

What to Watch Out For

Don't let common mistakes trip you up; remember the following when working on covalent bonding:

- ✔ The names of inorganic compounds normally consist of two words.
- ✔ All simple binary compounds have an *-ide* suffix on the second word in the name.
- ✔ The *-ite* and *-ate* suffixes normally don't occur in covalent compounds.
- ✔ The names of covalent compounds use multiplying prefixes, such as *di-, tri-, tetra-,* and *penta-*. The prefix *mono-* is rarely used anymore.
- ✔ Acids follow their own rules for naming.

Prefixes in Covalent-Compound Names

421–430 Review the prefixes associated with covalent compounds.

421. How many atoms does the prefix *di-* represent?

422. How many atoms does the prefix *hexa-* represent?

423. How many atoms does the prefix *hepta-* represent?

424. How many atoms does the prefix *tetra-* represent?

425. How many atoms does the prefix *nona-* represent?

426. How many atoms does the prefix *tri-* represent?

427. How many atoms does the prefix *mono-* represent?

428. How many atoms does the prefix *penta-* represent?

429. How many atoms does the prefix *octa-* represent?

430. How many atoms does the prefix *deca-* represent?

Naming Covalent Compounds

431–455 Provide the name of the specific covalent compound.

431. What is the name of CO?

432. What is the name of SBr_2?

433. What is the name of ICl?

434. What is the name of SO_2?

435. What is the name of PCl_5?

436. What is the name of XeF_2?

437. What is the name of SF_6?

438. What is the name of CBr_4?

439. What is the name of BCl_3?

440. What is the name of SiO_2?

441. What is the name of $AsCl_5$?

442. What is the name of $SbCl_3$?

443. What is the name of SiI_4?

444. What is the name of NF_3?

445. What is the name of CS_2?

446. What is the name of ClO_2?

447. What is the name of XeO_4?

448. What is the name of H_2O?

449. What is the name of SeF_6?

450. What is the name of S_2Cl_2?

451. What is the name of N_2O_3?

452. What is the name of P_4O_6?

453. What is the name of B_2Cl_4?

454. What is the name of BrF_3?

455. What is the name of S_2F_{10}?

Writing Formulas of Covalent Compounds

456–480 Give the chemical formulas of specific covalent compounds.

456. What is the chemical formula of silicon tetrabromide?

457. What is the chemical formula of nitrogen triiodide?

458. What is the chemical formula of carbon dioxide?

459. What is the chemical formula of arsenic pentafluoride?

460. What is the chemical formula of nitrogen monoxide?

461. What is the chemical formula of sulfur trioxide?

462. What is the chemical formula of chlorine monofluoride?

463. What is the chemical formula of xenon tetrafluoride?

464. What is the chemical formula of nitrogen dioxide?

465. What is the chemical formula of carbon tetrachloride?

466. What is the chemical formula of iodine heptafluoride?

467. What is the chemical formula of phosphorus tribromide?

468. What is the chemical formula of selenium tetrafluoride?

469. What is the chemical formula of chlorine dioxide?

470. What is the chemical formula of boron trifluoride?

471. What is the chemical formula of dinitrogen pentoxide?

472. What is the chemical formula of diphosphorus trioxide?

473. What is the chemical formula of dinitrogen dichloride?

474. What is the chemical formula of tetraphosphorus decoxide?

475. What is the chemical formula of arsenic trifluoride?

476. What is the chemical formula of dinitrogen oxide?

477. What is the chemical formula of tetraphosphorus trioxide?

478. What is the chemical formula of xenon trioxide?

479. What is the chemical formula of antimony pentachloride?

480. What is the chemical formula of dichlorine heptoxide?

Chapter 8

Molecular Geometry

• •

The difference in the electronegativity between two atoms is an indicator of the polarity of the bond between the elements. The difference ranges from 0 to about 3, and the bond ranges from nonpolar to polar to ionic.

The position of an element on the periodic table indicates the number of valence electrons on an atom. The key to a Lewis dot diagram is the sum of the valence electrons of all atoms in a species plus electrons indicated by the charge on an anion (negative ion) or minus electrons indicated by the charge on a cation (positive ion). A correct diagram must indicate all of these electrons. The bonds and lone pairs around an atom in a Lewis dot diagram indicate the shape and polarity of the species.

The Problems You'll Work On

In this chapter, you work with molecular geometry in the following ways:

✔ Counting electrons in Lewis dot diagrams

✔ Predicting bond types based on electronegativity values

✔ Recognizing molecular shapes

✔ Identifying polarity in bonds and molecules

Note: For access to the periodic table, see the Appendix.

What to Watch Out For

Remember the following when working on molecular geometry:

✔ The Lewis structure must account for all valence electrons plus or minus electrons gained or lost to form ions.

✔ Know the basic geometries and which geometries are inherently polar.

✔ The bond between two different atoms is likely to be polar. Some bonds are so polar they're ionic.

> ✔ The farther apart two nonmetals are on the periodic table, the more polar the bond. Hydrogen is an exception; its electronegativity falls between boron's and carbon's on the periodic table.

> ✔ Predicting the correct molecular shape relies on having a correct Lewis structure. Second-period elements never have more than an octet of electrons.

Valence Electrons

481–490 Answer the following questions concerning valence electrons.

481. How many electrons are in the Lewis dot diagram of a boron atom?

482. How many electrons are in the Lewis dot diagram of a barium atom?

483. How many electrons are in the Lewis dot diagram of a chlorine atom?

484. How many electrons are in the Lewis dot diagram of an O^{2-} ion?

485. How many electrons are in the Lewis dot diagram of a sodium ion?

486. Which group on the periodic table contains atoms that have the following Lewis dot diagram?

$$\cdot \dot{X}$$

487. Which group on the periodic table contains atoms that have the following Lewis dot diagram?

$$\dot{X}\colon$$

488. Which group on the periodic table contains atoms that have the following Lewis dot diagram?

$$\dot{X}\cdot$$

489. Which group on the periodic table contains atoms that have the following Lewis dot diagram?

$$\dot{X}\cdot$$

490. Which group on the periodic table contains atoms that have the following Lewis dot diagram?

Predicting Bond Types

491–501 Predict the type of bond.

491. What is the name for a measure of an atom's strength to attract a bonding pair of electrons?

492. What kind of bond forms between elements with the same or nearly the same electronegativity?

493. What kind of bond forms between elements when the electronegativity difference is moderate?

494. For the representative elements, metals are expected to have a _____ (*lower/ higher*) electronegativity, and nonmetals are expected to have a _____ (*lower/ higher*) electronegativity.

495. Based on their relative electronegativity values, what type of bond will form between carbon and bromine?

496. Based on their relative electronegativity values, what type of bond will form between carbon and fluorine?

497. Based on their relative electronegativity values, what type of bond will form between chlorine and fluorine?

498. Based on their relative electronegativity values, what type of bond will form between boron and hydrogen?

499. Based on their relative electronegativity values, what type of bond will form between carbon and hydrogen?

500. Based on their relative electronegativity values, what type of bond will form between phosphorus and fluorine?

501. Arrange the following bonds in order of increasing polarity.

Si–Cl, S–Cl, O–S

Basic Molecular Shapes

502–521 The following questions cover basic molecular shapes.

502. What term best describes the shape of the following molecule?

503. What term best describes the shape of the following molecule?

504. What term best describes the shape of the following molecule?

505. What term best describes the shape of the following molecule?

506. What term best describes the shape of the following molecule?

507. What term best describes the shape of the following molecule?

508. What molecular shape has bond angles that are approximately 180°?

509. What molecular shape has bond angles that are approximately 120°?

510. What molecular shape has bond angles that are approximately 109.5°?

511. What is the molecular shape of boron trichloride, BCl_3?

512. What is the molecular shape of bromine, Br_2?

513. What is the molecular shape of methane, CH_4?

514. What is the molecular shape of water, H_2O?

515. What is the molecular shape of ammonia, NH_3?

516. What is the molecular shape of carbon dioxide, CO_2?

517. Why does water, H_2O, have a bent molecular shape?

518. Why does nitrogen trifluoride, NF_3, have a trigonal pyramidal molecular shape?

519. Why does carbon dioxide, CO_2, have a linear molecular shape?

520. Why does carbon tetrachloride, CCl_4, have a tetrahedral molecular shape?

521. Why does boron trifluoride, BF_3, have a trigonal planar molecular shape?

Exceptional Molecular Shapes

522–536 The following questions deal with exceptional molecular shapes.

522. What term best describes the following molecular shape?

523. What term best describes the following molecular shape?

524. What term best describes the following molecular shape?

525. What term best describes the following molecular shape?

526. What term best describes the following molecular shape?

527. What term best describes the following molecular shape?

528. What molecular shape has bond angles between atoms that are 90°, 120°, and 180°?

529. Which molecular shape has 109.5° bond angles?

530. What is the molecular shape of sulfur tetra-fluoride, SF_4?

531. What is the molecular shape of phosphorus pentachloride, PCl_5?

532. What is the molecular shape of bromine pentafluoride, BrF_5?

533. What is the molecular shape of sulfur hexa-fluoride, SF_6?

534. What is the molecular shape of xenon tetra-fluoride, XeF_4?

535. What is the molecular shape of chlorine trifluoride, ClF_3?

536. What are the electron geometry and the molecular shape of xenon difluoride?

Polarity of Molecules

537–545 *These questions deal with the polarity of molecules.*

537. What do you need to know about a molecule to determine its polarity?

538. H_2 contains a _____ (*nonpolar/polar*) bond and is a _____ (*nonpolar/polar*) molecule.

539. HCl contains a _____ (*nonpolar/polar*) bond and is a _____ (*nonpolar/polar*) molecule.

540. CCl_4 contains _____ (*nonpolar/polar*) bonds and is a _____ (*nonpolar/polar*) molecule.

541. NH_3 contains _____ (*nonpolar/polar*) bonds and is a _____ (*nonpolar/polar*) molecule.

542. $SeCl_4$ contains _____ (*nonpolar/polar*) bonds and is a _____ (*nonpolar/polar*) molecule.

543. XeO_4 contains _____ (*nonpolar/polar*) bonds and is a _____ (*nonpolar/polar*) molecule.

544. ICl_3 contains _____ (*nonpolar/polar*) bonds and is a _____ (*nonpolar/polar*) molecule.

545. Which of the following molecules are polar?

CO_2, OF_2, BF_3, NBr_3, PCl_3, KrF_4, SnH_4, IF_5

Chapter 9

Chemical Reactions

*E*very chemical reaction begins with the formula of one or more substances (reactants) followed by an arrow and ends with the formula of one or more substances (products). There are many types of chemical reactions; however, only a few basic types appear in this chapter. Redox reactions tend to be more difficult to balance than the other reactions here, so you use a systematic approach to balance most redox reactions.

The Problems You'll Work On

In this chapter, you work with chemical reactions in the following ways:

- ✔ Identifying types of reactions
- ✔ Balancing chemical equations
- ✔ Balancing redox reactions

Note: See the Appendix if you need to check the periodic table.

What to Watch Out For

Remember the following when working on chemical reactions:

- ✔ When balancing equations, never change a chemical formula; change only the coefficients.
- ✔ There are only a few basic reaction types here, so don't invent new ones. But note that a reaction may fit into more than one category.
- ✔ Ionic equations contain ions.
- ✔ Acid-base reactions always produce a salt and usually produce water.
- ✔ Redox reactions must have an oxidation and a reduction.

Classifying Reactions from Chemical Equations

546–555 Given the chemical equation, name the type of reaction (combination, combustion, decomposition, single displacement, or double displacement).

546. What type of reaction is the following reaction?

$$2NaCl \rightarrow 2Na + Cl_2$$

547. What type of reaction is the following reaction?

$$2Na + 2HCl \rightarrow H_2 + 2NaCl$$

548. What type of reaction is the following reaction?

$$Ca + S \rightarrow CaS$$

549. What type of reaction is the following reaction?

$$2C_2H_2 + 5O_2 \rightarrow 4CO_2 + 2H_2O$$

550. What type of reaction is the following reaction?

$$KOH + HNO_3 \rightarrow KNO_3 + H_2O$$

551. What type of reaction is the following reaction?

$$2H_2 + O_2 \rightarrow 2H_2O$$

552. What type of reaction is the following reaction?

$$CH_4 + 2O_2 \rightarrow CO_2 + 2H_2O$$

553. What type of reaction is the following reaction?

$$N_2 + 3H_2 \rightarrow 2NH_3$$

554. What type of reaction is the following reaction?

$$2H_2O_2 \rightarrow 2H_2O + O_2$$

555. What type of reaction is the following reaction?

$$K + AgCl \rightarrow KCl + Ag$$

Classifying Reactions from Word Equations

556–565 *Given the word equation, name the type of chemical reaction (combination, combustion, decomposition, single displacement, or double displacement).*

556. What type of reaction is the following reaction?

aluminum + oxygen → aluminum oxide

557. What type of reaction is the following reaction?

sodium hydroxide + lead(II) nitrate →
sodium nitrate + lead(II) hydroxide

558. What type of reaction is the following reaction?

magnesium + hydrogen chloride →
magnesium chloride + hydrogen gas

559. What type of reaction is the following reaction?

mercury(II) oxide → mercury + oxygen gas

560. What type of reaction is the following reaction?

butane + oxygen gas →
carbon dioxide gas + water vapor

561. What type of reaction is the following reaction?

iron + copper(II) chloride →
iron(II) chloride + copper

562. What type of reaction is the following reaction?

sodium bicarbonate → sodium oxide +
water vapor + carbon dioxide gas

563. What type of reaction is the following reaction?

ethanol + oxygen gas →
carbon dioxide gas + water vapor

564. What type of reaction is the following reaction?

calcium oxide + water → calcium hydroxide

565. What type of reaction is the following reaction?

carbon monoxide + oxygen gas → carbon dioxide gas

Predicting Reactions

566–575 Given the reactants, predict the type of chemical reaction.

566. Predict the type of reaction these reactants would undergo:

$$Zn + H_2SO_4 \rightarrow$$

567. Predict the type of reaction this reactant would undergo when heated:

$$N_2O \rightarrow$$

568. Predict the type of reaction these reactants would undergo:

$$Ca(OH)_2 + H_2SO_4 \rightarrow$$

569. Predict the type of reaction these reactants would undergo:

$$Li + O_2 \rightarrow$$

570. Predict the type of reaction these reactants would undergo:

$$C_3H_8 + O_2 \rightarrow$$

571. Predict the type of reaction these reactants would undergo:

silver nitrate + sodium chloride →

572. Predict the type of reaction these reactants would undergo:

sodium + hydrogen gas →

573. Predict the type of reaction these reactants would undergo:

potassium bromide + chlorine gas →

574. Predict the type of reaction this reactant would undergo when heated:

copper(II) hydroxide →

575. Predict the type of reaction these reactants would undergo:

aluminum sulfate + calcium phosphate →

Balancing Chemical Reactions

576–585 *Balance the chemical reaction.*

576. What are the coefficients of the following equation when it's balanced?

$$\square N_2O_4 \rightarrow \square NO_2$$

577. What are the coefficients of the following equation when it's balanced?

$$\square CS_2 + \square Cl_2 \rightarrow \square CCl_4 + \square S_2Cl_2$$

578. What are the coefficients of the following equation when it's balanced?

$$\square Na_2CO_3 + \square Cd(NO_3)_2 \rightarrow \square CdCO_3 + \square NaNO_3$$

579. What are the coefficients of the following equation when it's balanced?

$$\square I_2 + \square Cl_2 \rightarrow \square ICl$$

580. What are the coefficients of the following equation when it's balanced?

$$\square KCl + \square F_2 \rightarrow \square KF + \square Cl_2$$

581. What are the coefficients of the following equation when it's balanced?

$$\square CH_4 + \square O_2 \rightarrow \square CO_2 + \square H_2O$$

582. What are the coefficients of the following equation when it's balanced?

$$\square Ca + \square O_2 \rightarrow \square CaO$$

583. What are the coefficients of the following equation when it's balanced?

$$\square N_2 + \square H_2 \rightarrow \square NH_3$$

584. What are the coefficients of the following equation when it's balanced?

$$\square SO_2 + \square O_2 \rightarrow \square SO_3$$

585. What are the coefficients of the following equation when it's balanced?

$$\square Mg(NO_3)_2 + \square K_3PO_4 \rightarrow \square Mg_3(PO_4)_2 + \square KNO_3$$

Balancing Reactions from Word Equations

586–595 Given the word equation, balance the chemical reaction.

586. What are the coefficients of the following reaction when it's balanced?

barium sulfite → barium oxide + sulfur dioxide gas

587. What are the coefficients of the following reaction when it's balanced?

silicon carbide + chlorine gas →
silicon tetrachloride + carbon

588. What are the coefficients of the following reaction when it's balanced?

silver sulfate + copper(II) chlorate →
silver chlorate + copper(II) sulfate

589. What are the coefficients of the following reaction when it's balanced?

calcium carbonate + hydrogen chloride →
calcium chloride + water + carbon dioxide

590. What are the coefficients of the following reaction when it's balanced?

boron + oxygen gas → diboron trioxide

591. What is the sum of the coefficients of the following reaction when it's balanced?

sodium + iodine → sodium iodide

592. What is the sum of the coefficients of the following reaction when it's balanced?

potassium chlorate →
potassium chloride + oxygen gas

593. What is the sum of the coefficients of the following reaction when it's balanced?

iron + oxygen gas → iron(III) oxide

594. What is the sum of the coefficients of the following reaction when it's balanced?

calcium nitride + water →
calcium hydroxide + ammonia

595. What are the coefficients of the following reaction when it's balanced?

octane $\left(C_8H_{18}\right)$ + oxygen gas →
carbon dioxide + water vapor

Predicting Products and Balancing Reactions

596–605 *Given the reactants, predict the products and balance the chemical reaction.*

596. When the following reaction is completed and balanced, what is the coefficient on the barium hydroxide?

iron(III) bromide + barium hydroxide →

597. When the following reaction is completed and balanced, what is the coefficient on the nickel(II) chloride?

potassium + nickel(II) chloride →

598. The compound in the following reaction is heated. What is the sum of the coefficients when the reaction is completed and balanced? (*Hint:* One of the products of heating this compound is another form of lead oxide.)

lead(IV) oxide →

599. When the following reaction is completed and balanced, what is the coefficient on the silver nitrate?

silver nitrate + aluminum chloride →

600. When the following reaction is completed and balanced, what is the sum of the coefficients? (*Hint:* The products are two elements.)

nitrogen trichloride →

601. Complete and balance the following reaction:

zinc + oxygen gas →

602. When the following reaction is completed and balanced, what is the sum of the coefficients? (*Hint:* One of the products is water.)

iron(III) oxide + hydrogen gas →

603. When the following reaction is completed and balanced, what is the coefficient on the nitrogen dioxide? (*Hint:* The products are two elements.)

nitrogen dioxide →

604. Complete and balance the following reaction:

silver nitrate + potassium dichromate →

605. When the following reaction is completed and balanced, what is the coefficient on the copper(II) sulfate?

aluminum + copper(II) sulfate →

Redox and Acid-Base Reactions

606–621 Answer the following questions on redox and acid-base (neutralization) reactions.

606. What is the oxidation number of Mn in $KMnO_4$?

607. What is the oxidation number of Br in $NaBrO_3$?

608. What is the oxidation number of Cr in $Cr_2O_7^{2-}$?

609. What is the oxidation number of S in $S_2O_3^{2-}$?

610. When a substance loses electrons during the course of a reaction, it goes through which process?

611. When a substance gains electrons during the course of a reaction, it goes through which process?

612. Which substance in the following forward reaction is oxidized?

$$Fe + Zn^{2+} \rightarrow Fe^{2+} + Zn$$

613. Which substance in the following forward reaction is reduced?

$$2Al + 3Fe^{2+} \rightarrow 2Al^{3+} + 3Fe$$

614. Which substance in the following forward reaction is the reducing agent?

$$Cu + 2AgNO_3 \rightarrow Cu(NO_3)_2 + 2Ag$$

615. Which substance in the following forward reaction is the oxidizing agent?

$$Br_2 + KOH \rightarrow KBrO_3 + KBr + H_2O$$

616. How many electrons are necessary to balance the following half-reaction that takes place in an acidic environment?

$$NO_3^- \rightarrow NO$$

617. How many electrons are necessary to balance the following half-reaction that takes place in a basic environment?

$$Br_2 \rightarrow BrO_3^-$$

618. How many hydrogen ions do you need to balance the following oxidation-reduction reaction?

$$Zn + H_3AsO_4 \rightarrow AsH_3 + Zn^{2+}$$

619. What is the coefficient of the OH^- when the following oxidation-reduction reaction is balanced?

$$Bi^{3+} + SnO_2^{2-} + OH^- \rightarrow Bi + SnO_3^{2-} + H_2O$$

620. What are the coefficients of the following reaction when it's balanced?

$$KIO_3 + KI + H_2SO_4 \rightarrow K_2SO_4 + I_2 + H_2O$$

621. What is the coefficient of ClO_4^- when the following reaction is balanced?

$$MnO_4^- + ClO_2^- \rightarrow MnO_2 + ClO_4^- \ (\text{in basic solution})$$

Chapter 10

Molar Calculations

. . . .

Moles are extremely important to many chemical calculations. Sometimes you need to find moles, and in other cases, moles are necessary to find something else. The molar mass is useful to convert mass to moles or to convert moles to mass. Converting moles to mass is useful in determining the percent composition and percent yield. Converting mass to moles is useful in determining empirical and molecular formulas and in finding the limiting reactant.

The Problems You'll Work On

In this chapter, you work with molar calculations in the following ways:

- ✔ Calculating molar mass
- ✔ Finding percent composition
- ✔ Writing empirical and molecular formulas
- ✔ Completing molar calculations
- ✔ Finding the percent yield
- ✔ Working with limiting reactants

Note: For access to the periodic table, see the Appendix.

What to Watch Out For

Remember the following when working on molar calculations:

- ✔ Reducing the empirical formula isn't possible.
- ✔ The molecular and empirical formulas may be the same.
- ✔ Moles may be the answer, but more often, moles are an intermediate. When in doubt, change to moles.
- ✔ A percent yield greater than 100 percent indicates that you (or the experimenter) made a mistake.
- ✔ The limiting reactant is the key for all subsequent calculations.
- ✔ The molar mass has units of grams/mole.

Calculating Molar Mass

622–636 Calculate the molar mass.

622. What is the molar mass of KCl rounded to two decimal places?

623. What is the molar mass of CaO rounded to two decimal places?

624. What is the molar mass of AlF_3 rounded to two decimal places?

625. What is the molar mass of B_2O_3 rounded to two decimal places?

626. What is the molar mass of CBr_4 rounded to two decimal places?

627. What is the molar mass of NH_4Cl rounded to two decimal places?

628. What is the molar mass of $Ba(NO_3)_2$ rounded to two decimal places?

629. What is the molar mass of $(NH_4)_2SO_4$ rounded to two decimal places?

630. What is the molar mass of silver sulfide rounded to two decimal places?

631. What is the molar mass of sodium carbonate rounded to two decimal places?

632. What is the molar mass of $Al_2(C_2O_4)_3$ rounded to two decimal places?

633. What is the molar mass of $Zn(NH_3)_4Cl_2$ rounded to two decimal places?

634. What is the molar mass of $CuSO_4 \cdot 5H_2O$ rounded to two decimal places?

635. What is the molar mass of manganese(II) nitrate hexahydrate rounded to two decimal places?

636. What is the molar mass of iron(II) phosphate octahydrate rounded to two decimal places?

Finding Mass Percent

637–651 *Find the percent composition of compounds.*

637. What is the mass percent of Na in NaBr rounded to two decimal places?

638. What is the mass percent of Sr in SrS rounded to two decimal places?

639. What is the mass percent of Cl in $KClO_3$ rounded to two decimal places?

640. What is the mass percent of O in CaC_2O_4 rounded to two decimal places?

641. What is the mass percent of S in $Na_2S_2O_3$ rounded to two decimal places?

642. What is the mass percent of N in NH_4NO_3 rounded to two decimal places?

643. What is the mass percent of Li in lithium hydrogen carbonate rounded to two decimal places?

644. What is the mass percent of silver in silver sulfide rounded to two decimal places?

645. What is the mass percent of each element in aluminum hydroxide rounded to two decimal places?

646. What is the mass percent of each element in zinc iodate rounded to two decimal places?

647. Rank the following compounds in order of increasing percent nitrogen content:

$$NO, NO_2, N_2O, N_2O_3, N_2O_5$$

648. Rank the following compounds in order of decreasing percent carbon content:

CH_4, CCl_4, C_2H_6, K_2CO_3, CaC_2O_4

649. Rank the following compounds in order of increasing percent sulfur content:

K_2S, K_2SO_3, K_2SO_4, $K_2S_2O_3$, $KSCN$

650. What is the mass percent of water in $CuSO_4 \cdot 5H_2O$ rounded to two decimal places?

651. What is the mass percent of oxygen in sodium phosphate dodecahydrate rounded to two decimal places? (**Hint:** The prefix *dodeca-* means 12.)

653. What is the empirical formula of a compound containing 85.62% C and 14.38% H?

654. What is the empirical formula of a compound containing 10.06 g of carbon, 0.84 g of hydrogen, and 89.09 g of chlorine?

655. A compound contains 1.4066 g of iron, 1.3096 g of chromium, and 1.6118 g of oxygen. What is the empirical formula of the compound?

656. A 100.00-g sample of a hydrate lost 18.73 g of water when heated. The anhydrous salt contains 43.82 g of cadmium and 12.50 g of sulfur, and the remaining mass is oxygen. What is the empirical formula of the anhydrous salt?

Empirical Formulas

652–656 Find the empirical formula.

652. Which of the following compounds could be correctly identified as an empirical formula?

C_2H_2, C_2H_4, C_2H_6, $C_2H_3O_2$, $C_4H_6O_4$

Molecular Formulas

657–661 Answer the questions about molecular formulas.

657. Which of the following pairs of compounds are correctly matched empirical and molecular formulas?

I. CH and CH_4

II. CH and C_2H_2

III. C_3H_4 and C_3H_8

658. Which of the following pairs of compounds are correctly matched empirical and molecular formulas?

    I.   CO and CO_2

    II. CH_2 and C_6H_{12}

    III. CH_2O and $C_6H_{12}O_6$

659. If the empirical formula for a compound is CH and the molar mass of the compound is 78.11 g/mol, what is the molecular formula?

660. What is the molecular formula of a compound with a molar mass of 174.20 g/mol and with a mass percent composition of 41.37% carbon, 8.10% hydrogen, 18.39% oxygen, and 32.16% nitrogen?

661. A compound contains 23.25% carbon, 1.95% hydrogen, 17.20% oxygen, and 3.01% nitrogen, and the rest is iodine. If the compound has a molar mass of approximately 465 g/mol, what are the empirical and molecular formulas of the compound?

Mole Calculations

662–711 *Complete the calculations involving the mole.*

662. How many moles of NaCl are in 10.0 g of NaCl?

663. How many moles of BH_3 are in 5.00 g of BH_3?

664. How many moles of Na_2CO_3 are in 275 g of Na_2CO_3?

665. How many moles of NH_4OH are in 400. g of NH_4OH?

666. How many moles are in 99 g of $KMnO_4$?

667. How many moles are in 150. g of beryllium sulfide?

668. How many moles are in 25.0 g of aluminum oxide?

669. How many moles are in 180. g of carbon tetrachloride?

670. How many moles are in 320. g of calcium phosphide?

671. How many moles are in 1.70 g of magnesium permanganate?

672. How many grams are in 5.25 mol of FeO?

673. How many grams are in 0.750 mol of S_2Cl_2?

674. How many grams are in 100. mol of KI?

675. How many grams are in 42.3 mol of $CaCO_3$?

676. How many grams are in 7.9 mol of $Al(NO_3)_3$?

677. How many grams are in 0.15 mol of lead(II) nitrate?

678. How many grams are in 12 mol of tetraphosphorus decoxide?

679. How many grams are in 0.25 mol of copper(I) oxide?

680. How many kilograms are in 248 mol of zinc hydroxide?

681. How many kilograms are in 367 mol of ammonium sulfate?

682. How many atoms are in 0.250 g of He?

683. How many atoms are in 30. g of Al?

684. How many molecules are in 200. g of carbon dioxide?

685. How many molecules are in 0.050 g of dinitrogen tetroxide?

686. How many atoms are in 250. g of sulfur, S_8?

687. How many grams are in 9.5×10^{26} molecules of Br_2?

688. How many grams are in 8.306×10^{21} atoms of Ni?

689. How many grams are in 3.00×10^{23} molecules of $C_6H_{12}O_6$?

690. What is the mass of 65,000,000 atoms of krypton gas?

691. What is the mass of 4.0×10^{22} formula units of AuF_3, gold(III) fluoride?

692. Given the following equation, how many moles of ammonia, NH_3, can be produced from 5.00 mol of nitrogen gas if there's an excess of hydrogen gas?

$$N_2(g) + 3H_2(g) \rightarrow 2NH_3(g)$$

693. Given the following balanced equation, how many moles of ammonia, NH_3, can be produced from 12.0 mol of hydrogen gas if there's an excess of nitrogen gas?

$$N_2(g) + 3H_2(g) \rightarrow 2NH_3(g)$$

694. Given the following balanced equation, how many moles of nitrogen gas would be necessary to make 24.0 mol of ammonia, NH_3, if there's an excess of hydrogen gas?

$$N_2(g) + 3H_2(g) \rightarrow 2NH_3(g)$$

695. Given the following balanced equation, how many moles of hydrogen gas would be necessary to make 36.0 mol of ammonia if there's an excess of nitrogen gas?

$$N_2(g) + 3H_2(g) \rightarrow 2NH_3(g)$$

696. If 3.00 mol of propane, C_3H_8, undergoes complete combustion according to the following reaction, how many moles of each product are produced?

$$C_3H_8(g) + 5O_2(g) \rightarrow 3CO_2(g) + 4H_2O(l)$$

697. According to the following reaction, what is the maximum number of grams of KCl that can be produced from 5.0 g of $KClO_3$?

$$2KClO_3(s) \rightarrow 2KCl(s) + 3O_2(g)$$

698. According to the following reaction, what is the maximum number of grams of O_2 that can be produced from 5.0 g of $KClO_3$?

$$2KClO_3(s) \rightarrow 2KCl(s) + 3O_2(g)$$

699. According to the following reaction, how many grams of $CaCO_3$ can be produced from 20.0 g of $Ca(OH)_2$?

$$Ca(OH)_2(aq) + CO_2(g) \rightarrow CaCO_3(s) + H_2O(l)$$

700. According to the following reaction, how many grams of NaCl can be produced from 80.0 g of NaOH?

$$HCl(aq) + NaOH(aq) \rightarrow NaCl(aq) + H_2O(l)$$

701. Potassium phosphate plus silver nitrate forms potassium nitrate and silver phosphate. How many grams of silver phosphate can be formed from 5.6 g of silver nitrate and an excess of potassium phosphate?

702. Sodium hydroxide plus copper(II) sulfate forms copper(II) hydroxide and sodium sulfate. With an excess of copper(II) sulfate, how many grams of sodium hydroxide are needed to make 10.0 g of copper(II) hydroxide?

703. When the following reaction is completed and balanced, how many grams of zinc are needed to react with 74.5 g of hydrogen chloride?

$$Zn + HCl \rightarrow$$

704. When the following reaction is completed and balanced, how many grams of potassium nitrate will be made when 220 g of calcium phosphate is produced?

$$Ca(NO_3)_2 + K_3PO_4 \rightarrow$$

705. When the following reaction is completed and balanced, what is the minimum number of grams of each reactant needed to make 0.25 g of barium dichromate?

barium chloride + potassium dichromate →

706. When the following reaction is completed and balanced, what is the maximum number of grams of product that can be made if there are 50.0 g of aluminum and 150. g of bromine? (*Hint:* The reaction has only one product.)

$$Al + Br_2 \rightarrow$$

707. When the following reaction is balanced, how many molecules of O_2 do you need to produce 8.02×10^{24} molecules of P_2O_5?

$$P + O_2 \rightarrow P_2O_5$$

708. According to the following reaction, how many molecules of H_2O can be produced from 1.69×10^{22} molecules of HNO_3?

$$C(s) + 4HNO_3(aq) \xrightarrow{\Delta} 4NO_2(g) + CO_2(g) + 2H_2O(l)$$

709. Barium sulfite decomposes to form barium oxide and sulfur dioxide gas. How many molecules of barium sulfite are needed to make 3.16×10^{21} molecules of sulfur dioxide?

710. Iodine gas plus chlorine gas makes iodine chloride gas. How many molecules of iodine are needed to make 5.28×10^{24} molecules of iodine chloride?

711. According to the following reaction, what is the minimum number of molecules of octane, C_8H_{18}, needed to make 4.52×10^{24} molecules of each product, assuming complete combustion of the octane?

$$C_8H_{18} + O_2 \rightarrow CO_2 + H_2O$$

Percent Yield

712–716 Complete the calculations related to percent yield.

712. If the theoretical yield of a product in a reaction was 1.358 g and the actual yield is 1.146 g, what is the percent yield of the product?

713. A student produced 5.15 g of a product during a chemistry experiment. The theoretical yield was 4.95 g. What is the percent yield of the product?

714. If 316 g of ammonia produces 1,225 g of ammonium bromide by reacting with excess hydrogen bromide, what is the percent yield of this reaction?

$$NH_3 + HBr \rightarrow NH_4Br$$

715. Ammonium bromide can be produced according to the following reaction:

$$3Br_2 + 8NH_3 \rightarrow 6NH_4Br + N_2$$

What is the percent yield of this reaction if 623 g of ammonia produces 2,341 g of ammonium bromide?

716. A student performed a precipitation reaction using 0.527 g of $BaCl_2$ dissolved in water with an excess of Na_2SO_4. If 0.551 g of $BaSO_4$ is recovered, what was the percent yield of $BaSO_4$?

Limiting Reactants

717–721 Perform calculations with limiting reactants.

717. According to the following balanced reaction, what is the maximum number of grams of SO_3 that can be prepared from 45.0 mol of SO_2 and 25.0 mol of O_2?

$$2SO_2 + O_2 \rightarrow 2SO_3$$

718. According to the following unbalanced reaction, what is the maximum amount of NaCl that can be prepared from 106 g of Cl_2 and 154 g of $NaClO_2$?

$$Cl_2 + NaClO_2 \rightarrow ClO_2 + NaCl$$

719. Aluminum combines with oxygen gas to form aluminum oxide, Al_2O_3. If 100.0 g of aluminum reacts with 100.0 g of oxygen, how many grams of aluminum oxide are produced, and which reactant is the limiting reactant?

720. A solution containing 155 g of KI is added to a solution containing 175 g of nitric acid and reacts according to the following equation:

$$6KI + 8HNO_3 \rightarrow 6KNO_3 + 2NO + 3I_2 + 4H_2O$$

How many grams of NO are produced, and which reactant is in excess?

721. A mixture containing 50.0 g of hydrogen gas and 50.0 g of oxygen gas is sparked and allowed to react. Which reactant is the limiting reactant, how much water is produced, and how much of the excess reactant remains?

Chapter 11

Thermochemistry

· ·

All processes involve energy. In this chapter, the key is the energy in chemical processes. Adding or removing heat may cause a phase change or a change in temperature. The specific heat of a substance is a useful factor relating the energy, mass, and temperature change. Calorimetry uses specific heat, temperature change, and mass to determine the energy involved in a chemical process. Heats of formation are a useful alternative to calorimetry for determining energy changes. Hess's law allows you to predict energy changes when calorimetry or heats of formation are inadequate.

The Problems You'll Work On

In this chapter, you work with thermochemistry in the following ways:

- ✔ Doing temperature conversions
- ✔ Interpreting energy and phase changes
- ✔ Completing specific heat and calorimetry problems
- ✔ Calculating with standard heats of formation
- ✔ Finding enthalpy changes with Hess's law

What to Watch Out For

Remember the following when working on thermochemistry:

- ✔ Dimensional analysis can be exceedingly useful in thermochemistry.
- ✔ Don't confuse exothermic and endothermic reactions.
- ✔ All phase changes involve energy. A phase change is exothermic in one direction and endothermic in the reverse direction.
- ✔ In calorimetry, the thermometer is part of the surroundings, not the system. When the thermometer shows a temperature increase, it's absorbing energy (endothermic). Endothermic to a thermometer is exothermic for the system.
- ✔ The heat of formation of an element in its standard state is always 0.

Converting Temperatures

722–731 Convert the temperatures between the Celsius and Kelvin scales.

722. 300°C = _____ K

723. 150°C = _____ K

724. –200°C = _____ K

725. –78°C = _____ K

726. 37°C = _____ K

727. 100 K = _____°C

728. 300 K = _____°C

729. 0 K = _____°C

730. 313 K = _____°C = _____°F

731. 233.15 K = _____°C = _____°F

Phase Changes and Energy

732–752 *Answer the questions on phase changes and the energy associated with the changes.*

732. Which line segment in the graph represents the process of melting?

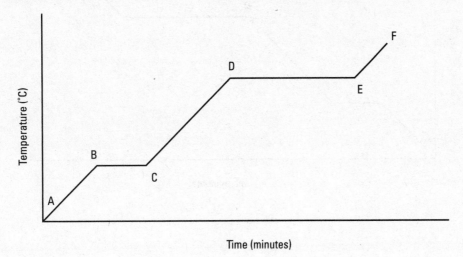

733. Which line segment in the graph represents the process of condensation?

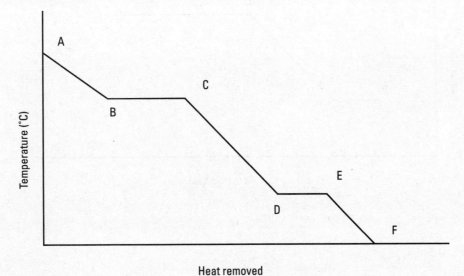

734. Which line segment in the graph represents the process of heating a liquid substance with no phase change?

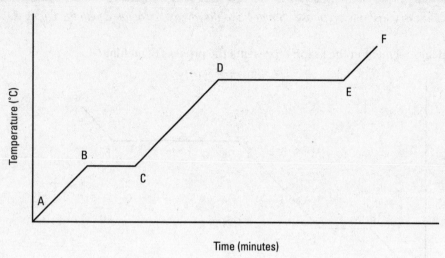

735. Which line segment in the graph represents the process of boiling?

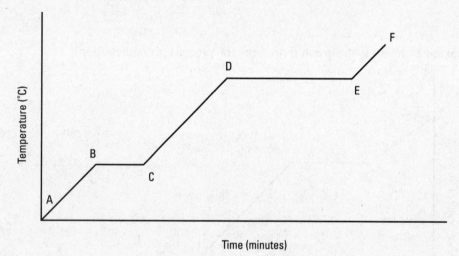

736. Which line segment in the graph represents the process of heating a solid substance with no phase change?

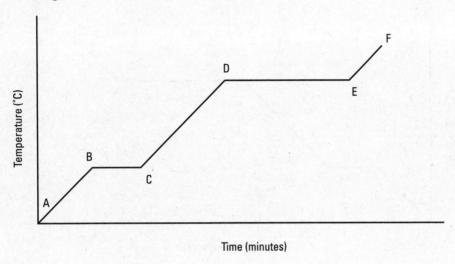

737. Which line segment in the graph represents the process of freezing?

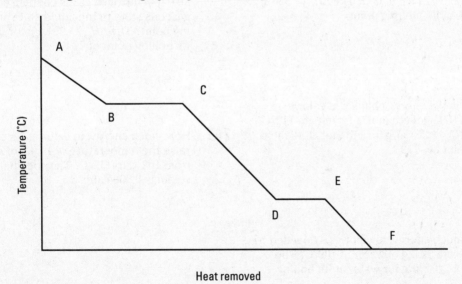

738. Which line segment in the graph represents the process of heating a gaseous substance with no phase change?

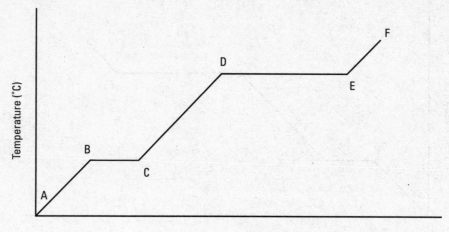

739. How much energy, in joules, does it take to melt 50.0 g of ice at 0°C? The ΔH_{fus} = 334 J/g for ice at its melting point.

740. What is the energy change, in calories, when 100. g of ethanol is frozen at –112°C? The ΔH_{fus} = 25 cal/g for ethanol at its melting point of –112°C.

741. How much energy, in calories, is needed to evaporate 50.0 g of water at 100°C? The ΔH_{vap} = 540 cal/g for water at its boiling point.

742. What is the energy change, in joules, when 20.0 g of nitrogen gas is condensed from the gaseous state to the liquid state at its boiling point? The ΔH_{vap} = 201 J/g for nitrogen at its boiling point of 77 K.

743. How much energy, in calories, is needed to raise the temperature of 10.0 g of steam from 102°C to 112°C? The specific heat of steam is 0.500 cal/g°C.

744. How much energy, in joules, is needed to raise the temperature of 75.0 g of water from 20.°C to 80.°C? The specific heat of water is 4.18 J/g°C.

745. How much energy, in calories, does it take to change 25.00 g of ice at −10.0°C to water at 0.0°C? The specific heat of ice is 0.500 cal/g°C, and ΔH_{fus} = 80.00 cal/g.

746. What is the energy change, in joules, when 125 g of water is cooled from 25.0°C to ice at 0.0°C? The specific heat of water is 4.18 J/g°C, and ΔH_{fus} = 334.0 J/g.

747. What is the energy change, in calories, when 40.0 g of steam is cooled from 120.0°C to 25.0°C? The specific heat of steam is 0.500 cal/g°C, ΔH_{vap} = 540.0 cal/g, and the specific heat of water is 1.00 cal/g°C.

748. How much energy, in joules, is necessary to raise the temperature of 200.0 g of water at 48.0°C to steam at 122.0°C? The specific heat of water is 4.18 J/g°C, ΔH_{vap} = 2,257 J/g, and the specific heat of steam is 2.09 J/g°C.

749. How much energy, in calories, is necessary to raise the temperature of 36.0 g of ice at −40.0°C to steam at 100.0°C? The specific heat of ice is 0.500 cal/g°C, the specific heat of water is 1.00 cal/g°C, ΔH_{fus} = 80.00 cal/g, and ΔH_{vap} = 540.0 cal/g.

750. What is the energy change, in joules, when 72.0 g of steam at 120.0°C changes to ice at 0.0°C? The specific heat of steam is 2.09 J/g°C, the specific heat of water is 4.18 J/g°C, ΔH_{fus} = 334.0 J/g, and ΔH_{vap} = 2,257 J/g.

751. How much energy, in joules, is necessary to raise the temperature of 4,536 g of ice at −78.0°C to steam at 105.0°C? The specific heat of ice and steam is 2.09 J/g°C, the specific heat of water is 4.18 J/g°C, ΔH_{fus} = 334.0 J/g, and ΔH_{vap} = 2,257 J/g.

752. What is the energy change, in calories, when 0.500 g of steam at 150.0°C changes to ice at −50.0°C? The specific heat of steam and ice is 0.500 cal/g°C, ΔH_{vap} = 540.0 cal/g, ΔH_{fus} = 80.00 cal/g, and the specific heat of water is 1.00 cal/g°C.

Specific Heat and Calorimetry

753–762 *Complete the calculations involving specific heat and calorimetry.*

753. What is the mass of a sample if the absorption of 172 calories results in a temperature increase of 5.00°C? The specific heat of the sample is 0.573 cal/g°C.

754. The addition of 197 J to a 22.0-g sample results in a temperature increase of 2.15°C. What is the specific heat of the sample?

755. What is the temperature change for a 38.1-g sample following the addition of 153 calories? The specific heat of the sample is 0.217 cal/g°C.

756. A 17.35-g sample was heated by 1,148 J. The specific heat of the sample is 2.17 J/g°C. What is the final temperature of the sample if the initial temperature was 15.5°C?

757. A 19.75-g sample was heated by 12.35 calories. The specific heat of the sample is 0.125 cal/g°C. What was the initial temperature of the sample if the final temperature is 37.0°C?

758. A 3.75-g sample with a specific heat of 0.986 cal/g°C experienced a 2.46°C increase in temperature. What was the sample's energy change in calories?

759. A 0.326-g sample with a specific heat of 0.896 J/g°C experienced a 1.37°C decrease in temperature. What was the sample's energy change in joules?

760. Two samples are placed in contact, and heat flows from the warmer sample to the cooler one. Sample 1, the cooler sample, has a specific heat of 2.18 cal/g°C and exhibits a 13.9°C increase in temperature. Sample 2 has a specific heat of 1.36 cal/g°C and exhibits a 18.8°C decrease in temperature. If Sample 1 has a mass of 1.35 g, what is the mass of Sample 2? Assume no heat is lost to the surroundings.

761. Two samples are placed in contact, and heat flows from the warmer sample to the cooler one. Sample 1, the cooler sample, exhibits a 20.1°C increase in temperature. Sample 2 has a specific heat of 2.15 J/g°C and exhibits a 12.8°C decrease in temperature. If Sample 1 has a mass of 1.42 g and Sample 2 has a mass of 2.70 g, what is the specific heat of Sample 1? Assume no heat is lost to the surroundings.

762. Two samples are placed in contact, and heat flows from the warmer sample to the cooler one. Sample 1, the warmer sample, exhibits a 15.0°C decrease in temperature. The initial temperature of Sample 2 was 25.1°C. Sample 1 has a specific heat of 0.581 cal/g°C, and Sample 2 has a specific heat of 0.381 cal/g°C. If Sample 1 has a mass of 5.13 g and Sample 2 has a mass of 4.19 g, what is the final temperature of Sample 2? Assume no heat is lost to the surroundings.

Heats of Formation

763–771 Complete the calculations related to heats of formation. Use the following table for these problems.

Standard Heats of Formation of Selected Substances

Substance	Formula	ΔH°_f (kJ/mol)
Ammonia	$NH_3(g)$	−46
Boron oxide	$B_2O_3(s)$	−1,274
Carbon dioxide	$CO_2(g)$	−394
Carbon monoxide	$CO(g)$	−111
Chlorine	$Cl_2(g)$	0*
Diamond	$C(dia)$	2
Ethyl alcohol	$C_2H_5OH(l)$	−277
Glucose	$C_6H_{12}O_6(s)$	−1,273
Graphite	$C(gr)$	0*
Nitrogen oxide	$NO(g)$	91
Oxygen	$O_2(g)$	0*
Water	$H_2O(l)$	−286
Water vapor	$H_2O(g)$	−242

The heat of formation for all elements in their standard state is, by definition, exactly zero.

763. Use standard heats of formation to determine the heat of reaction (ΔH°_{rxn}) for the following reaction:

$$C(gr) \rightarrow C(dia)$$

764. Use standard heats of formation to determine the heat of reaction (ΔH°_{rxn}) for the following reaction:

$$C(dia) + O_2(g) \rightarrow CO_2(g)$$

765. Use standard heats of formation to determine the heat of reaction (ΔH°_{rxn}) for the following reaction:

$$2CO(g) + O_2(g) \rightarrow 2CO_2(g)$$

766. Use standard heats of formation to determine the heat of reaction (ΔH°_{rxn}) for the combustion of ammonia:

$$4NH_3(g) + 5O_2(g) \rightarrow 4NO(g) + 6H_2O(g)$$

767. Use standard heats of formation to determine the heat of reaction (ΔH°_{rxn}) for the combustion of ethyl alcohol:

$$C_2H_5OH(l) + 3O_2(g) \rightarrow 2CO_2(g) + 3H_2O(l)$$

768. Use standard heats of formation to determine the heat of reaction ($\Delta H°_{rxn}$) for the combustion of glucose:

$$C_6H_{12}O_6(s)+6O_2(g)\rightarrow 6CO_2(g)+6H_2O(l)$$

769. The heat of reaction for the following reaction is −76 kJ:

$$2NO(g)+Cl_2(g)\rightarrow 2NOCl(g)$$

Determine the standard heat of formation for nitrosyl chloride (NOCl).

770. The heat of reaction for the combustion of propane is −2,045 kJ. This reaction is

$$C_3H_8(g)+5O_2(g)\rightarrow 3CO_2(g)+4H_2O(g)$$

Determine the standard heat of formation of propane.

771. If the heat of reaction for the following reaction is −9,090. kJ, what is the standard heat of formation of pentaborane-9 (B_5H_9)?

$$2B_5H_9(g)+12O_2(g)\rightarrow 5B_2O_3(s)+9H_2O(l)$$

Enthalpy Changes with Hess's Law

772–781 Practice using Hess's law. This law says that the overall enthalpy change of a process is the sum of the individual enthalpy changes of the steps involved in the process.

772. Determine the heat of reaction for the following reaction using Hess's Law:

$$CuCl_2(s)+Cu(s)\rightarrow 2CuCl(s)$$

Use these thermochemical equations to find the answer:

$$2Cu(s)+Cl_2(g)\rightarrow 2CuCl(s) \qquad \Delta H=-36 \text{ kJ}$$
$$CuCl_2(s)\rightarrow Cu(s)+Cl_2(g) \qquad \Delta H=206 \text{ kJ}$$

773. Determine the heat of reaction for the following reaction using Hess's law:

$$2H_2O(l)+2F_2(g)\rightarrow O_2(g)+4HF(g)$$

Use these thermochemical equations to find the answer:

$$H_2(g)+F_2(g)\rightarrow 2HF(g) \qquad \Delta H=-542 \text{ kJ}$$
$$2H_2O(l)\rightarrow 2H_2(g)+O_2(g) \qquad \Delta H=572 \text{ kJ}$$

774. Determine the heat of reaction for the following reaction using Hess's law:

$$2CO(g) + O_2(g) \rightarrow 2CO_2(g)$$

Use these thermochemical equations to find the answer:

$$C(gr) + O_2(g) \rightarrow CO_2(g) \qquad \Delta H = -394 \text{ kJ}$$
$$2CO(g) \rightarrow 2C(gr) + O_2(g) \qquad \Delta H = 222 \text{ kJ}$$

775. Determine the heat of reaction for the following reaction using Hess's law:

$$C_2H_6(g) + \frac{7}{2}O_2(g) \rightarrow 2CO_2(g) + 3H_2O(l)$$

Use these thermochemical equations to find the answer:

$$C_2H_6(g) \rightarrow C_2H_4(g) + H_2(g) \qquad \Delta H = 136 \text{ kJ}$$
$$H_2(g) + \frac{1}{2}O_2(g) \rightarrow H_2O(l) \qquad \Delta H = -286 \text{ kJ}$$
$$C_2H_4(g) + 3O_2(g) \rightarrow 2CO_2(g) + 2H_2O(l) \qquad \Delta H = -1,411 \text{ kJ}$$

776. Determine the heat of reaction for the following reaction using Hess's law:

$$Mg(s) + H_2(g) + O_2(g) \rightarrow Mg(OH)_2(s)$$

Use these thermochemical equations to find the answer:

$$MgO(s) + H_2O(l) \rightarrow Mg(OH)_2(s) \qquad \Delta H = -37 \text{ kJ}$$
$$2Mg(s) + O_2(g) \rightarrow 2MgO(s) \qquad \Delta H = -1,204 \text{ kJ}$$
$$2H_2O(l) \rightarrow 2H_2(g) + O_2(g) \qquad \Delta H = 572 \text{ kJ}$$

777. Determine the heat of reaction for the following reaction using Hess's law:

$$4NH_3(g) + 5O_2(g) \rightarrow 4NO(g) + 6H_2O(g)$$

Use these thermochemical equations to find the answer:

$$N_2(g) + 3H_2(g) \rightarrow 2NH_3(g) \qquad \Delta H = -92 \text{ kJ}$$
$$2H_2(g) + O_2(g) \rightarrow 2H_2O(g) \qquad \Delta H = -484 \text{ kJ}$$
$$N_2(g) + O_2(g) \rightarrow 2NO(g) \qquad \Delta H = 181 \text{ kJ}$$

778. Determine the heat of reaction for the following reaction using Hess's law:

$$CH_4(g) + NH_3(g) \rightarrow HCN(g) + 3H_2(g)$$

Use the following thermochemical equations to find the answer:

$$H_2(g) + 2C(gr) + N_2(g) \rightarrow 2HCN(g) \qquad \Delta H = 271 \text{ kJ}$$
$$C(gr) + 2H_2(g) \rightarrow CH_4(g) \qquad \Delta H = -75 \text{ kJ}$$
$$N_2(g) + 3H_2(g) \rightarrow 2NH_3(g) \qquad \Delta H = -92 \text{ kJ}$$

779. Determine the heat of reaction for the following reaction using Hess's law:

$$CH_4(g) + 4Cl_2(g) \rightarrow CCl_4(g) + 4HCl(g)$$

Use the following thermochemical equations to find the answer:

$$H_2(g) + Cl_2(g) \rightarrow 2HCl(g) \qquad \Delta H = -92 \text{ kJ}$$
$$C(gr) + 2Cl_2(g) \rightarrow CCl_4(g) \qquad \Delta H = -96 \text{ kJ}$$
$$C(gr) + 2H_2(g) \rightarrow CH_4(g) \qquad \Delta H = -75 \text{ kJ}$$

780. Determine the heat of reaction for the following reaction using Hess's law:

$$CO(g) + 2H_2(g) \rightarrow CH_3OH(l)$$

Use the following thermochemical equations to find the answer:

$$H_2(g) + \frac{1}{2}O_2(g) \rightarrow H_2O(l) \qquad \qquad \Delta H = -286 \text{ kJ}$$

$$C(gr) + \frac{1}{2}O_2(g) \rightarrow CO(g) \qquad \qquad \Delta H = -111 \text{ kJ}$$

$$C(gr) + O_2(g) \rightarrow CO_2(g) \qquad \qquad \Delta H = -394 \text{ kJ}$$

$$CH_3OH(l) + \frac{3}{2}O_2(g) \rightarrow CO_2(g) + 2H_2O(l) \qquad \qquad \Delta H = -727 \text{ kJ}$$

781. Determine the heat of reaction for the following reaction using Hess's law:

$$3NO_2(g) + H_2O(l) \rightarrow 2HNO_3(aq) + NO(g)$$

Use these thermochemical equations to find the answer:

$$2NO(g) + O_2(g) \rightarrow 2NO_2(g) \qquad \qquad \Delta H = -173 \text{ kJ}$$

$$2N_2(g) + 5O_2(g) + 2H_2O(l) \rightarrow 4HNO_3(aq) \qquad \qquad \Delta H = -255 \text{ kJ}$$

$$N_2(g) + O_2(g) \rightarrow 2NO(g) \qquad \qquad \Delta H = 181 \text{ kJ}$$

Chapter 12

Gases

....

Most of the gas laws depend upon various combinations of volume *(V)*, temperature *(T)*, pressure *(P)*, and moles *(n)*. In some cases, all four are important; in others, one or two are held constant. Although a number of pressure or volume units will work, all temperatures used in the calculations must be in kelvins. Graham's law is the only law in this chapter that doesn't depend upon *V*, *T*, *P*, or *n*.

The Problems You'll Work On

In this chapter, you work with gases in the following ways:

- ✔ Converting units of pressure
- ✔ Using Boyle's law, Charles's law, and Gay-Lussac's law
- ✔ Finding pressure, volume, and temperature with the combined gas law
- ✔ Relating volume to moles with Avogadro's law
- ✔ Working with the ideal gas law
- ✔ Finding partial pressures with Dalton's law
- ✔ Exploring effusion rates with Graham's law
- ✔ Doing stoichiometry calculations

Note: The Appendix includes a periodic table for reference.

What to Watch Out For

Remember the following when working on gases:

- ✔ Know the basic forms of the gas laws. Know which gas laws are direct relationships and which are inverse relationships.
- ✔ All temperatures used in calculations must be in kelvins. A negative kelvin temperature is impossible.
- ✔ The conversion 22.4 L/mol only works for gases at standard temperature and pressure (STP).

Converting Pressure Units

782–789 Convert between the pressure units.

782. How many torr are in 2.5 atm of pressure?

783. How many kilopascals are in 0.75 atm of pressure?

784. How many atmospheres are in 528 mm Hg?

785. The pressure in a car tire is 38.7 psi. What is the pressure in atmospheres?

786. A vessel is pressurized to 5.00 atm. What is the pressure in millimeters of mercury?

787. How many millimeters of mercury are in 1,050 torr?

788. How many torr are in 3.11 kPa?

789. The pressure exerted by a gas in a bicycle tire is 50.0 psi. How many kilopascals is that?

Boyle's Law

790–799 Perform the calculations using Boyle's law. This law may be implemented as $P_1V_1 = P_2V_2$ when the gas's temperature is held constant.

790. The pressure of 5.0 L of a gas changes from 3.0 atm to 10.0 atm while the temperature remains constant. What is the gas's new volume in liters?

791. The original pressure of a gas was 765 torr, and the volume changed from 17.5 L to 12.5 L while the temperature remained constant. What is the gas's new pressure in torr?

792. The pressure of 44.8 L of a gas changes from 0.75 atm to 0.25 atm while the temperature remains constant. What is the gas's new volume in liters?

793. As the volume of a gas changes from 547 mL to 861 mL, the temperature remains constant. If the original pressure was 1.75 atm, what is the gas's new pressure in atmospheres?

794. The pressure of a gas changes from 95.0 kPa to 211 kPa, and the volume changes to 45.0 mL. What was the gas's original volume in milliliters if the temperature remained constant during the process?

795. A gas's volume changes from 2,645 mL to 379 mL while the temperature remains constant. If the final pressure is 5.10 atm, what was the gas's original pressure in torr?

796. The pressure of a gas increased from 1,020 torr to 7,660 torr while the temperature was held constant. If the final volume of the gas is 0.210 L, what was the original volume in milliliters?

797. The volume of a gas decreased from 1.00 L to 600. mL while the temperature remained constant. If the final pressure is 720. torr, what was the gas's initial pressure in atmospheres?

798. The pressure of a gas changes from 1.50 atm to 540 torr, and the volume changes to 730 mL. What was the gas's initial volume in liters if the temperature remained constant?

799. The pressure of a gas changes from 920 mm Hg to 2.5 atm while the temperature remains constant. If the final volume is 0.34 L, what was the gas's original volume in milliliters?

Charles's Law

800–809 Complete the calculations using Charles's law. You can implement this law as $\frac{V_1}{T_1} = \frac{V_2}{T_2}$ (where T is in kelvins and pressure is held constant).

800. If a 4.2-L sample of a gas is heated from 200. K to 400. K while the pressure remains constant, what is the new volume in liters?

801. A gas sample has a volume of 473 mL after being heated from 20. K to 40. K at constant pressure. What was the gas's initial volume in milliliters?

802. A gas is cooled from 323 K to 223 K while the pressure is held constant. If the final volume is 2.50 L, what was the gas's original volume in liters?

803. A 20.0-L gas sample is held at a constant pressure while the temperature drops from 300. K to 75.0 K. What is the gas's new volume in liters?

804. While the pressure is held constant, a 50.0-mL sample of a gas is cooled from 100.0°C to 50.0°C. What is the gas's new volume in milliliters?

805. While the pressure is held constant, the volume of a gas sample changes from 350. mL to 0.100 L. What is the final temperature in kelvins if the original temperature was 127°C?

806. If 981 mL of a gas is heated from 335 K at constant pressure and the volume increases to 1,520 mL, what will be the new temperature in kelvins?

807. The volume of a gas decreased from 2.63 L to 627 mL, and the final temperature is 275 K. If the pressure remained constant, what was the initial temperature of the gas in kelvins?

808. The volume of a gas changed from 90.0 mL to 10.0 mL while the pressure was held constant. If the final temperature of the gas is –35°C, what was the initial temperature in kelvins?

809. The volume of a gas increases from 750 mL to 3.0 L. If the pressure was held constant and the starting temperature was 95°C, what will be the new temperature in kelvins?

Gay-Lussac's Law

810–819 Complete the calculations using Gay-Lussac's law. You can implement this law as $\frac{P_1}{T_1} = \frac{P_2}{T_2}$ (where T is in kelvins and the volume is held constant).

810. The temperature of a gas rises from 300. K to 450. K while the volume remains constant. If the original pressure is 900. torr, what will the final pressure be in torr?

811. While the temperature of a gas changes from 500. K to 300. K, the volume doesn't change. If the initial pressure was 2.5 atm, what is the gas's final pressure in atmospheres?

812. The temperature of a gas sample decreases from 425 K to 225 K, and the final pressure is 675 torr. If the volume is held constant, what was the gas's initial pressure in torr?

813. The temperature of a gas sample increases from 315 K to 505 K, and the final pressure is 8.10 atm. If the volume is held constant, what was the gas's initial pressure in atmospheres?

814. A gas sample is cooled from 310 K to 280 K while the volume is held constant. What will be the final pressure in kilopascals if the original pressure is 470 kPa?

815. A gas starts at 801°C with a pressure of 780. mm Hg and reaches a pressure of 1,280 mm Hg. If the volume remains constant, what is the final temperature in kelvins?

816. A gas starts at 100.°C with a pressure of 1.00 atm and reaches a pressure of 2.50 atm. If the volume is held constant, what is the final temperature of the gas in degrees Celsius?

817. The final temperature of a gas sample is 521 K, and the pressure changes from 428 torr to 1.25 atm. If the volume remains constant, what was the original temperature in kelvins?

818. The final temperature of a gas sample is 175 K, and the pressure changes from 44.1 psi to 2,028 torr. If the volume is held constant, what was the original temperature in degrees Celsius?

819. The initial temperature of a gas sample was 75°C, and the pressure changes from 0.612 atm to 34.9 in. Hg. If the volume is held constant, what is the final temperature in degrees Celsius?

The Combined Gas Law

820–829 Complete the calculations using the combined gas law, which may be represented as $\frac{P_1 V_1}{T_1} = \frac{P_2 V_2}{T_2}$ (where T is in kelvins and the number of moles is constant).

820. Solve for the missing value:

P_1 = 2.00 atm	P_2 = ? atm
V_1 = 75.0 mL	V_2 = 125 mL
T_1 = 223 K	T_2 = 273 K

821. Solve for the missing value:

P_1 = 970 torr	P_2 = 760 torr
V_1 = 4.1 L	V_2 = ? L
T_1 = 273 K	T_2 = 400. K

822. Solve for the missing value:

$P_1 = ?$ atm	$P_2 = 5.1$ atm
$V_1 = 327$ mL	$V_2 = 188$ mL
$T_1 = 323$ K	$T_2 = 544$ K

823. Solve for the missing value:

$P_1 = 280$ mm Hg	$P_2 = 760$ mm Hg
$V_1 = ?$ mL	$V_2 = 50.$ mL
$T_1 = 520$ K	$T_2 = 220$ K

824. A gas sample that has a pressure of 4.23 atm, a volume of 1,870 mL, and a temperature of 293 K is allowed to expand to a volume of 6.01 L with a final temperature of 373 K. What is the final pressure of the gas in atmospheres?

825. A sample of a gas has a volume of 10.0 L at a pressure of 450 torr and a temperature of 773 K. What will the final volume be in liters if the conditions are changed to standard temperature and pressure (STP)?

826. The original volume of a gas sample was 0.75 L at a temperature of 25°C. The final volume, temperature, and pressure are 2,647 mL, 100.°C, and 5.5 atm, respectively. What was the original pressure in atmospheres?

827. The original pressure of a gas sample was 3.00 atm at a temperature of 300.°C. The final volume, temperature, and pressure are 594 mL, 200.°C, and 3,862 mm Hg, respectively. What was the original volume in milliliters?

828. A gas sample had an initial pressure, volume, and temperature of 970 torr, 220. mL, and 50. K, respectively. Its final pressure is 0.78 atm, and its final volume is 2.4 L. What is the gas's final temperature in kelvins?

829. A gas sample had an initial pressure of 35.3 psi and an initial volume of 10.0 L. Its final pressure is 6.18 atm, its final volume is 4,290 mL, and its final temperature is 572 K. What was the initial temperature in kelvins?

Avogadro's Law

*830–835 Complete the calculations using Avogadro's law, which states that the volume of a gas is directly proportional to the number of moles: $V \propto n$. Using Avogadro's law, you can determine that at standard temperature and pressure (273 K and 1 atm), 1 mol of gas will occupy 22.4 L. **Note:** Assume that this volume is an exact value for the following problems.*

830. In liters, what is the volume of a sample of helium gas that has a mass of 100. g at standard temperature and pressure?

831. In liters, what is the volume of a sample of $CO_2(g)$ that has a mass of 25.0 g at standard temperature and pressure?

832. How many molecules of nitrogen gas are present in 9.14 L of $N_2(g)$ at standard temperature and pressure?

833. How many grams are in a 37.89-L sample of methane, CH_4, at standard temperature and pressure?

834. A 55.0-L gas sample contains 2.10 mol of dinitrogen tetroxide (N_2O_4) at 1.0 atm of pressure and 298 K. If the sample completely decomposes according to the following reaction, how many liters of NO_2 are produced?

$$N_2O_4(g) \rightarrow 2NO_2(g)$$

835. At a certain temperature and pressure, a 12.4-L sample containing 0.296 mol of ozone (O_3) reacts completely according to the following reaction. How many liters of O_2 are produced?

$$2O_3(g) \rightarrow 3O_2(g)$$

The Ideal Gas Law

836–845 Complete the calculations using the ideal gas law, which says PV = nRT (where P is pressure in atmospheres, V is volume in liters, n is the number of moles, R is a constant equal to 0.0821 L·atm/ K·mol, and T is temperature in kelvins).

836. What is the pressure in atmospheres of a 5.00-mol gas sample that occupies 12.5 L at 400. K?

837. What is the volume, in liters, of 8.91 mol of a gas at a pressure of 0.747 atm at 200. K?

838. How many moles of helium are in a gas sample that has a volume of 500.0 mL at a pressure of 3.00 atm and a temperature of 298 K?

839. What is the temperature, in kelvins, of 0.60 mol of a gas that occupies 2.8 L at a pressure of 4.1 atm?

840. How much space, in liters, would 40.6 mol of carbon dioxide gas occupy at a temperature of 205°C and a pressure of 1.50 atm?

841. What is the temperature, in kelvins, of a 0.333-mol sample of a gas that has a pressure of 6.98 psi and a volume of 11.2 L?

842. How many moles of oxygen gas occupy 3,050 mL at a pressure of 672 torr and 75°C?

843. What is the pressure, in atmospheres, of 0.618 g of chlorine gas (Cl_2) if it occupies a volume of 75.0 mL at a temperature of 255 K?

844. What is the molar mass of a gas if 261 g of the gas occupies 555 L at a pressure of 0.90 atm and a temperature of 373 K?

845. What is the molar mass of a gas that has a mass of 12.15 g at a pressure of 1.05 atm, a volume of 11.3 L, and a temperature of 523 K?

Dalton's Law of Partial Pressures

846–854 Complete the calculations using Dalton's law of partial pressures, which says that the total pressure is the sum of the individual partial pressures of the components of a gas mixture.

846. What is the total pressure of a mixture of gases containing N_2, O_2, and He with the following partial pressures?

$$P_{N_2} = 255 \text{ torr}$$
$$P_{O_2} = 491 \text{ torr}$$
$$P_{He} = 101 \text{ torr}$$

847. Samples of Ar, Cl_2, and F_2 are mixed together. If their individual pressures are as follows, what is the total pressure, in atmospheres, exerted by the gases?

$$P_{Ar} = 1.52 \text{ atm}$$
$$P_{Cl_2} = 567 \text{ torr}$$
$$P_{F_2} = 843 \text{ torr}$$

848. Gases P_x, P_y, and P_z exist in a ratio of 7:3:2 in a gas sample. If the total pressure is 5.12 atm, what is the partial pressure of P_x in atmospheres?

849. A mixture of SO_3, SO_2, and O_2 has a total pressure of 4.42 atm. The partial pressure of SO_3 is 1.77 atm, and the partial pressure of SO_2 is 1.02 atm. What is the partial pressure of O_2 in atmospheres?

850. A mixture of gases contains butane, ethane, and propane with a total pressure of 1,482 torr. The mixture contains 0.822 mol of butane, 0.282 mol of ethane, and 0.550 mol of propane at the same temperature. What is the partial pressure of the propane in torr?

851. When 15.0 g of N_2 gas and 20.0 g of O_2 gas are added to a 10.0-L container at 25°C, what is the partial pressure of the N_2 gas in atmospheres?

852. When 15.0 g of N_2 gas and 20.0 g of O_2 gas are added to a 10.0-L container at 25°C, what is the total pressure, in atmospheres, that the gases exert on the container?

853. When a sample of ammonium nitrite is heated, it decomposes according to the following reaction to form 4.25 L of nitrogen gas (collected over water) at 26°C and 757.0 torr. What is the pressure due to the nitrogen gas, and how many grams of ammonium nitrite were used?

$$NH_4NO_2(g) \rightarrow N_2(g) + 2H_2O(g)$$

Note: The vapor pressure of water at 26°C is 25.21 torr.

854. Oxygen gas is collected over water during the thermal decomposition of potassium chlorate at a pressure of 762.0 torr and a temperature of 22°C. What is the pressure, in torr, due to the oxygen if 242 mL of oxygen gas is formed? And how many grams of potassium chloride remain after the reaction is complete? **Note:** At 22°C, the vapor pressure of water is 19.83 torr.

Graham's Law

855–857 Answer the questions on Graham's law.

855. At the same temperature and pressure, which gas, Ar or O_2, will diffuse faster and why?

856. At the same temperature and pressure, which gas, NH_3 or CO_2, will diffuse more slowly and why?

857. Given equal amounts of SO_2 and ClO_2 at the same temperature and pressure, which gas effuses faster and by how much?

Gas Stoichiometry

858–861 Complete the calculations using gas stoichiometry.

858. Excess chlorine gas reacts with 60.9 L of hydrogen gas to form hydrogen chloride gas according to the following equation:

$$H_2(g) + Cl_2(g) \rightarrow 2HCl(g)$$

What is the maximum volume of hydrogen chloride gas that can be produced at standard temperature and pressure?

859. When mercury(II) oxide is heated, liquid mercury and oxygen gas are the products. What is the maximum mass of mercury that can be produced if 100. L of oxygen gas is collected at standard temperature and pressure?

860. Liquid hydrogen peroxide decomposes to form liquid water and oxygen gas. How many liters of oxygen gas can be produced if 862 g of hydrogen peroxide completely decomposes at standard temperature and pressure?

861. Nitrogen gas reacts with hydrogen gas to form ammonia. What is the maximum volume of ammonia gas at standard temperature and pressure that can be produced from 25.0 L of nitrogen gas and 50.0 L of hydrogen gas?

Chapter 13

Solutions (The Chemistry Kind)

· ·

Solutions contain a solvent and one or more solutes. The *concentration* is the quantity of solute in a given amount of solution. Concentrations are important when considering colligative properties, which depend on the number of solute particles present, not on their identities. Freezing point depression, boiling point elevation, and osmotic pressure are examples of colligative properties.

The Problems You'll Work On

In this chapter, you work with solutions in the following ways:

- ✔ Identifying solution components
- ✔ Calculating small concentrations (ppt, ppm, ppb)
- ✔ Diluting solutions
- ✔ Finding molarity and molality
- ✔ Determining freezing point depression and boiling point elevation
- ✔ Finding osmotic pressure

Note: For reference, you can find the periodic table in the Appendix.

What to Watch Out For

Remember the following when working on solutions:

- ✔ Water is a very important solvent, but it isn't the only solvent.
- ✔ Conversion from one concentration unit to another is easier through dimensional analysis.
- ✔ Molality is the only concentration unit that has the solute in the denominator.
- ✔ Molality *(m)*, not molarity *(M)*, is used for freezing point depression and boiling point elevation.

✔ All colligative properties depend on the total particles present in the solution, not their identities.

✔ The freezing point of a solution is the normal freezing point of the solvent minus the depression value. The boiling point of a solution is the normal boiling point of the solvent plus the elevation value.

Solutions, Solvents, and Solutes

862–871 The following questions concern terms associated with solutions.

862. For all solutions, what is a solute?

863. For all solutions, what is a solvent?

864. Air is a solution. What is the solvent?

865. High-carbon steel is a solution. What is the solvent?

866. Vodka is a solution. A sample of 80-proof vodka is 40% alcohol. What is the solvent?

867. Some sodium chloride is added to 100 mL of water. The NaCl completely dissolves. Additional NaCl is added until no more of the solid dissolves. The solution precipitates out the undissolved solid. What is the final type of solution formed?

868. Some sodium bromide is added to 250 mL of water. The NaBr completely dissolves. Additional NaBr is added until no more of the solid dissolves. The solution is separated from the undissolved solid and 250 mL of water is added to the solution. What is the final type of solution formed?

869. You're given a sodium acetate ($NaC_2H_3O_2$) solution. A friend drops a small crystal of sodium acetate into the solution, and immediately the solution turns white as a large amount of precipitate forms. What type of solution was the original solution?

870. If you know the molarity of a solution, what additional information is necessary to determine the moles of solute present?

871. If you know the molality of a solution, what additional information is necessary to determine the moles of solute present?

Concentration Calculations

872–880 *Calculate the concentration of the solution.*

872. A solution contains 1.5×10^{-2} g of lead(II) ions, Pb^{2+}, in 1,275 mL of solution. What is the concentration of lead in parts per thousand (ppt)? Assume the density of the solution is 1.00 g/mL.

873. A solution contains 5.5×10^{-3} g of the pesticide DDT in 2,125 mL of solution. What is the concentration of DDT in parts per million (ppm)? Assume the density of the solution is 1.00 g/mL.

874. A solution contains 8.5×10^{-8} mol of mercury in 1.5 L of solution. What is the concentration of mercury in parts per billion (ppb)? Assume the density of the solution is 1.00 g/mL.

875. A 0.200-mol sample of sodium nitrate, $NaNO_3$, is dissolved in sufficient water to produce 2.5 L of solution. What is the molarity of the solution?

876. A 0.1250-mol sample of sodium hydroxide, NaOH, is dissolved in sufficient water to produce 1,725 mL of solution. What is the molarity of the solution?

877. What is the molarity of a solution formed by dissolving 26 g of NaCl (molar mass = 58 g/mol) in sufficient water to produce 1.5 L of solution?

878. How many moles of lithium chloride, LiCl, are in 3.5 L of a 0.16-M LiCl solution?

879. How many liters of a 0.25-M sulfuric acid (H_2SO_4) solution are needed to supply 15 g of sulfuric acid? The molar mass of sulfuric acid is 98 g/mol.

880. How many grams of magnesium chloride, $MgCl_2$, are in 1.75 L of a 0.564-M $MgCl_2$ solution? The molar mass of magnesium chloride is 95.2 g/mol.

Dilution

881–888 Answer the dilution problems using $M_1V_1 = M_2V_2$

881. A student has 1.0 L of 12-M hydrochloric acid, HCl. She wants to dilute the solution to 1.5 M. How many liters of solution can she make?

882. How many liters of 2.0-M sodium hydroxide, NaOH, can you prepare from 0.50 L of 15-M NaOH?

883. What is the molarity of a solution formed by mixing 1.5 L of 6.0-M nitric acid, HNO_3, with water to get a final volume of 6.5 L?

884. What is the molarity of a solution formed by adding 15 mL of 6.0-M phosphoric acid, H_3PO_4, to 135 mL of water?

885. How many liters of a 12-M hydrochloric acid (HCl) solution are necessary to prepare 5.0 L of 1.0-M HCl?

886. How many milliliters of 14-M nitric acid, HNO_3, are necessary to prepare 875 mL of 3.0-M nitric acid?

887. Normal saline solution has a sodium chloride (NaCl) concentration of 0.154 M. How many liters of saline solution can you prepare from 1.00 L of 1.00-M NaCl?

888. How many liters of 0.250-M calcium chloride ($CaCl_2$) solution can you prepare from 5.00 L of 1.00-M $CaCl_2$ solution?

Molality

889–894 Do the following molality problems.

889. What is the molality of a solution containing 0.25 mol of potassium bromide, KBr, in 1.25 kg of water?

890. What is the molality of a solution containing 0.014 mol of calcium nitrate, $Ca(NO_3)_2$, in 1,775 g of water?

891. What is the molality of a solution containing 10.5 g of hydrogen chloride, HCl, in 0.750 kg of water? The molar mass of HCl is 36.5 g/mol.

892. How many moles of zinc chloride, $ZnCl_2$, are in a 0.16-m solution containing 1.5 kg of water?

893. How many grams of cadmium chloride, $CdCl_2$, are in a 0.015-m cadmium chloride solution if 2.5 kg of solvent is present? The molar mass of cadmium chloride is 183.317 g/mol.

894. A solution of glucose, $C_6H_{12}O_6$, was prepared by adding some of this nonelectrolyte to 175 g of water. Based on a freezing point depression experiment, the concentration of the solution was 0.125 m. How many grams of glucose were in the solution? The molar mass of glucose is 180.2 g/mol.

Colligative Properties

895–911 Answer the questions on colligative properties.

895. Calcium nitrate, $Ca(NO_3)_2$, is a strong electrolyte. What is the van't Hoff factor, *i*, for a calcium nitrate solution?

896. Chloric acid, $HClO_3$, is a strong electrolyte, and chlorous acid, $HClO_2$, is a weak electrolyte. The van't Hoff factor of a dilute chloric acid solution is 2. How does the van't Hoff factor of a dilute chlorous acid solution compare?

897. What is the freezing point depression (ΔT) of a 1.50-m solution of sucrose in water? Sucrose is a nonelectrolyte, and the freezing point depression constant of water is 1.86°C/m.

898. What is the freezing point depression (ΔT) of a 12.6-m solution of ethylene glycol (antifreeze) in water? Ethylene glycol is a nonelectrolyte, and the freezing point depression constant of water is 1.86°C/m.

899. Isopropyl alcohol, C_3H_7OH, is a nonelectrolyte. A 0.200-mol sample of isopropyl alcohol is dissolved in 1.00 kg of water. If the boiling point elevation constant (K_b) of water is 0.512°C/m, what is the boiling point elevation (ΔT) of the solution?

900. Propyl alcohol, C_3H_7OH, is a nonelectrolyte. A 0.370-mol sample of propyl alcohol is dissolved in 1.75 kg of water. If the boiling point elevation constant (K_b) of water is 0.512°C/m, what is the boiling point elevation (ΔT) of the solution?

901. Ethyl alcohol, C_2H_5OH, is a water-soluble nonelectrolyte. A solution is prepared by adding 17.0 g of ethyl alcohol to 0.750 kg of water. What is the freezing point depression (ΔT) for this solution? The molar mass of ethyl alcohol is 46.1 g/mol, and the freezing point depression constant (K_f) for water is 1.86°C/m.

902. Propylene glycol, $C_3H_6(OH)_2$, is a water-soluble nonelectrolyte used in some antifreezes. A 125-g sample of propylene glycol is dissolved in 1.25 kg of water. What is the boiling point of the solution? The molar mass of propylene glycol is 76.1 g/mol, and the boiling point elevation constant (K_b) of water is 0.512°C/m.

903. Hydrogen bromide, HBr, is a strong electrolyte in water. What is the boiling point elevation (ΔT) of a solution containing 125 g of HBr in 1.75 kg of water? The molar mass of HBr is 80.9 g/mol, and the boiling point elevation constant (K_b) of water is 0.512°C/m.

904. A 25.0-g sample of methyl alcohol, CH_3OH, is dissolved in 1.50 kg of water. What is the freezing point of the solution? The molar mass of methyl alcohol (a nonelectrolyte) is 32.0 g/mol, and the freezing point depression constant (K_f) for water is 1.86°C/m.

905. Rubidium chloride, RbCl, is a strong electrolyte. What is the freezing point of a solution formed by dissolving 15.0 g of RbCl in 0.125 kg of water? The molar mass of RbCl is 121 g/mol, and the freezing point depression constant (K_f) for water is 1.86°C/m.

906. Hexane, C_6H_{14}, is a nonelectrolyte. A solution contains 125 g of hexane in 1.50 kg of cyclohexane, C_6H_{12}. What is the boiling point of the solution? The molar mass of hexane is 86.2 g/mol, and the boiling point elevation constant (K_b) of cyclohexane is 2.79°C/m. The normal boiling point of cyclohexane is 80.70°C.

907. What is the osmotic pressure (Π) of a solution containing 0.125 mol of a nonelectrolyte in 0.500 L of solution at 25°C? $R = 0.0821$ L·atm/mol·K.

908. What is the osmotic pressure (Π) of a 1.25-M solution of a nonelectrolyte solution at 25°C? $R = 0.0821$ L·atm/mol·K.

909. Unlike many metallic compounds, mercury(II) chloride, $HgCl_2$, is a nonelectrolyte. What is the osmotic pressure (Π) of a solution containing 5.00 g of mercury(II) chloride in 0.275 L of solution at 45°C? The molar mass of mercury(II) chloride is 272 g/mol, and $R = 0.0821$ L·atm/mol·K.

910. An aqueous solution of a polymer contains 0.410 g of this compound in 1.00 L of solution. At 37°C, the osmotic pressure (Π) of the solution is 3.00×10^{-3} atm. What is the molar mass of the polymer? $R = 0.0821$ L·atm/mol·K.

911. Calcium nitrate, $Ca(NO_3)_2$, is a strong electrolyte. A solution contains 4.75 g of calcium nitrate in 125 mL of solution. What is the osmotic pressure (Π) of this solution at 37°C? The molar mass of calcium nitrate is 164 g/mol, and $R = 0.0821$ L·atm/mol·K.

Chapter 14

Acids and Bases

· ·

Simply knowing the concentration of an acid or base is often insufficient to understand its actions. The pH relates to the concentration of hydrogen ions in a solution, and the pOH relates to the concentration of the hydroxide ions in the solution. Acids and bases may be strong or weak. There are only a few strong acids and bases, and calculations involving them are always simpler than calculations involving weak acids or bases. A buffer solution provides a means to control the pH of a solution.

The Problems You'll Work On

In this chapter, you work with acids and bases in the following ways:

- ✔ Describing acids and bases
- ✔ Identifying conjugate acids and conjugate bases
- ✔ Determining pH and pOH for strong and weak acid or base solutions
- ✔ Doing stoichiometry of titrations
- ✔ Completing buffer problems

Note: See the Appendix if you need to check the periodic table.

What to Watch Out For

Remember the following when working on acids and bases:

- ✔ Know the strong acids and strong bases.
- ✔ Dimensional analysis helps in calculating the concentration.
- ✔ Use the stoichiometry when doing calculations.
- ✔ At the equivalence point, the concentrations of the reactants are zero (other than through hydrolysis).
- ✔ Calculations involving weak acids or bases always involve an equilibrium constant, K_a or K_b. Strong acids and bases do not have K's.

✔ Either a K or the Henderson-Hasselbalch equation works for a buffer solution.

✔ Concentrations have units; pH and K's do not.

✔ There's only one way to write a K_a or a K_b expression; any variation is wrong.

Identifying Acids and Bases

912–917 Identify acids and bases.

912. Which member of the following set is a strong acid?

HNO_2, HNO_3, NH_3, KOH, CH_3OH

913. Which member of the following set is a strong base?

HNO_2, HNO_3, NH_3, KOH, CH_3OH

914. Which member of the following set is a weak acid?

HNO_2, HNO_3, NH_3, KOH, CH_3OH

915. Which member of the following set is a weak base?

HNO_2, HNO_3, NH_3, KOH, CH_3OH

916. Which member of the following set is neither an acid nor a base?

HNO_2, HNO_3, NH_3, KOH, CH_3OH

917. Which member of the following set is a weak base?

HNO_2, HNO_3, CH_3NH_2, $CsOH$, C_2H_5OH

Conjugate Acids and Bases

918–924 Identify conjugate acids and conjugate bases.

918. What is the conjugate base of HCl?

919. What is the conjugate acid of $C_2H_3O_2^-$?

920. What is the conjugate acid of NH_3?

921. What is the conjugate base of NH_3?

922. What is the conjugate base of H_3PO_4?

923. What is the conjugate acid of H_2SO_4?

924. Which of the following substances may serve both as a conjugate acid and as a conjugate base?

Cl^-, HPO_4^{2-}, Fe^{3+}, AsO_4^{3-}, Mg^{2+}

Finding pH and pOH of Strong Acids and Bases

925–932 Determine the pH or pOH of a strong acid or base solution.

925. What is the pH of a 0.01-M HNO_3 solution?

926. What is the pOH of a 0.001-M KOH solution?

927. What is the pH of a 0.015-M NaOH

solution?

928. What is the pOH of a 0.0025-M HCl solution?

929. What is the pH of a 0.003-M $Ba(OH)_2$ solution?

930. What is the pH of a 1.0×10^{-3} M $Sr(OH)_2$ solution?

931. What is the pH of a 1.0-M HBr solution?

932. What is the pH of a 15-M NaOH solution?

Finding pH and pOH of Weak Acids and Bases

933–947 Determine the pH or pOH of the weak acid or weak base solution.

933. What is the pH of a 1.0-M acetic acid ($HC_2H_3O_2$) solution? $K_a = 1.7 \times 10^{-5}$.

934. What is the pH of a 1.0-M nitrous acid (HNO_2) solution? $K_a = 5.0 \times 10^{-4}$.

935. What is the pOH of a 1.0-M ammonia (NH_3) solution? $K_b = 1.8 \times 10^{-5}$.

936. What is the pH of a 0.100-M hydrocyanic acid (HCN) solution? $K_a = 6.2 \times 10^{-10}$.

937. What is the pH of a 0.20-M acetic acid ($HC_2H_3O_2$) solution? $K_a = 1.7 \times 10^{-5}$.

938. What is the pH of a 0.0015-M periodic acid (HIO_4) solution? $K_a = 2.8 \times 10^{-2}$.

939. What is the pH of a 0.0025-M chlorous acid ($HClO_2$) solution? $K_a = 1.1 \times 10^{-2}$.

940. What is the pOH of a 1.0-M pyridine (C_5H_5N) solution? $K_b = 1.7 \times 10^{-9}$.

941. What is the pH of a 0.15-M ammonia (NH_3) solution? $K_b = 1.8 \times 10^{-5}$.

942. What is the pH of a 2.5-M methylamine (CH_3NH_2) solution? $K_b = 5.2 \times 10^{-4}$.

943. What is the pH of a 0.20-M chlorous acid ($HClO_2$) solution? $K_a = 1.1 \times 10^{-2}$.

944. What is the pH of a 0.015-M cyanic acid (HOCN) solution? $pK_a = 3.46$.

945. What is the pOH of a 1.0-M methylamine (CH_3NH_2), solution? $pK_b = 3.28$.

946. What is the pH of a 0.25-M sodium hydrogen sulfate ($NaHSO_4$) solution? The K_a of HSO_4^- is 1.1×10^{-2}.

947. What is the pH of a 0.50-M calcium acetate ($Ca(C_2H_3O_2)_2$) solution? The K_a of $HC_2H_3O_2$ is 1.7×10^{-5}.

Stoichiometry of Titrations

948–959 *Answer the questions on the stoichiometry of titrations.*

948. How many moles of hydrochloric acid (HCl) will react with 25.00 mL of 0.1000-M sodium hydroxide (NaOH) solution?

The reaction is

$$HCl(aq) + NaOH(aq) \rightarrow NaCl(aq) + H_2O(aq)$$

949. What is the molarity of a hypochlorous acid (HClO) solution if 25.00 mL of this solution reacts completely with 35.42 mL of a 0.1250-M potassium hydroxide (KOH) solution?

The reaction is

$$HClO(aq) + KOH(aq) \rightarrow KClO(aq) + H_2O(aq)$$

950. What is the molarity of a chlorous acid ($HClO_2$) solution if 25.00 mL of this solution reacts completely with 39.32 mL of a 0.1350-M sodium hydroxide (NaOH) solution?

The reaction is

$$HClO_2(aq) + NaOH(aq) \rightarrow NaOH(aq) + H_2O(aq)$$

951. How many moles of lithium hydroxide, LiOH, will react with 25.00 mL of 0.2500-M sulfuric acid (H_2SO_4) solution?

The reaction is

$$H_2SO_4(aq) + 2LiOH(aq) \rightarrow Li_2SO_4(aq) + 2H_2O(aq)$$

952. What is the molarity of a calcium hydroxide ($Ca(OH)_2$) solution if 25.00 mL of this solution reacts completely with 37.24 mL of a 0.1275-M nitric acid (HNO_3) solution?

The reaction is

$$2HNO_3(aq) + Ca(OH)_2(aq) \rightarrow$$
$$Ca(NO_3)_2(aq) + 2H_2O(aq)$$

953. What is the molarity of strontium hydroxide ($Sr(OH)_2$) solution if 25.00 mL of the solution reacts with 39.26 mL of a 0.2500-M phosphoric acid (H_3PO_4) solution?

The reaction is

$$2H_3PO_4(aq) + 3Sr(OH)_2(aq) \rightarrow$$
$$Sr_3(PO_4)_2(aq) + 6H_2O(aq)$$

954. How many moles of nitrous acid (HNO_2) will react with 35.00 mL of 0.1000-M potassium hydroxide (KOH)?

955. What is the molarity of a rubidium hydroxide (RbOH) solution if 25.00 mL of this solution reacts completely with 43.29 mL of a 0.1235-M sulfuric acid (H_2SO_4) solution?

956. What is the molarity of a phosphoric acid (H_3PO_4) solution if 100.0 mL of this solution titrates with 27.98 mL of a 0.1000-M calcium hydroxide ($Ca(OH)_2$) solution?

957. What is the molarity of an ammonia (NH_3) solution if 50.00 mL of the solution reacts with 47.98 mL of a 0.3215-M sulfuric acid (H_2SO_4) solution?

958. How many grams of acetic acid ($HC_2H_3O_2$) are in a vinegar sample if the sample requires 39.95 mL of 0.09527-M sodium hydroxide (NaOH) to reach the endpoint of a titration? The molar mass of acetic acid is 60.52 g/mol.

959. How many grams of ammonia (NH_3) are in a sample of household cleaner if the sample requires 37.86 mL of 0.08271-M hydrochloric acid (HCl) to reach the endpoint of a titration? The molar mass of ammonia is 17.031 g/mol.

Buffer Solutions

960–970 *The following questions deal with buffer solutions.*

960. Which of the following compounds would create a buffer when added to a sodium nitrite ($NaNO_2$) solution?

CH_3OH, KNO_2, NaOH, HNO_2, or KCl

961. Which of the following compounds would create a buffer when added to an ammonia (NH_3) solution?

CH_3OH, $(NH_4)_2SO_4$, NaOH, CH_3NH_2, or KCl

962. Which of the following compounds would create a buffer when added to a sodium carbonate (Na_2CO_3) solution?

CH_3OH, $KHCO_3$, NaOH, $CaCO_3$, or NaCl

963. What is the pH of a buffer solution that is 0.75 M in acetic acid ($HC_2H_3O_2$) and 0.50 M in sodium acetate? The pK_a of acetic acid is 4.76.

964. What is the pOH of a buffer solution that is 0.25 M in ammonia (NH_3) and 0.35 M in ammonium chloride (NH_4Cl)? The pK_b of ammonia is 4.75.

965. What is the pH of a buffer solution that is 0.25 M in hydrofluoric acid (HF) and 0.35 M in sodium fluoride (NaF)? The pK_a of hydrofluoric acid is 3.17.

966. What is the pOH of a buffer solution that is 0.75 M in methylamine (CH_3NH_2) and 0.50 M in methylammonium chloride (CH_3NH_3Cl)? The pK_b of methylamine is 3.28.

967. What is the pH of a buffer solution that is 0.35 M in ammonia (NH_3) and 0.25 M in ammonium chloride (NH_4Cl)? The pK_b of ammonia is 4.75.

968. What is the pH of a buffer solution that is 0.50 M in methylamine (CH_3NH_2) and 0.25 M in methylammonium chloride (CH_3NH_3Cl)? The pK_b of methylamine is 3.28.

969. What is the pH of a buffer solution formed by mixing 1.00 mol of hydrofluoric acid (HF) with 0.25 mol of sodium hydroxide (NaOH) in enough water to make a liter of solution? The pK_a of hydrofluoric acid is 3.17.

970. What is the pH of a buffer solution formed by mixing 0.75 mol of ammonia (NH_3) and 0.25 mol of hydrochloric acid (HCl) in enough water to make a liter of solution? The pK_b of ammonia is 4.75.

Titrations and pH Changes

971–980 Answer the questions on pH changes during titrations.

971. A student prepares to do a titration by adding 0.1000-M sodium hydroxide (NaOH) to 25.00 mL of a 0.08750-M acetic acid ($HC_2H_3O_2$) solution. What was the initial pH of the acetic acid solution? The pK_a of acetic acid is 4.76, and the reaction is

$$NaOH(aq) + HC_2H_3O_2(aq) \rightarrow$$
$$Na^+(aq) + C_2H_3O_2^-(aq) + H_2O(l)$$

972. A student prepares to do a titration by adding 0.1000-M hydrochloric acid (HCl) to 25.00 mL of a 0.08750-M ammonia (NH$_3$) solution. What was the initial pH of the ammonia solution? The pK_b of ammonia is 4.75, and the reaction is

$$NH_3(aq) + HCl(aq) \rightarrow NH_4^+(aq) + Cl^-(aq)$$

973. What is the pH at the equivalence point for the titration of 50.00 mL of a 0.5000-M sodium hydroxide (NaOH) solution with 0.7500-M hydrochloric acid (HCl)? The reaction is

$$NaOH(aq) + HCl(aq) \rightarrow Na^+(aq) + Cl^-(aq) + H_2O(l)$$

974. A student prepares to do a titration by adding 0.05000-M barium hydroxide (Ba(OH)$_2$) to 25.00 mL of a 0.08800-M acetic acid (HC$_2$H$_3$O$_2$) solution. What was the pH of the solution after he added 11.00 mL of base? The pK_a of acetic acid is 4.76, and the reaction is

$$Ba(OH)_2(aq) + 2HC_2H_3O_2(aq) \rightarrow$$
$$Ba^{2+}(aq) + 2C_2H_3O_2^-(aq) + 2H_2O(l)$$

975. A student prepares to do a titration by adding 0.05000-M sulfuric acid (H$_2$SO$_4$) to 25.00 mL of a 0.08800-M ammonia (NH$_3$) solution. What was the pH of the solution after she added 11.00 mL acid? The pK_b of ammonia is 4.75, and the reaction is as follows. Ignore hydrolysis of the sulfate ion.

$$2NH_3(aq) + H_2SO_4(aq) \rightarrow 2NH_4^+(aq) + SO_4^{2-}(aq)$$

976. A student prepares to do a titration by adding 0.05000-M barium hydroxide (Ba(OH)$_2$) to 25.00 mL of a 0.08800-M acetic acid (HC$_2$H$_3$O$_2$) solution. What was the pH of the solution after she added 20.00 mL of base? The pK_a of acetic acid is 4.76, and the reaction is

$$Ba(OH)_2(aq) + 2HC_2H_3O_2(aq) \rightarrow$$
$$Ba^{2+}(aq) + 2C_2H_3O_2^-(aq) + 2H_2O(l)$$

977. A student prepares to do a titration by adding 0.05000-M sulfuric acid (H$_2$SO$_4$) to 25.00 mL of a 0.08800-M ammonia (NH$_3$) solution. What is the pH of the solution after he adds 17.00 mL of acid? The pK_b of ammonia is 4.75, and the reaction is as follows. Ignore hydrolysis of the sulfate ion.

$$2NH_3(aq) + H_2SO_4(aq) \rightarrow 2NH_4^+(aq) + SO_4^{2-}(aq)$$

978. A student prepares to do a titration by adding 0.05000-M barium hydroxide $(Ba(OH)_2)$ to 25.00 mL of a 0.08800-M acetic acid $(HC_2H_3O_2)$ solution. What was the pH of the solution at the equivalence point? The pK_a of acetic acid is 4.76.

979. A student prepares to do a titration by adding 0.4800-M barium hydroxide $(Ba(OH)_2)$ to 25.00 mL of a 0.7600-M acetic acid $(HC_2H_3O_2)$ solution. What was the pH of the solution at the equivalence point? The pK_a of acetic acid is 4.76.

980. A student prepares to do a titration by adding 0.06000-M sulfuric acid (H_2SO_4) to 25.00 mL of a 0.09800-M ammonia (NH_3) solution. What was the pH of the solution at the equivalence point? The pK_b of ammonia is 4.75. Ignore hydrolysis of the sulfate ion.

Chapter 15

Graphing Basics

● ●

Graphs are a visual means of summarizing data. A good graph gives more information than a simple table of x and y values. Bad graphs can be very misleading. Scientists need to be able to transmit information in the form of good graphs.

The Problems You'll Work On

In this chapter, you work with graphing basics in the following ways:

- ✔ Identifying parts of a graph
- ✔ Determining graph scales
- ✔ Drawing and interpreting graphs
- ✔ Predicting values

What to Watch Out For

Remember the following when working on graphing:

- ✔ Very few graphs simply connect the data points.
- ✔ The more neatly a graph is drawn, the easier it is to read.
- ✔ The graph should occupy most, if not all, of the space available.
- ✔ The scales should place the highest values near the top or right of the graph. The lowest values should appear near the bottom or left of the graph.

Graphing

981–1,001 Answer the following questions on graphing. When data appears in a two-column format, the first column contains the data for the x-axis, and the second column contains the data for the y-axis. You can sketch the graphs by hand or use a program such as Excel.

981. What is the name of the position on the graph denoted by $x = 0$ and $y = 0$?

982. What do you call the process of finding a value on a graph that falls within the range of the plotted points?

983. What do you call the process of estimating a value on a graph that falls outside the range of the plotted points?

984. A plot of mass (y-axis) versus volume (x-axis) gives you a straight-line plot. What does the intercept of this line with the x-axis represent?

985. A plot of volume (y-axis) versus temperature (x-axis) gives you a straight-line plot. What does the intercept of this line with the x-axis represent?

986. In a graph of distance versus time, what does the slope of the line represent?

987. In a graph of mass versus volume, what does the slope of the line represent?

988. In a graph of heat energy versus temperature, what does the slope of the line represent?

989. According to Charles's law, the volume of an ideal gas is directly proportional to the temperature of the gas. What does a graph of volume versus temperature look like?

990. According to Boyle's law, the pressure of an ideal gas is inversely proportional to the volume of the gas. What does a graph of volume versus pressure look like?

991. What does a graph of the volume (y-axis) versus the edge of a cube (x-axis) look like?

992. Graph the following data with a line of best fit. Use your graph to estimate how many pennies would be in a stack that has a mass of 84.1 g.

Number of Pennies	Mass (g)
5	15.2
10	31.5
15	46.2
20	60.4
25	77.3
30	91.7

993. In a graph depicting the following information, what does the slope of the line represent?

Number of Pennies	Mass (g)
5	15.2
10	31.5
15	46.2
20	60.4
25	77.3
30	91.7

994. Graph the following data with a line of best fit. Then use the graph to estimate the volume of 31.0 g of the substance.

Volume (mL)	Mass (g)
2.2	11.9
3.9	28.2
5.5	37.7
7.1	48.9
9.1	60.1
10.6	76.0

995. Graph the following data with a line of best fit. What is the mass of 10.0 mL of the substance?

Volume (mL)	Mass (g)
2.2	11.9
3.9	28.2
5.5	37.7
7.1	48.9
9.1	60.1
10.6	76.0

996. Graph the following data. Then use the graph to estimate the concentration at an absorbance of 0.318.

Concentration (M)	Absorbance
0.100	0.385
0.075	0.300
0.050	0.187
0.025	0.099

997. Graph the following data. Then use the graph to estimate the absorbance at a concentration of 0.125 M.

Concentration (M)	Absorbance
0.100	0.385
0.075	0.300
0.050	0.187
0.025	0.099

998. What is the *x*-intercept of the best-fit line for the graph of the following data? (**Hint:** The *x*-axis should go from –300°C to +200°C.)

Temperature (°C)	Volume (L)
0.0	1.00
50.0	1.18
100.0	1.37
150.0	1.55

999. Describe the shape of the graph of the following data.

Volume (mL)	pH
0.00	3.09
5.00	4.07
10.00	4.49
13.00	4.70
16.00	4.92
19.00	5.18
22.00	5.58
23.00	5.82
24.00	6.30
24.25	6.51
24.50	7.04
24.75	9.46
24.90	10.23
25.00	10.31
25.10	10.47
25.25	10.61
25.50	10.78
25.75	10.93
26.00	11.01
27.00	11.28
28.00	11.43
31.03	11.56

1,000. Graph the following data for the titration of a weak acid with a strong base. Then use the graph to estimate the pH at 12.25 mL.

Volume (mL)	pH
0.00	3.09
5.00	4.07
10.00	4.49
13.00	4.70
16.00	4.92
19.00	5.18
22.00	5.58
23.00	5.82
24.00	6.30
24.25	6.51
24.50	7.04
24.75	9.46
24.90	10.23
25.00	10.31
25.10	10.47
25.25	10.61
25.50	10.78
25.75	10.93
26.00	11.01
27.00	11.28
28.00	11.43
31.03	11.56

1,001. A student conducts a kinetic experiment for a second-order reaction. Graphing which two columns in the following table will produce a straight line?

Time (min)	[A] (M)	ln[A]	$\frac{1}{[A]}(M^{-1})$
0	0.2000	−1.61	5.0
5	0.0282	−3.57	35.5
10	0.0156	−4.17	64.1
15	0.0106	−4.55	94.3
20	0.0080	−4.83	125

Part II
The Answers

Go to www.dummies.com/cheatsheet/1001chemistry to access the Cheat Sheet created specifically for *1,001 Chemistry Practice Problems For Dummies.*

In this part . . .

Here you get answers and explanations for all 1,001 problems. As you read the solutions, you may realize that you need a little more instruction. Fortunately, the *For Dummies* series offers several excellent resources. We highly recommend the following titles (all published by Wiley):

- *Chemistry Essentials For Dummies* by John T. Moore

- *Chemistry For Dummies* by John T. Moore

- *Chemistry Workbook For Dummies* by Peter J. Mikulecky, Katherine Brutlag, Michelle Rose Gilman, and Brian Peterson

When you're ready to step up to more advanced chemistry courses, you'll find the help you need in these titles:

- *Chemistry II For Dummies* by John T. Moore

- *Inorganic Chemistry For Dummies* by Michael Matson and Alvin W. Orbaek

- *Organic Chemistry I For Dummies* by Arthur Winter

- *Organic Chemistry II For Dummies* by John T. Moore and Richard H. Langley

- *Biochemistry For Dummies* by John T. Moore and Richard H. Langley

Visit www.dummies.com for more information.

Chapter 16

Answers and Explanations

Here are the answer explanations for all 1,001 chemistry questions in this book. For reference, you can find the periodic table in the Appendix.

1. **a gram**

A gram (g) is a common metric unit of mass used in the laboratory.

2. **a centimeter**

A centimeter (cm) is a common metric unit of length used to measure small objects in the laboratory. You can also measure small objects in millimeters (mm).

3. **a milliliter**

A milliliter (mL) is a common metric unit of volume used in the laboratory. You can also measure volume in cubic centimeters (cm^3 or cc), but that unit is less common in the lab.

4. **a millimeter of mercury**

A millimeter of mercury (mm Hg), also known as a *torr,* is a common metric unit of pressure used in the laboratory when dealing with gases. Other units of pressure are pascals, atmospheres, and bars; however, they aren't as common in the chemistry lab.

5. **a joule**

The joule (J) is a basic unit of energy or work in the metric system. It's equal to a newton-meter (N·m).

6. **kilo-**

Kilo- is the metric prefix that represents 1,000, or 10^3.

7. **milli-**

Milli- is the metric prefix that represents $\frac{1}{1,000}$, or 10^{-3}.

8. **centi-**

Centi- is the metric prefix that represents $\frac{1}{100}$, or 10^{-2}.

9. **nano-**

Nano- is the metric prefix that represents 10^{-9}, which equals $\frac{1}{1,000,000,000}$, or 1 billionth.

10. **mega-**

Mega- is the metric prefix that represents 10^6, which equals 1,000,000, or 1 million.

11. **kilogram**

The kilogram (kg) is the most appropriate unit to express the mass of an adult human. A person who weighs 150 lb. (pounds) on Earth has a mass of approximately 68 kg.

12. **cubic centimeters**

To measure a small regularly shaped object like a child's wooden block, you'd likely use centimeters (cm). After measuring the length, width, and height, you multiply these values together ($V = lwh$) and record the volume in cubic centimeters (cm^3).

13. **degrees Celsius**

Scientists would most likely use the metric unit degrees Celsius (°C) to report the temperature on a warm autumn day. They could also use kelvins (K), the SI unit of temperature.

14. **milligram**

The milligram (mg) is the most common metric measurement for small doses of solid medication.

15. **kelvin or ampere**

The SI base unit of temperature, the *kelvin,* was named after Lord Kelvin, who was also known as William Thomson. The *ampere,* named after André-Marie Ampère, is another base unit in SI; it measures electrical current.

Other units named after people — newtons, joules, watts, and so on — are derived units, not base units.

16. **a quart**

One liter is about 1.06 qt. (quarts).

17. **a yard**

A meter is a little over 39 in. (inches), which is closest to a yard (36 in.).

18. **8 fl. oz.**

One cup contains 8 fl. oz. (fluid ounces).

19. **yard**

The yard is an English unit for length that's comparable to a meter.

20. **centimeters**

Measuring the width of the board in centimeters (cm) is appropriate.

21. **100 mg**

To convert from decigrams to milligrams, use the relationship 1 dg = 100 mg. Start with the given value (1 dg) over 1 and then multiply by 100 mg/1 dg so that the decigrams cancel:

$$\frac{1\,\text{dg}}{1} \times \frac{100\,\text{mg}}{1\,\text{dg}} = 100\,\text{mg}$$

Another option is to convert decigrams to grams and grams to milligrams:

$$\frac{1\,\text{dg}}{1} \times \frac{1\,\text{g}}{10\,\text{dg}} \times \frac{1,000\,\text{mg}}{1\,\text{g}} = 100\,\text{mg}$$

22. **10 dL**

To convert from liters to deciliters, use the conversion factor 10 dL/1 L. Cancel the units that are the same in the numerator and denominator to make sure you end up with deciliters:

$$\frac{1\,\text{L}}{1} \times \frac{10\,\text{dL}}{1\,\text{L}} = 10\,\text{dL}$$

23. **0.001 km**

To convert from meters to kilometers, recall that 1 km = 1,000 m. Divide 1 by 1,000 to get 0.001. The meters in the numerator and denominator cancel, leaving kilometers as the final unit:

$$\frac{1\,\text{m}}{1} \times \frac{1\,\text{km}}{1,000\,\text{m}} = 0.001\,\text{km}$$

24. **100 cm**

To determine how many centimeters are in a meter, recall that the prefix *centi-* represents 1/100. You can use 1 cm/0.01 m or 100 cm/1 m as your conversion factor. Generally, it's easier to avoid the conversions that use decimals. Multiply 1 m by 100 cm/1 m. The meters cancel, giving you 100 cm:

$$\frac{1\,\cancel{m}}{1} \times \frac{100\text{ cm}}{1\,\cancel{m}} = 100\text{ cm}$$

25. **100 g**

To find out how many grams are in a hectogram, recall that the prefix *hecto-* represents 100. Multiply 1 hg by 100 g/1 hg. The hectograms cancel, giving you 100 g:

$$\frac{1\,\cancel{hg}}{1} \times \frac{100\text{ g}}{1\,\cancel{hg}} = 100\text{ g}$$

26. **25,000 mL**

You can determine how many milliliters are in a dekaliter in two steps. Convert dekaliters to liters (1 daL = 10 L) and then convert liters to milliliters (1 L = 1,000 mL): daL → L → mL. Here's the answer:

$$\frac{2.5\,\cancel{daL}}{1} \times \frac{10\,\cancel{L}}{1\,\cancel{daL}} \times \frac{1,000\text{ mL}}{1\,\cancel{L}} = 25,000\text{ mL}$$

In scientific notation, the answer is 2.5×10^4 mL.

27. **4,900,000 cg**

You can find the number of centigrams that are in 49 kg in two steps. Convert kilograms to grams (1 kg = 1,000 g) and then convert grams to centigrams (1 g = 100 cg): kg → g → cg:

$$\frac{49\,\cancel{kg}}{1} \times \frac{1,000\,\cancel{g}}{1\,\cancel{kg}} \times \frac{100\text{ cg}}{1\,\cancel{g}} = 4,900,000\text{ cg}$$

In scientific notation, the answer is 4.9×10^6 cg.

28. **0.00037 GW**

To get from watts to gigawatts, you need to know that *giga-* means 1×10^9, or 1,000,000,000, of something. Multiply 370,000 W by 1 GW/1,000,000,000 W. The watts cancel, giving you the answer in gigawatts:

$$\frac{370,000\,\cancel{W}}{1} \times \frac{1\text{ GW}}{1,000,000,000\,\cancel{W}} = 0.00037\text{ GW}$$

Alternatively, you can move the decimal point in 370,000 W nine places to the left to convert to watts. In scientific notation, the answer is 3.7×10^{-4} GW.

29.

126,000,000,000 µg

Going from very large units to very small units can be challenging. Be sure to double-check your conversions and the number of zeros. Convert megagrams to grams (1 Mg = 1,000,000 g) and convert grams to micrograms (1 g = 1,000,000 µg): Mg → g → µg:

$$\frac{0.126 \, \cancel{Mg}}{1} \times \frac{1,000,000 \, \cancel{g}}{1 \, \cancel{Mg}} \times \frac{1,000,000 \, \mu g}{1 \, \cancel{g}}$$

$$= 126,000,000,000 \, \mu g$$

$$= 1.26 \times 10^{11} \, \mu g$$

30.

0.000000000000080 km

Converting from very small units to very large units can be challenging. Think of the steps, picometers to meters (1×10^{12} pm = 1 m) and then meters to kilometers (1,000 m = 1 km): pm → m → km:

$$\frac{80 \, \cancel{pm}}{1} \times \frac{1 \, \cancel{m}}{1,000,000,000,000 \, \cancel{pm}} \times \frac{1 \, km}{1,000 \, \cancel{m}}$$

$$= 0.000000000000080 \, km$$

$$= 8.0 \times 10^{-14} \, km$$

31.

0.002 m³

Conversions that involve cubic units are often challenging because people aren't used to thinking of conversion factors in three dimensions.

To get from units used for measuring liquids (liters) to units used to measure solids (cubic meters), remember that 1 mL = 1 cm³. Your plan to get from liters to cubic meters may look like this: L → mL → cm³ → m³:

$$\frac{2 \, \cancel{L}}{1} \times \frac{1,000 \, \cancel{mL}}{1 \, \cancel{L}} \times \frac{1 \, cm^3}{1 \, \cancel{mL}} \times \frac{(1 \, m)^3}{(100 \, cm)^3}$$

$$= \frac{2 \, \cancel{L}}{1} \times \frac{1,000 \, \cancel{mL}}{1 \, \cancel{L}} \times \frac{1 \, \cancel{cm^3}}{1 \, \cancel{mL}} \times \frac{1 \, m^3}{(100 \times 100 \times 100) \, \cancel{cm^3}}$$

$$= 0.002 \, m^3$$

In scientific notation, the answer is 2×10^{-3} m³.

32.

640,000 mL

Conversions that involve cubic units are often challenging because people aren't used to thinking of conversion factors in three dimensions.

To convert from cubic meters to milliliters, you may choose to use the relationship 1 mL = 1 cm³ as follows: m³ → cm³ → mL:

$$\frac{0.64 \text{ m}^3}{1} \times \frac{(100 \text{ cm})^3}{(1 \text{ m})^3} \times \frac{1 \text{ mL}}{1 \text{ cm}^3}$$

$$= \frac{0.64 \text{ m}^3}{1} \times \frac{(100 \times 100 \times 100) \text{ cm}^3}{1 \text{ m}^3} \times \frac{1 \text{ mL}}{1 \text{ cm}^3}$$

$$= 640{,}000 \text{ mL}$$

In scientific notation, the answer is 6.4×10^5 mL.

---

33. **22 mi.**

Converting between metric units and English units mainly requires using the right conversion factor. To go from kilometers to miles, you can use the conversion factor 1 mi./1.61 km. If you set up the problem correctly, the kilometers cancel out, leaving you with miles:

$$\frac{35 \text{ km}}{1} \times \frac{1 \text{ mi.}}{1.61 \text{ km}} \approx 22 \text{ mi.}$$

The given measurement has two significant figures, so you round the answer to two significant figures as well.

---

34. **0.079 in.**

To convert between centimeters and inches, use the conversion factor 1 in./2.54 cm. The centimeters cancel out, giving you the answer in inches:

$$\frac{0.20 \text{ cm}}{1} \times \frac{1 \text{ in.}}{2.54 \text{ cm}} \approx 0.079 \text{ in.}$$

The given measurement has two significant figures, so you round the answer to two significant figures as well.

---

35. **221 yd.**

When converting between meters and yards, start with the given (202 m) and use the conversion factor 1 yd./0.914 m. The meters cancel out, giving you the answer in yards:

$$\frac{202 \text{ m}}{1} \times \frac{1 \text{ yd.}}{0.914 \text{ m}} \approx 221 \text{ yd.}$$

The given measurement has three significant figures, so you round the answer to three significant figures as well.

---

36. **130 lb.**

To convert from kilograms to pounds, you can use the conversion factors 1,000 g/1 kg and 1 lb./454 g:

$$\left(\frac{58 \text{ kg}}{1}\right)\left(\frac{1{,}000 \text{ g}}{1 \text{ kg}}\right)\left(\frac{1 \text{ lb.}}{454 \text{ g}}\right) \approx 130 \text{ lb.}$$

The given measurement has two significant figures, so you round the answer to two significant figures as well.

Note: You have to make some assumptions to do this conversion, because the kilogram is a unit of mass and the pound is a unit of weight (force). Dividing a weight by the acceleration due to gravity yields a mass. On Earth, the force of gravity is nearly constant, allowing the pound to function as a unit of mass. For clarity, if *pound* is a mass, the designation should be *pound-mass;* when *pound* is a weight, the designation is *pound-force.*

37.   **7.97 qt.**

Liters and quarts are similar in size, so be careful with the conversion. Use the conversion factor 1 qt./0.946 L. The liters cancel out, giving you the answer in quarts:

$$\frac{7.54\ \cancel{L}}{1} \times \frac{1\ \text{qt.}}{0.946\ \cancel{L}} \approx 7.97\ \text{qt.}$$

The given measurement has three significant figures, so you round the answer to three significant figures as well.

38.   **0.22 cm**

To convert from inches to centimeters, multiply the inches by 2.54 cm/1 in.:

$$\frac{0.087\ \cancel{\text{in.}}}{1} \times \frac{2.54\ \text{cm}}{1\ \cancel{\text{in.}}} \approx 0.22\ \text{cm}$$

The given measurement has two significant figures, so you round the answer to two significant figures as well.

39.   **745 km**

To find how many kilometers are in a certain number of miles, simply multiply the miles by 1.61 km/1 mi.:

$$\frac{463\ \cancel{\text{mi.}}}{1} \times \frac{1.61\ \text{km}}{1\ \cancel{\text{mi.}}} \approx 745\ \text{km}$$

The given measurement has three significant figures, so you round the answer to three significant figures as well.

40.   **41,000 g**

To convert from pounds to grams, take the number of pounds and multiply it by 454 g/1 lb.:

$$\frac{91\ \cancel{\text{lb.}}}{1} \times \frac{454\ \text{g}}{1\ \cancel{\text{lb.}}} \approx 41,000\ \text{g}$$

The given measurement has two significant figures, so you round the answer to two significant figures as well. In scientific notation, the answer is 4.1×10^4 g.

41. **1,980 L**

There are 3.78 L in a gallon, so simply multiply the number of gallons, 525 gal., by 3.78 L/1 gal.:

$$\frac{525 \text{ gal.}}{1} \times \frac{3.78 \text{ L}}{1 \text{ gal.}} \approx 1{,}980 \text{ L}$$

The given measurement has three significant figures, so you round the answer to three significant figures as well. In scientific notation, the answer is 1.98×10^3 L.

42. **3.0 atm**

To find the number of atmospheres in 44 psi (pounds per square inch), use the appropriate pressure conversion. There are 14.7 psi in 1 atm, so multiply 44 psi by 1 atm/14.7 psi:

$$\frac{44 \text{ psi}}{1} \times \frac{1 \text{ atm}}{14.7 \text{ psi}} \approx 3.0 \text{ atm}$$

The given measurement has two significant figures, so you round the answer to two significant figures as well.

43. **8.46 c.**

Writing a short plan for the conversions may be helpful. Converting from liters to quarts and then from quarts to cups gets you through this problem: L → qt. → c. Always write the units to make sure they cancel correctly; then do the math:

$$\frac{2.00 \text{ L}}{1} \times \frac{1 \text{ qt.}}{0.946 \text{ L}} \times \frac{4 \text{ c.}}{1 \text{ qt.}} \approx 8.46 \text{ c.}$$

The given measurement has three significant figures, so you round the answer to three significant figures as well.

44. **36.1 lb.**

To convert from hectograms to pounds, you can either go from hectograms to grams to pounds or from hectograms to kilograms to pounds. Here's the hg → g → lb. conversion:

$$\frac{164 \text{ hg}}{1} \times \frac{100 \text{ g}}{1 \text{ hg}} \times \frac{1 \text{ lb.}}{454 \text{ g}} \approx 36.1 \text{ lb.}$$

The given measurement has three significant figures, so you round the answer to three significant figures as well.

45. **0.155 gal.**

When converting from milliliters to gallons, I find it easiest to convert milliliters to liters and then convert liters to gallons: mL → L → gal.:

$$\frac{587 \text{ mL}}{1} \times \frac{1 \text{ L}}{1{,}000 \text{ mL}} \times \frac{1 \text{ gal.}}{3.78 \text{ L}} \approx 0.155 \text{ gal.}$$

The given measurement has three significant figures, so you round the answer to three significant figures as well.

46. 969,000 cm

A slightly longer problem like this requires a plan. Changing miles to kilometers and then kilometers to centimeters is the shorter way to do this. Or you can change the kilometers to meters and then the meters to centimeters, as follows: mi. $\rightarrow$ km $\rightarrow$ m $\rightarrow$ cm:

$$\frac{6.02 \text{ mi.}}{1} \times \frac{1.61 \text{ km}}{1 \text{ mi.}} \times \frac{1,000 \text{ m}}{1 \text{ km}} \times \frac{100 \text{ cm}}{1 \text{ m}} \approx 969,000 \text{ cm}$$

The given measurement has three significant figures, so you round the answer to three significant figures as well. In scientific notation, the answer is 9.69×10^5 cm.

47. 1,020,000 dg

You can do this conversion by converting pounds to grams and grams to decigrams: lb. $\rightarrow$ g $\rightarrow$ dg. Multiply the number of pounds by 454 g/lb. and then convert the grams to decigrams by multiplying by 10 dg/1 g:

$$\frac{225 \text{ lb.}}{1} \times \frac{454 \text{ g}}{1 \text{ lb.}} \times \frac{10 \text{ dg}}{1 \text{ g}} \approx 1,020,000 \text{ dg}$$

The given measurement has three significant figures, so you round the answer to three significant figures as well. In scientific notation, the answer is 1.02×10^6 dg.

48. 6,400 mL

Convert quarts to liters and liters to milliliters: qt. $\rightarrow$ L $\rightarrow$ mL. Multiply the number of quarts by 0.946 L/1 qt.; then convert the liters to milliliters by multiplying by 1,000 mL/1 L:

$$\frac{6.8 \text{ qt.}}{1} \times \frac{0.946 \text{ L}}{1 \text{ qt.}} \times \frac{1,000 \text{ mL}}{1 \text{ L}} \approx 6,400 \text{ mL}$$

The given measurement has two significant figures, so you round the answer to two significant figures as well. In scientific notation, the answer is 6.4×10^3 mL.

49. 466 cm

Convert feet into inches and then convert inches into centimeters: ft. $\rightarrow$ in. $\rightarrow$ cm. Multiply the 15.3 ft. by 12 in./1 ft. to convert to inches; then multiply by 2.54 cm/1 in. to convert to centimeters:

$$\frac{15.3 \text{ ft.}}{1} \times \frac{12 \text{ in.}}{1 \text{ ft.}} \times \frac{2.54 \text{ cm}}{1 \text{ in.}} \approx 466 \text{ cm}$$

The given measurement has three significant figures, so you round the answer to three significant figures as well.

50. 47 L

Convert pints to quarts and then convert quarts to liters: pt. → qt. → L. Multiply 99 pt. by 1 qt./2 pt. to convert to quarts; then multiply the quarts by 0.946 L/1 qt. to convert to liters. The pints and quarts cancel out, giving you the answer in liters:

$$\frac{99 \text{ pt.}}{1} \times \frac{1 \text{ qt.}}{2 \text{ pt.}} \times \frac{0.946 \text{ L}}{1 \text{ qt.}} \approx 47 \text{ L}$$

The given measurement has two significant figures, so you round the answer to two significant figures as well.

51. 908 kg

To convert short tons to pounds, multiply the number of tons by 2,000 lb./1 ton. To get from pounds to kilograms (kg), multiply by 0.454 kg/1 lb.:

$$\frac{1.00 \text{ ton}}{1} \times \frac{2,000 \text{ lb.}}{1 \text{ ton}} \times \frac{0.454 \text{ kg}}{1 \text{ lb.}} \approx 908 \text{ kg}$$

The given measurement has three significant figures, so you round the answer to three significant figures as well.

52. 552 cm

You may be able to find a conversion factor that goes straight from yards to centimeters. If so, this conversion takes only one step. Otherwise, you may remember that 3 ft. = 1 yd., 12 in. = 1 ft., and 2.54 cm = 1 in. Set up the problem, starting with 6.04 yd. over 1, and then line up the conversions so that the units cancel correctly. Here's the overall plan: yd. → ft. → in. → cm. And here are the calculations:

$$\frac{6.04 \text{ yd.}}{1} \times \frac{3 \text{ ft.}}{1 \text{ yd.}} \times \frac{12 \text{ in.}}{1 \text{ ft.}} \times \frac{2.54 \text{ cm}}{1 \text{ in.}} \approx 552 \text{ cm}$$

The given measurement has three significant figures, so you round the answer to three significant figures as well.

53. 0.063 c.

First, remember that cc represents cubic centimeters (cm^3). Cubic centimeters easily convert to milliliters because they're in a 1:1 ratio. Then you can convert milliliters to fluid ounces and fluid ounces to cups: cc → mL → fl. oz. → c.:

$$\frac{15 \text{ cc}}{1} \times \frac{1 \text{ mL}}{1 \text{ cc}} \times \frac{1 \text{ fl. oz.}}{29.6 \text{ mL}} \times \frac{1 \text{ c.}}{8 \text{ fl. oz.}} \approx 0.063 \text{ c.}$$

The given measurement has two significant figures, so you round the answer to two significant figures as well.

54. 1,610,000 mm

You can approach this conversion in many ways. One option is to convert yards to inches, inches to centimeters, and centimeters to millimeters: yd. → in. → cm → mm:

$$\frac{1,760 \text{ yd.}}{1} \times \frac{36 \text{ in.}}{1 \text{ yd.}} \times \frac{2.54 \text{ cm}}{1 \text{ in.}} \times \frac{10 \text{ mm}}{1 \text{ cm}} \approx 1,610,000 \text{ mm}$$

The given measurement has three significant figures, so you round the answer to three significant figures as well. In scientific notation, the answer is 1.61×10^6 mm.

---

55.   **53,000 pt.**

Converting hectoliters to pints involves many steps, so put together a plan. You can go from hectoliters to liters to gallons to quarts to pints: hL → L → gal. → qt. → pt. The only division in the plan occurs when you convert liters to gallons, so you can multiply all the values in the numerator ($250 \times 100 \times 4 \times 2$) and then divide by 3.78. The units cancel, giving you the answer in pints:

$$\frac{250 \text{ hL}}{1} \times \frac{100 \text{ L}}{1 \text{ hL}} \times \frac{1 \text{ gal.}}{3.78 \text{ L}} \times \frac{4 \text{ qt.}}{1 \text{ gal.}} \times \frac{2 \text{ pt.}}{1 \text{ qt.}} \approx 53,000 \text{ pt.}$$

The given measurement has two significant figures, so you round the answer to two significant figures as well. In scientific notation, the answer is 5.3×10^4 pt.

---

56.   **5,100 g**

A slug is an English mass unit that you may not encounter except when working conversion problems. First convert slugs to pounds and then convert to grams: slugs → lb. → g. Converting to grams isn't difficult if you remember that there are 454 g in 1 lb. Here's the conversion:

$$\frac{0.35 \text{ slugs}}{1} \times \frac{32.2 \text{ lb.}}{1 \text{ slug}} \times \frac{454 \text{ g}}{1 \text{ lb.}} \approx 5,100 \text{ g}$$

The given measurement has two significant figures, so you round the answer to two significant figures as well. In scientific notation, the answer is 5.1×10^3 g.

The conversion takes an extra step if you convert pounds to kilograms and then convert the kilograms to grams, but you should get the same answer either way.

---

57.   **0.2540 km**

You can find the number of kilometers in 9,999 in. by converting inches to centimeters (by multiplying by 2.54 cm/1 in.) and then changing centimeters to meters (multiplying by 1 m/100 cm) and meters to kilometers (multiplying by 1 km/1,000 m): in. → cm → m → km.

$$\frac{9,999 \text{ in.}}{1} \times \frac{2.54 \text{ cm}}{1 \text{ in.}} \times \frac{1 \text{ m}}{100 \text{ cm}} \times \frac{1 \text{ km}}{1,000 \text{ m}} \approx 0.2540 \text{ km}$$

The given measurement has four significant figures, so you round the answer to four significant figures as well.

---

58.   **25.9 oz.**

Converting from kilograms to ounces requires one metric conversion (kilograms to grams), one metric-English conversion (grams to pounds), and one English conversion (pounds to ounces): kg → g → lb. → oz. The units cancel out, leaving you with ounces:

$$\frac{0.734 \text{ kg}}{1} \times \frac{1,000 \text{ g}}{1 \text{ kg}} \times \frac{1 \text{ lb.}}{454 \text{ g}} \times \frac{16 \text{ oz.}}{1 \text{ lb.}} \approx 25.9 \text{ oz.}$$

The given measurement has three significant figures, so you round the answer to three significant figures as well.

59. **1,600,000 μL**

Converting from ounces to microliters requires one English conversion (ounces to quarts), one English-metric conversion (quarts to liters), and one metric conversion (liters to microliters): oz. → qt. → L → μL:

$$\frac{55 \text{ oz.}}{1} \times \frac{1 \text{ qt.}}{32 \text{ oz.}} \times \frac{0.946 \text{ L}}{1 \text{ qt.}} \times \frac{1,000,000 \text{ μL}}{1 \text{ L}} \approx 1,600,000 \text{ μL}$$

The given measurement has two significant figures, so you round the answer to two significant figures as well. In scientific notation, the answer is 1.6×10^6 μL.

60. **1,498.4 dozen**

Finding the number of dozen eggs in 17,981 eggs uses the conversion factor 1 dozen/12 eggs:

$$\frac{17,981 \text{ eggs}}{1} \times \frac{1 \text{ dozen}}{12 \text{ eggs}} \approx 1,498.4 \text{ dozen}$$

The given egg count has five significant figures, so you round the answer to five significant figures as well.

61. **17.1 yr.**

When dealing with a number of days that would exceed four years (1,461 days), account for the extra day in a leap year by using the relationship 365.25 days = 1 yr. Take the number of days and multiply by 1 yr./365.25 days. The days cancel out, leaving you with years:

$$\frac{6,250 \text{ days}}{1} \times \frac{1 \text{ yr.}}{365.25 \text{ days}} \approx 17.1 \text{ yr.}$$

The given length of time has three significant figures, so you round the answer to three significant figures as well.

62. **13,000 weeks**

To do this conversion, you need to know that a century is 100 years. Multiply the 2.5 centuries by the number of years in a century (100); then multiply the years by the number of weeks in a year (52) to get the number of weeks: centuries → years → weeks:

$$\frac{2.5 \text{ centuries}}{1} \times \frac{100 \text{ yr.}}{1 \text{ century}} \times \frac{52 \text{ weeks}}{1 \text{ yr.}} = 13,000 \text{ weeks}$$

The given measurement has two significant figures, so the answer has two significant figures as well.

63. | **$2,870**

First, recognize that 1 penny = 3.16 g is a conversion factor that you'll use in your calculations. The question asks you to find the mass of 1.00 ton of pennies, so 1.00 ton is your starting point. Convert tons to pounds, pounds to grams, grams to pennies, and pennies to dollars: tons → lb. → g → pennies → $:

$$\frac{1.00 \text{ ton}}{1} \times \frac{2,000 \text{ lb.}}{1 \text{ ton}} \times \frac{454 \text{ g}}{1 \text{ lb.}} \times \frac{1 \text{ penny}}{3.16 \text{ g}} \times \frac{\$1}{100 \text{ pennies}} \approx \$2,870$$

The given average mass has three significant figures, so you round the answer to three significant figures as well.

64. | **43.8 s**

The tricky part is sorting out which of the given numbers to use where. Starting with the total distance in meters, you can multiply by 1 yd./0.914 m and then use 100 yd. = 10.0 s to get to the time unit in the numerator: m → yd. → s:

$$\frac{400 \text{ m}}{1} \times \frac{1 \text{ yd.}}{0.914 \text{ m}} \times \frac{10.0 \text{ s}}{100 \text{ yd.}} \approx 43.8 \text{ s}$$

The race distances (400 m and 100 yd.) are considered exact measures, so you can ignore those numbers when looking at significant figures. The given time has three significant figures, so the answer does as well.

65. | **18 L**

This problem is a bit like a conversion from fluid ounces to liters, but you have to take into consideration that 60 guests are each drinking 10. fl. oz. of soda. The first step is to figure out how many ounces of soda that is; then you can convert from ounces to milliliters by multiplying by 29.6 mL/1 fl. oz. Last, you convert from milliliters to liters by multiplying by 1 L/1,000 mL. Here's the conversion plan: guests → fl. oz. → mL → L:

$$\frac{60 \text{ guests}}{1} \times \frac{10. \text{ fl. oz.}}{1 \text{ guest}} \times \frac{29.6 \text{ mL}}{1 \text{ fl. oz.}} \times \frac{1 \text{ L}}{1,000 \text{ mL}} \approx 18 \text{ L}$$

The given volume measurement has two significant figures, so you round the answer to two significant figures as well.

66. | **8.3 sandwiches**

The question gives you two relationships to use: 1 guest eats 25.4 cm of sandwich, and 1 sandwich is 6.0 ft. long. Take the amount of sandwich that each person will eat and multiply it by 60 guests to find out how many centimeters of sandwich you need. Next, you can convert centimeters to inches (by multiplying by 1 in./2.54 cm), inches to feet (by multiplying by 1 ft./12 in.), and feet to sandwiches (by multiplying by 1 sandwich/6.0 ft.): guests → cm → in. → ft. → sandwiches:

$$\frac{60 \text{ guests}}{1} \times \frac{25.4 \text{ cm}}{1 \text{ guest}} \times \frac{1 \text{ in.}}{2.54 \text{ cm}} \times \frac{1 \text{ ft.}}{12 \text{ in.}} \times \frac{1 \text{ sandwich}}{6.0 \text{ ft.}} \approx 8.3 \text{ sandwiches}$$

The given sub measurement has two significant figures, so you round the answer to two significant figures as well.

67. **3,780 cm³**

You find the volume of a rectangular solid by multiplying the length times the width times the height. You can convert measurements before multiplying:

$$l = \frac{230. \text{ mm}}{1} \times \frac{1 \text{ cm}}{10 \text{ mm}} = 23.0 \text{ cm}$$

$$w = \frac{274 \text{ mm}}{1} \times \frac{1 \text{ cm}}{10 \text{ mm}} = 27.4 \text{ cm}$$

$$h = \frac{60.0 \text{ mm}}{1} \times \frac{1 \text{ cm}}{10 \text{ mm}} = 6.00 \text{ cm}$$

$$V = lwh = 23.0 \text{ cm} \times 27.4 \text{ cm} \times 6.00 \text{ cm} \approx 3,780 \text{ cm}^3$$

The given measurements have three significant figures, so you round the answer to three significant figures as well.

You can also find the volume in cubic millimeters and then do the conversion:

$$\frac{230. \text{ mm} \times 274 \text{ mm} \times 60.0 \text{ mm}}{1} \times \frac{(1 \text{ cm})^3}{(10 \text{ mm})^3}$$

$$= \frac{(230. \times 274 \times 60.0) \text{ mm}^3}{1} \times \frac{1 \text{ cm}^3}{(10 \times 10 \times 10) \text{ mm}^3}$$

$$\approx 3,780 \text{ cm}^3$$

68. **0.0630 m²**

The surface area of the front cover of the textbook equals to the length of the book times the width of the book: $A = lw$. You can convert from millimeters to meters before multiplying them together:

$$l = \frac{230. \text{ mm}}{1} \times \frac{1 \text{ m}}{1,000 \text{ mm}} = 0.230 \text{ m}$$

$$w = \frac{274 \text{ mm}}{1} \times \frac{1 \text{ m}}{1,000 \text{ mm}} = 0.274 \text{ m}$$

$$A = lw = 0.230 \text{ m} \times 0.274 \text{ m} \approx 0.0630 \text{ m}^2$$

The given measurements have three significant figures, so you round the answer to three significant figures as well.

Another option is to find the surface area before converting the measurements:

$$\frac{230. \text{ mm} \times 274 \text{ mm}}{1} \times \frac{(1 \text{ m})^2}{(1,000 \text{ mm})^2}$$

$$= \frac{(230. \times 274) \text{ mm}^2}{1} \times \frac{1 \text{ m}^2}{(1,000 \times 1,000) \text{ mm}^2}$$

$$\approx 0.0630 \text{ m}^2$$

69. **72 tiles**

One approach is to first find the area of the hallway in square feet. Then convert the square feet to square inches and divide by the area of one tile.

$$\frac{(10.0 \text{ ft.} \times 5.0 \text{ ft.})}{1} \times \frac{(12 \text{ in.})^2}{(1 \text{ ft.})^2} \times \frac{1 \text{ tile}}{(10.0 \text{ in.})^2}$$

$$= \frac{(10.0 \times 5.0) \text{ ft.}^2}{1} \times \frac{(12 \times 12) \text{ in.}^2}{1 \text{ ft.}^2} \times \frac{1 \text{ tile}}{(10.0 \times 10.0) \text{ in.}^2}$$

$$= 72 \text{ tiles}$$

The given 5.0-ft. measurement has only two significant figures, so the answer has two significant figures as well.

Another option is to find the area of the hallway in square inches and divide by the area of a tile, also in square inches.

Or you can determine the number of tiles that fit across the 5.0-ft. hallway (by dividing the width of the hallway by the width of a tile) and multiply that number by the number of tiles that fit down the 10.0-ft. hallway:

Across: $\dfrac{5.0 \text{ ft.}}{1} \times \dfrac{12 \text{ in.}}{1 \text{ ft.}} \times \dfrac{1 \text{ tile}}{10 \text{ in.}} = 6 \text{ tiles}$

Down: $\dfrac{10.0 \text{ ft.}}{1} \times \dfrac{12 \text{ in.}}{1 \text{ ft.}} \times \dfrac{1 \text{ tile}}{10 \text{ in.}} = 12 \text{ tiles}$

Total: $6 \times 12 = 72 \text{ tiles}$

70. 54,000 cm/min.

This problem is challenging because you have to convert both the numerator and the denominator. Keeping track of the units is really important in this kind of problem. You can convert the numerator or the denominator first — the order doesn't matter. The following equation converts hours to minutes first by multiplying by 1 hr./60 min. Then it converts miles to kilometers (by multiplying by 1.61 km/1 mi.), kilometers to meters (by multiplying by 1,000 m/1 km), and then meters to centimeters (by multiplying by 100 cm/ 1 m). Here's the conversion plan: mi./hr. → mi./min. → km/min. → m/min. → cm/min.:

$$\frac{20. \text{ mi.}}{1 \text{ hr.}} \times \frac{1 \text{ hr.}}{60 \text{ min.}} \times \frac{1.61 \text{ km}}{1 \text{ mi.}} \times \frac{1,000 \text{ m}}{1 \text{ km}} \times \frac{100 \text{ cm}}{1 \text{ m}} \approx 54,000 \text{ cm/min.}$$

The given measurement has two significant figures, so you round the answer to two significant figures as well. In scientific notation, the answer is 5.4×10^4 cm/min.

As you go through a problem like this, make sure you cross out the units that cancel. If you end up with the correct units in your answer — in this case, cm/min. — you know you've probably set up the problem correctly.

71. 85 lb.

In this problem, you're given a volume and a conversion factor of $1.00 \text{ cm}^3 = 19.3$ g. So to start, this is a conversion problem from liters to cubic centimeters. You can convert liters to milliliters by multiplying by 1,000 mL/1 L. Milliliters and cubic centimeters are equivalent, so you can multiply by 1 cm³/1 mL. Next, multiply by 19.3 g/1.00 cm³ to get to mass. Last, convert grams pounds by multiplying by 1 lb./454 g. Here's the overall plan: L → mL → cm³ → g → lb.:

$$\frac{2.0\,L}{1} \times \frac{1{,}000\,mL}{1\,L} \times \frac{1\,cm^3}{1\,mL} \times \frac{19.3\,g}{1.00\,cm^3} \times \frac{1\,lb.}{454\,g} = 85\,lb.$$

The given volume measurement has only two significant figures, so you round the answer to two significant figures as well.

72. **2.5 min.**

To find the number of minutes the horse takes to run a distance of 12 furlongs, first convert the distance to miles in order to use the given speed of 385.3 mph. Then convert the time from hours to minutes: furlongs → rods → yd. → mi. → hr. → min.:

$$\frac{12\,furlongs}{1} \times \frac{40\,rods}{1\,furlong} \times \frac{5.5\,yd.}{1\,rod} \times \frac{1\,mi.}{1{,}760\,yd.} \times \frac{1\,hr.}{35.3\,mi.} \times \frac{60\,min.}{1\,hr.} = 2.5\,min.$$

The given length measurement has only two significant figures, so you round the answer to two significant figures as well.

If you don't know how many yards are in a mile (1 mi. = 1,760 yd.), you may need to take the extra step of converting yards to feet (3 yd. = 1 ft.) and then converting feet to miles (5,280 ft. = 1 mi.), but you should get the same answer.

73. **0.43 s**

Finding the number of seconds the pitch will take to travel from the pitcher to the batter requires using the distance between the two locations and the speed of the ball. Convert the distance from feet to miles by multiplying by 1 mi./5,280 ft. Next, divide by the speed (96 mi./1 hr.) and then convert the hours to minutes (by multiplying by 60 min./1 hr.) and the minutes to seconds (by multiplying by 60 s/1 min.): ft. → mi. → hr. → min. → s:

$$\frac{60.5\,ft.}{1} \times \frac{1\,mi.}{5{,}280\,ft.} \times \frac{1\,hr.}{96\,mi.} \times \frac{60\,min.}{1\,hr.} \times \frac{60\,s}{1\,min.} \approx 0.43\,s$$

The given speed measurement has only two significant figures, so you round the answer to two significant figures as well.

74. **0.00072 mm**

This problem requires that you think in three dimensions. You're given a mass (grams), a conversion between mass and volume (grams per cubic centimeter), and a surface area (cubic meters). Volume equals length × width × height, and surface area = length × width, so you can find the height by dividing volume by surface area:

$$\frac{\text{Volume}}{\text{Surface Area}} = \frac{l \times w \times h}{l \times w}$$

Now for the unit conversions. Change kilograms to grams by multiplying by 1,000 g/1 kg. Multiply by 1 cm³/19.3 g to get cubic centimeters. Next, convert the cubic centimeters to cubic millimeters. That takes care of the numerator. You have to convert the square meters in the denominator to square millimeters before you can divide. Here's the conversion plan: (kg → g → cm³ → mm³) ÷ (m² → mm²) = mm³/mm² = mm:

$$\frac{\dfrac{25\ \cancel{kg}}{1}\times\dfrac{1{,}000\ \cancel{g}}{1\ \cancel{kg}}\times\dfrac{1\ cm^3}{19.3\ \cancel{g}}\times\dfrac{(10\ mm)^3}{(1\ cm)^3}}{\dfrac{1{,}810\ m^2}{1}\times\dfrac{(1{,}000\ mm)^2}{(1\ m)^2}}$$

$$=\frac{\dfrac{25\ \cancel{kg}}{1}\times\dfrac{1{,}000\ \cancel{g}}{1\ \cancel{kg}}\times\dfrac{1\ \cancel{cm^3}}{19.3\ \cancel{g}}\times\dfrac{(10\ \cancel{mm}\times10\ \cancel{mm}\times10\ mm)}{1\ \cancel{cm^3}}}{\dfrac{1{,}810\ \cancel{m^2}}{1}\times\dfrac{(1{,}000\ \cancel{mm}\times1{,}000\ \cancel{mm})}{1\ \cancel{m^2}}}$$

$$\approx 0.00072\ mm$$

The given mass measurement has only two significant figures, so you round the answer to two significant figures as well. In scientific notation, the answer is 7.2×10^{-4} mm.

Alternatively, you can leave the numerator in cubic centimeters and convert the denominator from square meters to square centimeters. That gives you the answer in centimeters, which you can convert to millimeters by multiplying by 10 mm/1 cm.

---

75. **2.57 s**

You're waiting for a reply, so you have to account for the time for a signal to get to the moon and back, which means that the signal has to travel $2 \times 239{,}000$ miles. Convert miles to kilometers (multiply by 1.61 km/1 m) and kilometers to meters (multiply by 1,000 m/1 km). Then to get the time, divide by the speed:

$$\frac{2(239{,}000\ \cancel{mi.})}{1}\times\frac{1.61\ \cancel{km}}{1\ \cancel{mi.}}\times\frac{1{,}000\ \cancel{m}}{1\ \cancel{km}}\times\frac{1\ s}{300{,}000{,}000\ \cancel{m}}\approx 2.57\ s$$

The given distance measurement has three significant figures, so you round the answer to three significant figures as well. The speed of the radio waves is considered exact, so you can disregard it when figuring out significant figures.

---

76. 8.76×10^2

Scientific notation expresses numbers with one digit to the left of the decimal point and any number of significant digits to the right of the decimal. For numbers greater than 1, the exponent on the 10 is positive. The number 876 equals 8.76×100, and 100 is the same as 10^2, so 876 is 8.76×10^2.

---

77. 4.000001×10^6

For numbers greater than 1, the exponent on the 10 is positive in scientific notation. The number 4,000,001 equals $4.000001 \times 1{,}000{,}000$, and 1,000,000 is the same as 10^6, so 4,000,001 is 4.000001×10^6.

---

78. 5.10×10^{-4}

When a number is less than 1, the exponent on the 10 is negative in scientific notation. Move the decimal point in 0.000510 until you have just one digit to the left of the decimal point. The number of times you move the decimal becomes the exponent on the 10. You move the decimal point four places to the right, so 0.000510 become 5.10×10^{-4}.

79. 9×10^6

The number 900 is the same as 9×100, or 9×10^2, so the problem becomes $9 \times 10^2 \times 10^4$. To multiply $10^2 \times 10^4$, you just need to add the exponents on the 10s: You end up with $9 \times 10^2 \times 10^4 = 9 \times 10^{2+4} = 9 \times 10^6$.

80. 1×10^1

Although writing 10 in scientific notation doesn't have much practical value, you can do it. The decimal moves one place to the left, and the 10 becomes 1 (writing .0 after the 1 isn't necessary unless you have a decimal point after 10 in the original number). You then multiply that 1 by a power of ten. The exponent on the 10 is 1, giving you 1×10^1.

81. 200

The exponent on the 10 is positive, so the decimal point in 2.00 moves two places to the right. The answer is 200, because 2.00×10^2 is the same as 2.00×100.

82. 0.09

The exponent on the 10 is –2, so the value is less than 1. Move the understood decimal point in 9 two places to the left, giving you a number between 0 and 1. The answer is 0.09.

83. 4,795.2

With an exponent of +3 on the 10, the decimal point in 4.7952 moves three places to the right. Moving the decimal point three places to the right is the same as multiplying by 1,000, giving you 4,795.2 as the answer.

84. 0.0000164

With an exponent of –5 on the 10, the decimal point in 1.64 moves five places to the left, giving you 0.0000164.

Tip: When you have a large negative exponent, the easiest way to check your answer is to count the 0s to the left of your nonzero digit, including the 0 to the left of the decimal point. The number of 0s should be the same as the number of the exponent.

85. 0.083

The number 0.83×10^{-1} looks like it's in scientific notation, but it isn't, because you don't have a nonzero digit in front of the decimal point. Regardless, the negative exponent on the 10 indicates that the decimal point in 0.83 needs to move one place to the left, so the number is 0.083 in standard form.

86. 5.97×10^3

When you're adding numbers in scientific notation and the exponents on the 10s are the same, you can just add the numbers in front of the powers of ten: $1.26 + 4.71 = 5.97$. The 5.97 falls between 1 and 10 — it has one nonzero digit to the left of the decimal point — so you can finish giving the answer in scientific notation by writing the $\times 10^3$. The answer is 5.97×10^3.

Note: This approach works because you're essentially factoring out the 10^3 before adding the decimal numbers: $(1.26 \times 10^3) + (4.71 \times 10^3) = 10^3(1.26 + 4.71) = 10^3(5.97)$.

87. 6.0×10^{-1}

The exponents on the 10s are the same, so you can just add the numbers in front of the powers of ten: $3.9 + 2.1 = 6.0$. The sum is between 1 and 10, so you retain the power $(\times 10^{-1})$ from the original numbers. The answer is 6.0×10^{-1}.

Note: This approach works because you're essentially factoring out 10^{-1} before adding the decimal numbers: $(3.9 \times 10^{-1}) + (2.1 \times 10^{-1}) = 10^{-1}(3.9 + 2.1) = 10^{-1}(6.0)$.

88. 8.6×10^2

Because the exponents on the 10s don't match, it's easiest to take both numbers out of scientific notation, do the subtraction, and then put the answer back into scientific notation. Round the decimal number to the tenths place to give the answer the right number of significant figures.

$$8.9 \times 10^2 = 890$$
$$\underline{-3.3 \times 10^1 = -33}$$
$$= 857$$
$$\approx 8.6 \times 10^2$$

89. -5.1×10^1

The exponents on the 10s don't match, so take the numbers out of scientific notation, do the subtraction, and then put the answer back into scientific notation. Round the decimal number to the tenths place to give the answer the right number of significant figures.

$$7.4 \times 10^{-1} = 0.74$$
$$\underline{-5.2 \times 10^1 = -52.}$$
$$= -51.26$$
$$\approx -5.1 \times 10^1$$

90. 1.2031×10^3

The exponents on the 10s are the same, so you can add the numbers in front of the powers of ten: $8.240 + 3.791 = 12.031$. Because 12.031 is greater than 10 — it has two digits to the left of the decimal point — change it into scientific notation, giving you 1.2031×10^1.

Next, multiply that number by 10^2, the power of ten from the original problem; you get $1.2031 \times 10^1 \times 10^2$. To multiply $10^1 \times 10^2$, add the exponents, giving you $10^{1+2} = 10^3$. The final answer in scientific notation is 1.2031×10^3.

91. 9.5×10^6

The exponents aren't the same, so take the numbers out of scientific notation, do the subtraction, and then put the answer back into scientific notation. Round the decimal number to the tenths place to give the answer the right number of significant figures.

$$
\begin{aligned}
1.0 \times 10^7 &= 10,000,000 \\
-5.2 \times 10^5 &= -520,000 \\
\hline
&= 9,480,000 \\
&\approx 9.5 \times 10^6
\end{aligned}
$$

92. 6.04×10^{-3}

Take the numbers out of scientific notation, do the addition, and then write the sum in scientific notation. Round the decimal number to two decimal places to give the answer the right number of significant figures.

$$
\begin{aligned}
5.42 \times 10^{-3} &= 0.00542 \\
+6.19 \times 10^{-4} &= +0.000619 \\
\hline
&= 0.006039 \\
&\approx 6.04 \times 10^{-3}
\end{aligned}
$$

93. 8.41×10^6

In this problem, each number in scientific notation has a different exponent. Take the numbers out of scientific notation, do the addition and subtraction, and put the answer back into scientific notation. Round the decimal number to the hundredths place to give the answer the right number of significant figures.

$$
\begin{aligned}
8.20 \times 10^6 &= 8,200,000 \\
-7.31 \times 10^4 &= -73,100 \\
+2.846 \times 10^5 &= +284,600 \\
\hline
&= 8,411,500 \\
&\approx 8.41 \times 10^6
\end{aligned}
$$

94. 4.5×10^{-2}

To multiply numbers that are in scientific notation, you can multiply the numbers in front of the $\times 10$ (here, $1.0 \times 4.5 = 4.5$) and then add the exponents from the powers of ten ($10^{-7} \times 10^5 = 10^{-7+5} = 10^{-2}$). The answer is 4.5×10^{-2}.

95. 1.0×10^1

When dividing numbers that are in scientific notation, divide the numbers in front of the $\times 10$ (here, $1.0 \div 1.0 = 1.0$) and subtract the exponents on the powers of ten. Keep in mind that subtracting a negative number is the same as adding a positive number: $10^{-3} \div 10^{-4} = 10^{-3-(-4)} = 10^{-3+4} = 10^1$. The answer is 1.0×10^1.

96. 6.3×10^{15}

Multiply the numbers in front of the $\times 10$ (here, $3.15 \times 2.0 = 6.3$) and then add the exponents ($10^{12} \times 10^3 = 10^{12+3} = 10^{15}$) to give you 6.3×10^{15}.

97. 4.9×10^4

First, divide the numbers in front of the $\times 10$. In this case, $4.7 \div 9.6 = 0.49$. Because this number is less than 1, simply subtracting the exponents on the 10s isn't enough to find the exponent in the answer. You need to account for 0.49 being less than 1 by moving the decimal one place to the right (0.49 becomes 4.9) and by subtracting another 1 from the exponent.

Dividing the powers of 10 in the problem gives you $10^{-2} \div 10^{-7} = 10^{-2-(-7)} = 10^{-2+7} = 10^5$. Subtracting 1 from the exponent (to account for moving the decimal point to the right) gives you $10^{5-1} = 10^4$, so the complete answer is 4.9×10^4.

98. 1.68×10^{11}

To multiply numbers that are in scientific notation, you can multiply the numbers in front of the $\times 10$ and then add the exponents from the powers of ten. When you multiply 8.40 by 2.00, you get 16.8. To put this in scientific notation, you need to move the decimal point one place to the left, so 16.8 becomes 1.68×10^1.

To multiply powers of ten, add the exponents together: $10^{15} \times 10^{-5} = 10^{15+(-5)} = 10^{10}$. Next, add another 1 to the exponent to account for moving the decimal point in 16.8 to the left: $10^{10+1} = 10^{11}$. The final answer is 1.68×10^{11}.

99. 3.1×10^5

To divide numbers that are in scientific notation, you can divide the numbers in front of the $\times 10$ and then subtract the exponents from the powers of ten. When you divide 1.0 by 3.2, you get 0.31, a number less than 1. To put the number in scientific notation, you need to move the decimal one place to the right (0.31 becomes 3.1) and decrease the exponent on the 10 by 1.

You're dividing in this problem, so subtract the exponents from the powers of ten: $10^8 \div 10^2 = 10^{8-2} = 10^6$. Then subtract another 1 from the exponent to account for moving the decimal point in 0.31 one place to the right: $10^{6-1} = 10^5$. The answer becomes 3.1×10^5.

100. 1.9×10^{-6}

Start by doing calculations with the numbers in front of the $\times$ 10: $9.76 \times 3.55 \div 1.8 = 19.25$. Round to two digits to give the answer the right number of significant figures: That means 19.25 becomes 1.9×10^1.

Now deal with the powers of ten from the problem. You add exponents to multiply powers of ten and subtract exponents to divide by powers of ten: $10^{-9} \times 10^{-3} \div 10^{-5} = 10^{-9 + (-3) - (-5)} = 10^{-9 - 3 + 5} = 10^{-7}$. Finally, multiply the 1.9×10^1 by the 10^{-7} to get the answer: $1.9 \times 10^1 \times 10^{-7} = 1.9 \times 10^{1 + (-7)} = 1.9 \times 10^{-6}$.

101. 1.1×10^{-1}

Start by doing the math with the numbers in front of the $\times$ 10: $2.48 \times 4.756 \times 9.1 = 107.33$. Round to two digits to give the answer the right number of significant figures: That means 107.33 becomes 110, or 1.1×10^2.

Now add the exponents to multiply the powers of ten: $10^3 \times 10^{-4} \times 10^{-2} = 10^{3 + (-4) + (-2)} = 10^{-3}$. Multiply the 1.1×10^2 by the 10^{-3}, and you get $1.1 \times 10^2 \times 10^{-3} = 1.1 \times 10^{2 - 3} = 1.1 \times 10^{-1}$.

102. 3.6×10^{-4}

Using the rules for order of operations, complete the multiplication portion first. To multiply numbers that are in scientific notation, you can multiply the numbers in front of the $\times$ 10 and then add the exponents from the powers of ten:

$$\left(6.27 \times 10^{-2}\right) \times \left(2.9 \times 10^{-3}\right) = \left(6.27 \times 2.9\right)\left(10^{-2} \times 10^{-3}\right)$$

$$= 18.183\left(10^{-2 + (-3)}\right)$$

$$= 18.183 \times 10^{-5}$$

$$= 1.8183 \times 10^{-4}$$

Then do the addition. The powers of ten are the same, so you can simply add the numbers in front of the $\times$ 10. (Mathematically, you're factoring out 10^{-4}.) Round the answer to one decimal place to give the answer the right number of significant figures.

$$\left(1.8 \times 10^{-4}\right) + \left(1.8183 \times 10^{-4}\right) = \left(1.8 + 1.8183\right)10^{-4}$$

$$= 3.6183 \times 10^{-4}$$

$$\approx 3.6 \times 10^{-4}$$

103. 1.526×10^{-5}

Here, the order of operations applies. Start with the division. To divide numbers that are in scientific notation, you can divide the numbers in front of the $\times$ 10 and then subtract the exponents from the powers of ten:

$$\left(9.189 \times 10^{-19}\right) \div \left(0.6021 \times 10^{-13}\right) = \left(9.189 \div 0.6021\right)\left(10^{-19} \div 10^{-13}\right)$$

$$= 15.262\left(10^{-19 - (-13)}\right)$$

$$= 15.262 \times 10^{-6}$$

$$= 1.5262 \times 10^{-5}$$

Answers
101–200

Then do the addition. The powers of ten are different, so take the numbers out of scientific notation, do the addition, and then put the answer back in scientific notation. Round the answer to three decimal places to give the answer the right number of significant figures:

$$1.5262 \times 10^{-5} = 0.000015262$$

$$\underline{+4.5 \times 10^{-11} = +0.000000000045}$$

$$= 0.000015262045$$

$$\approx 1.526 \times 10^{-5}$$

The answer has four significant figures because the –11 exponent makes the second value too small to make a significant difference in the first value. It would be important only if it affected one of the significant figures in the first number.

104. 4.48×10^2

Based on order of operations, do the division from left to right before doing the addition. Start with $(1.1 \times 10^1) \div (3.68 \times 10^{-6})$. To divide numbers that are in scientific notation, divide the numbers in front of the $\times$ 10 and then subtract the exponents on the powers of ten:

$$\left(1.1 \times 10^1\right) \div \left(3.68 \times 10^{-6}\right) = \left(1.1 \div 3.68\right)\left(10^1 \div 10^{-6}\right)$$

$$= 0.2989\left(10^{1-(-6)}\right)$$

$$= 0.2989 \times 10^7$$

Then divide that answer by 8.2×10^4:

$$\left(0.2989 \times 10^7\right) \div \left(8.2 \times 10^4\right) = \left(0.2989 \div 8.2\right)\left(10^7 \div 10^4\right)$$

$$= 0.03645\left(10^{7-4}\right)$$

$$= 0.03645 \times 10^3$$

$$= 36.45$$

Now add $4.115 \times 10^2 = 411.5$ and put the answer back into scientific notation (round to 448 to give the answer the right number of significant figures):

$$36.45 + 411.5 = 447.95$$

$$\approx 448 = 4.48 \times 10^2$$

All the digits in 411.5 are significant; however, due to the division by a two-significant-figure value (8.2×10^4), only the first two digits in 36.45 are significant. Because 36.45 is understood to have no significant figures after the ones place, neither can the sum 448.

105. 5×10^{-2}

In this problem, follow the order of operations. Solve the numerator and the denominator before dividing.

In the numerator, do the multiplication before doing the addition. In the denominator, take the numbers out of scientific notation before subtracting. Here are the first couple of steps:

$$\frac{\left(4.6\times10^2\right)+\left(6.97\times10^9\right)\times\left(3\times10^{-7}\right)}{\left(5.18\times10^4\right)-\left(2.00\times10^3\right)}$$

$$=\frac{\left(4.6\times10^2\right)+\left(6.97\times10^9\times3\times10^{-7}\right)}{51,800-2,000}$$

Remember that to multiply powers of ten by each other, you add the exponents:

$$=\frac{\left(4.6\times10^2\right)+\left(6.97\times3\right)\left(10^9\times10^{-7}\right)}{49,800}$$

$$=\frac{\left(4.6\times10^2\right)+\left(20.91\right)\left(10^{9+(-7)}\right)}{49,800}$$

$$=\frac{\left(4.6\times10^2\right)+\left(20.91\times10^2\right)}{49,800}$$

To do the addition in the numerator, you can take the numbers out of scientific notation:

$$=\frac{460+2,091}{49,800}$$

$$=\frac{2,551}{49,800}$$

Now you're ready to divide. You can put the numbers back into scientific notation to help you with the calculations:

$$\frac{2,551}{49,800}=\frac{2.551\times10^3}{4.98\times10^4}$$

$$=\left(2.551\div4.98\right)\left(10^3\div10^4\right)$$

$$=\left(0.5122\right)\left(10^{3-4}\right)$$

$$=0.5122\times10^{-1}$$

$$=5.122\times10^{-2}$$

$$\approx5\times10^{-2}$$

The answer has one significant figure. Why? Multiplying by (3×10^{-4}) in the first step would give you a one-significant-figure answer; when you then add (4.6×10^2) to the product, the sum would likewise have one significant figure. Finally, dividing a one-significant-figure number by a three-significant-figure number yields a one-significant-figure answer.

106. 3

All nonzero digits are significant, so 343 has three significant figures.

107. 4

All nonzero digits are significant, so 0.4592 has four significant figures.

108. 6

All nonzero digits and zeros that are sandwiched between two nonzero digits are significant. In 705,204, the 7, 5, 2, and 4 are nonzero digits. The zeros trapped between the 7 and 5 as well as the 2 and 4 are also significant, so 705,204 has a total of six significant figures.

109. 2

Leading zeros (zeros to the left of a nonzero digit and to the right of the decimal point) aren't significant — they're just placeholders. The 7 and 5 are significant, so 0.0075 has two significant figures.

110. 3

Zeros to the right of nonzero digits but before an understood (not written) decimal point are not significant. The 2, 4, and 8 are significant, so 248,000 has three significant figures.

111. 5

The nonzero digits and the trapped zeros are significant, but the trailing zeros are not. So in 9,400,300, the five digits from the 9 to the 3 are significant.

112. 5

The nonzero digits, the trapped zeros, and the trailing zero are all significant. Zeros to the right of a nonzero digit and to the right of the decimal point are significant. So in 1.0070, all five digits are significant.

113. 8

Zeros to the right of a nonzero digit and to the right of the decimal point are significant. This traps the zeros between the 3 and the final 0, making all eight digits in 3,000,000.0 significant.

114. 5

The nonzero digits, the trapped zeros, and the trailing zeros are all significant. The leading zeros are not significant. In 0.0040800, the five digits from the 4 to the final 0 are significant.

115. 3

The nonzero digits and the trailing zero are significant; the leading zero before the decimal point is not. The number 0.870 has three significant figures: the 8, 7, and final 0.

116. 5,500

Adding 5,379 + 100 gives you 5,479. After adding, round the final answer with the least-accurate decimal place in mind. The first number (5,379) ends with the ones place, and the second number (100) is significant only to the hundreds place, so round 5,479 to the hundreds place: 5,500. (You can also write the answer in scientific notation: 5.5×10^3.)

117. 13.0

First, do the addition: 12.4 + 0.59 = 12.99. Now determine the least-accurate decimal place: 12.4 ends with the tenths place, and 0.59 ends with the hundredths place. The tenths place is less accurate than the hundredths place, so round 12.99 to 13.0. (You can also write the answer in scientific notation: 1.30×10^1.)

118. 27.56

Do the subtraction first: 61.035 – 33.48 = 27.555. Round the answer with the least-accurate decimal place in mind. In this question, the first number (61.035) ends with the thousandths place, and the second number (33.48) ends with the hundredths place, so round the answer to the hundredths place: $27.555 \approx 27.56$. (You can also write the answer in scientific notation: 2.756×10^1.)

119. 96

Do the math and then round the answer with the least-accurate decimal place in mind. The addition gives you 71 + 24.87 + 0.0003 = 95.8703. The ones place from the 71 is the least accurate for the three numbers, so round the answer to 96. (You can also write the answer in scientific notation: 9.6×10^1.)

120. –467

Doing the subtraction gives you 0.387 – 467 = –466.613. When subtracting, round the final answer to the least-accurate decimal place that both numbers have in common. In this case, the first number is accurate to the thousandths place and the second number, 467, is accurate only to the ones place, so round the answer to the ones place. The answer is –467. (You can also write the answer in scientific notation: -4.67×10^2.)

121. 0.0467

When adding, you want to round your final answer to the least-accurate decimal place. Sometimes it's easier to see which digit to round your answer to if you line up the numbers vertically:

$$0.005689$$
$$\underline{+0.0410}$$
$$0.046689$$

The least-accurate place is the 0 in the ten-thousandths place of 0.0410, so round the final answer to the ten-thousandths place, giving you 0.0467. (You can also write the answer in scientific notation: 4.67×10^{-2}.)

122. **–60.405**

First, line up the numbers based on the location of the decimal point. Then do the addition and subtraction and round the answer to the least-accurate decimal place.

$$60.0080$$
$$-128.35429$$
$$+7.941$$
$$\overline{-60.40529}$$

The least-accurate decimal place is the thousandths place in 7.941, so round the answer to –60.405. (You can also write the answer in scientific notation: -6.0405×10^1.)

123. **5,100**

Doing the addition gives you 130 + 4,600 + 395.2 = 5,125.2. When adding, round the answer to the least-accurate decimal place. The least-accurate place in 130 is the tens, the least-accurate place in 4,600 is the hundreds, and the least-accurate place in 395.2 is the tenths. Of the three, the hundreds place is the least accurate, so round the answer, 5,125.2, to the hundreds place: 5,100. (You can also write the answer in scientific notation: 5.1×10^3.)

124. **210**

The division gives you $0.0074 \div 0.000035 \approx 211.4285714$. When dividing, the answer should have the same number of significant figures as the number with the fewest significant figures, regardless of where the decimal point is located. In this problem, each number has only two significant figures, so the answer should have two significant figures. You round the answer to 210. (You can also write the answer in scientific notation: 2.1×10^2.)

125. **26,000**

The multiplication gives you $75 \times 349 = 26,175$. In multiplication problems, identify the number that has the fewest significant figures to determine how many significant figures the answer should have. The number 75 has two significant figures, and 349 has three significant figures. The final answer should have the smaller number of significant figures, which is two, so round the answer to 26,000. (You can also write the answer in scientific notation: 2.6×10^4.)

126. **41.6**

The multiplication gives you $7.98 \times 5.21 = 41.5758$. For a multiplication problem, your answer should have the same number of significant figures as the given number with the fewest significant figures. Both 7.98 and 5.21 have three significant figures, so round the answer to 41.6. (You can also write the answer in scientific notation: 4.16×10^1.)

127. **2.0×10^3**

The division gives you $5.00 \div 0.0025 = 2,000$. With division, the answer should have the same number of significant figures as the number with the fewest significant figures. In this question, 5.00 has three significant figures and 0.0025 has two, so the answer should have two significant figures. The easiest way to express 2,000 with two significant figures is to put it into scientific notation, with one zero after the decimal point: 2.0×10^3.

128. **50 cm²**

The multiplication gives you 7.0 cm × 7 cm = 49 cm². Each measurement has a different number of significant figures: 7.0 has two, and 7 has one. One is the lower number, so the answer should have one significant figure. Round 49 to 50. (You can also write the answer in scientific notation: 5×10^1 cm².)

129. **0.0334**

The division gives you 6.48 ÷ 194.21 ≈ 0.033365944. When dividing, you record the answer with the same number of significant figures as the number with the fewest number of significant figures. In this case, you have only three significant figures coming from 6.48, so round the answer to 0.0334. (You can also write the answer in scientific notation: 3.34×10^{-2}.)

130. 2.2×10^{-13}

The multiplication gives you 0.000000029 × 0.00000745 = 0.00000000000021605 = 2.1605 × 10^{-13}. The first number you're multiplying has two significant figures, and the second number has three. Because the answer should have the lower number of significant figures, 2.1605 rounds to 2.2, and the 10^{-13} remains. The answer is 2.2×10^{-13}.

131. **0.20**

For a multiplication and division problem, the answer should have the same number of significant figures as the given number with the fewest number of significant figures. The calculations give you the following:

$$\frac{0.0034 \times 518.27}{9.00} \approx 0.195790889$$

Here, 0.0034 has two significant figures, 518.27 has five significant figures, and 9.00 has three significant figures. The fewest number of significant figures is two, so round the answer to 0.20. (You can also write the answer in scientific notation: 2.0×10^{-1}.)

132. **2,070**

Complete the multiplication first: 2,300.00 × 0.854 = 1,964.2. Then add 110, giving you 2,074.2, and round to the least-accurate decimal place, which is the tens place in 110. The answer is 2,070. (You can also write the answer in scientific notation: 2.07×10^3.)

133. **9.0 g/mL**

Do the subtraction first to see how many significant figures would be in the denominator: 25.0 mL − 23.8 mL = 1.2 mL. Then do the division: 10.78 ÷ 1.2 ≈ 8.983333333. The final answer should have the same number of significant figures as the 1.2 mL, so you round the answer to 9.0 g/mL.

134. 4.39

At first, you may think this answer should have two significant figures. But think about doing the addition in the numerator: 8.1 + 2.32 + 0.741 = 11.161. If you were to round this answer before dividing by the denominator, it would have three significant figures, because with addition, you round to the least-accurate decimal place: in this case, the tenths place. When you're ready to divide, both the numerator and denominator would have three significant figures, so the final answer should have three significant figures as well.

The usual practice is to do all the calculations and round at the end to avoid cumulative rounding errors. So here are the calculations:

$$\frac{8.1+2.32+0.741}{2.54} = \frac{11.161}{2.54} \approx 4.394094488$$

Rounded to three significant figures, the answer is 4.39.

Note that if you round and then divide, your answer will be slightly different: 4.4094, which rounds to 4.41.

135. 55

Do the math first:

$$\frac{250+12}{2.0} \times \frac{1.0}{3.57-1.2} = \frac{262}{2.0} \times \frac{1.0}{2.37} \approx 55.2742616$$

After doing the addition and subtraction, you can determine the number of significant figures that need to be in the answer.

The numerator in the first fraction is accurate only to the tens place (due to the 250), and the denominator in the second fraction is accurate only to the tenths place (due to the 1.2). If you were to round 262 and 2.37 at this point, the resulting numbers (260 and 2.4) would have two significant figures.

The numbers you're multiplying and dividing by also have two significant figures, so the final answer should have two significant figures as well. The final answer is 55. (You can also write the answer in scientific notation: 5.5×10^1.)

136. gas

Think of air in a balloon. Gases take the shape of their containers, and they're easily compressed or expanded (they don't have a definite volume).

137. solid

Think of a solid like ice or a block of wood. A solid has a definite shape and a definite volume.

138. liquid

Matter that takes the shape of its container but has a definite volume is a liquid. Think of room-temperature water. One liter of water is 1 liter of water, whether it's in a vase or spilled all over the kitchen table — same volume, different shape.

139. freezing

When a liquid (such as water) is becoming solid (ice), it's going through a phase change called *freezing*.

140. condensation

When a gas cools, it changes into a liquid. This phase change is called *condensation*.

141. vaporization

When you add enough energy to a liquid that it changes in phase to a gas, it's said to *vaporize* or go through *vaporization*. This phase change is called *evaporation* when it occurs below the liquid's boiling point.

142. sublimation

When matter goes from a solid phase to a gas phase without becoming a liquid, it's said to *sublime*. You can observe sublimation in substances such as dry ice, moth balls, and iodine crystals.

143. deposition

Deposition is the opposite of sublimation. Deposition occurs when a substance goes from the gaseous phase to the solid phase without becoming a liquid in between. You can see the result of deposition as frost in a freezer; the water vapor in the air turns to ice.

144. pure substance; element

Gold (Au), a pure substance, is element number 79 on the periodic table of the elements.

145. pure substance; compound

Table sugar is a pure substance made of the compound sucrose. *Sucrose* is a carbohydrate that contains fixed amounts of carbon, hydrogen, and oxygen atoms bonded together.

146. mixture; homogeneous

Fresh air is a mixture of colorless gases containing mostly nitrogen gas and oxygen gas. Each breath of air contains the same composition of gases, so the mixture is homogeneous.

147. pure substance; element

Oxygen (O), a pure substance, is element number 8 on the periodic table of the elements.

148. mixture; heterogeneous

Vegetable soup is a mixture of water, salt, and assorted vegetables. Each bowl you serve has a slightly different composition of ingredients — that's what makes it heterogeneous.

149. mixture; heterogeneous

Each serving of fruit salad that you take from a serving bowl generally has a different number of each kind of fruit. This variety in each sample makes fruit salad heterogeneous.

150. pure substance; element

Calcium (Ca), a pure substance, is element number 20 on the periodic table of the elements.

151. mixture; heterogeneous

Concrete is made of a variety of rocks and *cement,* which holds the concrete together. The variety of rocks depends on where you live or what's in the mix. You can see the rock differences — although they're sometimes subtle — when you break off a piece of concrete. These differences are what make concrete a heterogeneous mixture.

152. mixture; heterogeneous

Smog is air that contains a variety of particulates (dirt, dust, and so on) and pollutants. Smog can be different shades of color, depending on the types and concentrations of contaminants. This variety of substances mixed together is rarely homogeneous.

153. intensive

An *intensive* property is independent of the amount of the substance that's present. Intensive properties are helpful in identifying an unknown substance.

The boiling point of water is an intensive property. If you have a large pot and a small pot full of boiling water on the stove, the water will be boiling at the same temperature in both.

154. physical

A *physical* change involves a change in the physical properties of a substance, not a change in chemical composition. The shape of a substance is a physical property.

155. chemical

In a *chemical* change, a substance changes into a new substance. The burning of a match illustrates a chemical change, and the match's ability to burn is a chemical property of the match.

156. extensive

An *extensive* property depends on exactly how much of a substance is present. Mass is an example of an extensive property.

157. intensive physical

Density is the ratio of an object's mass to its volume. This physical property can help chemists identify a substance, especially because density is also an intensive property (one that stays the same no matter how much of the substance you have).

158. extensive physical

Length is a physical property that describes size. Length is extensive because it depends on exactly how much of a substance you have.

159. intensive physical

Color is a physical property that's intensive. A piece of paper that's green is still green if you cut the paper in half; the color is independent of the size of the paper.

160. chemical

Flammability is a chemical property because if a flammable substance catches fire, the substance changes into new substances. For example, as wood burns, carbon dioxide and water escape, and a residue of ash remains behind.

161. extensive physical

Mass is a physical property that describes how much of a substance you have. Mass is extensive because it depends on exactly how much is present — more matter, more mass.

162. intensive physical

Odor is a physical property that's intensive. A small orange has a distinctive odor that's essentially the same as the odor of a large orange.

163. intensive physical

Ductility is a physical property that describes how a substance can be drawn into thin wires. A small block of copper is just as ductile as a large block of copper. Ductility is based on the identity of the substance, not how much is present, so it's an intensive property.

164.

intensive physical

Conductivity is a physical property that's intensive. A piece of silver metal will conduct electricity whether it's 10 cm long or 30 cm long. The fact that silver is a good conductor is based on its identity.

165.

intensive physical

Solubility is a physical property because you can evaporate away the solvent and be left with the solute. Solubility is intensive; whether something dissolves or not depends more on the nature of the substance and the solvent than on the amount.

166.

12 g/cm³

Density is the ratio of the mass of a substance to its volume. Enter the numbers in the formula and do the math:

$$\text{density} = \frac{\text{mass}}{\text{volume}}$$
$$= \frac{57.5 \text{ g}}{5.0 \text{ cm}^3}$$
$$= 11.5 \text{ g/cm}^3$$

The 5.0 cm³ measurement limits the answer to two significant figures. Rounded to two significant figures, the answer is 12 g/cm³.

167.

0.884 g/mL

To find the density in grams per milliliter, divide the mass (in grams) by the volume (in milliliters):

$$\text{density} = \frac{\text{mass}}{\text{volume}}$$
$$= \frac{22.1 \text{ g}}{25.0 \text{ mL}}$$
$$= 0.884 \text{ g/mL}$$

Both the mass and the volume measurements have three significant figures, so the answer has three significant figures as well.

168.

1.98 kg/m³

First, convert the number of grams to kilograms; you can do this by multiplying 3,960 g by 1 kg/1,000 g:

$$\frac{3,960 \text{ g}}{1} \times \frac{1 \text{ kg}}{1,000 \text{ g}} = 3.96 \text{ kg}$$

To get the density, divide the mass by the volume:

$$density = \frac{mass}{volume}$$

$$= \frac{3.96 \text{ kg}}{2.00 \text{ m}^3}$$

$$= 1.98 \text{ kg/m}^3$$

Both the mass and the volume measurements have three significant figures, so the answer has three significant figures as well.

169.   240 g

Because density = mass/volume, you can multiply density by volume to find the mass. Here's the initial setup:

$$density = \frac{mass}{volume}$$

$$mass = density \times volume$$

$$= \frac{1.2 \text{ g}}{1 \text{ mL}} \times 0.200 \text{ L}$$

Before you do the math, you need to make the volume units — milliliters and liters — consistent so they'll cancel out, leaving you with the mass in grams. Because 1 L = 1,000 mL, you can multiply by the conversion factor 1,000 mL/1 L. Here are the calculations, with the conversion factor in parentheses:

$$mass = density \times volume$$

$$= \frac{1.2 \text{ g}}{1 \text{ mL}} \times \left(\frac{1,000 \text{ mL}}{1 \text{ L}} \right) \times \frac{0.200 \text{ L}}{1}$$

$$= 240 \text{ g}$$

The density measurement, 1.2 g/mL, has only two significant figures, limiting your answer to two significant figures. The answer is 240 g.

170.   73 g

Because density = mass/volume, you can multiply density by volume to find the mass:

$$density = \frac{mass}{volume}$$

$$mass = density \times volume$$

You have a cube with a side that's 3.00 cm long. The volume of a cube is equal to length × width × height, or side³, so the cube's volume is $V = (3.00 \text{ cm})^3 = 27.0 \text{ cm}^3$. To find the mass, multiply the density by the volume:

$$mass = density \times volume$$

$$= \frac{2.7 \text{ g}}{1 \text{ cm}^3} \times \frac{27.0 \text{ cm}^3}{1}$$

$$= 72.9 \text{ g}$$

The density measurement (2.7 g/cm³) has only two significant figures, limiting the answer to two significant figures, so round the answer to 73 g.

171.

4.0 g/cm³

Density is the mass divided by the volume. You need both the mass of the block and its volume, so first solve for the volume of the block:

$$\text{volume} = \text{length} \times \text{width} \times \text{height}$$
$$= 5.0 \text{ cm} \times 3.0 \text{ cm} \times 2.0 \text{ cm}$$
$$= 30. \text{ cm}^3$$

Then enter the mass and volume in the density formula:

$$\text{density} = \frac{\text{mass}}{\text{volume}} = \frac{120. \text{ g}}{30. \text{ cm}^3} = 4.0 \text{ g/cm}^3$$

The given volume measurement (30. cm³) has only two significant figures, so you include two significant figures in your answer.

172.

29 kg

You need to convert several units in this problem, so make a plan. Start by putting the quantity that has only one unit (the 1.5 L) over 1; writing the number this way keeps it from falling into the denominator.

Note that the density gives you a conversion factor (19.3 g = 1 cm³) that lets you relate the given volume to the mass you want to find. Your plan for the conversions in this problem may look something like this: L → mL → cm³ → g → kg. Here are the calculations with the appropriate units canceled out:

$$\frac{1.5 \cancel{L}}{1} \times \frac{1,000 \cancel{\text{mL}}}{1 \cancel{L}} \times \frac{1 \cancel{\text{cm}^3}}{1 \cancel{\text{mL}}} \times \frac{19.3 \cancel{g}}{1 \cancel{\text{cm}^3}} \times \frac{1 \text{ kg}}{1,000 \cancel{g}} = 28.95 \text{ kg}$$

The given volume (1.5 L) has only two significant figures, so your answer should have two significant figures as well. Rounding to two significant figures gives you an answer of 29 kg.

173.

110 mL

Density is the mass divided by the volume, and you want to find the volume. You can start by solving the density equation for volume:

$$\text{density} = \frac{\text{mass}}{\text{volume}}$$
$$\text{density} \times \text{volume} = \text{mass}$$
$$\text{volume} = \frac{\text{mass}}{\text{density}}$$

Now just enter the numbers and do the math:

$$\text{volume} = \frac{\text{mass}}{\text{density}}$$
$$= \frac{77.0 \cancel{g}}{0.71 \cancel{g}/\text{mL}} \approx 108.45 \text{ mL}$$

The given density measurement (0.71 g/mL) has only two significant figures, so the answer should have two significant figures as well; 108.45 mL rounds to 110 mL.

174.   **4.00 cm**

You're given the cube's mass and density. Density is the mass divided by the volume, so you can solve for the volume of the cube using the density formula:

$$\text{density} = \frac{\text{mass}}{\text{volume}}$$

$$\text{density} \times \text{volume} = \text{mass}$$

$$\text{volume} = \frac{\text{mass}}{\text{density}}$$

$$\text{volume} = \frac{672 \text{ g}}{10.5 \text{ g/cm}^3} = 64.0 \text{ cm}^3$$

Then solve for the length of a side of the cube. The volume, V, of a cube is side $\times$ side $\times$ side, or s^3. Enter the volume in the formula and solve for s:

$$V = s^3$$

$$\sqrt[3]{V} = s$$

$$\sqrt[3]{64.0 \text{ cm}^3} = s$$

$$s = 4.00 \text{ cm}$$

All the measurements in this problem have three significant figures, so the answer also has three significant figures.

175.   **calorie**

A *calorie* is the amount of energy required to raise the temperature of 1 g of water by 1°C. A calorie equals 4.184 joules (J).

176.   **kilojoules/mole**

The heat content of chemicals is measured in kilojoules per mole (kJ/mol).

177.   **Calorie (kilocalorie)**

The food calorie is really a kilocalorie (kcal), or 1,000 calories. You can also indicate a food calorie by writing *Calorie* with a capital *C*.

A *calorie* (small *c*) is the amount of heat necessary to raise the temperature of 1 g of water by 1°C.

178.   **potential; position**

Objects stored above ground level have energy based on their position, which is potential energy.

179.   **kinetic; motion**

A rolling ball is moving, so it has kinetic energy, the energy of motion.

Note: If the ball is rolling downhill, it has a combination of potential and kinetic energy until it reaches the bottom of the hill. At the bottom, all the potential energy (energy of position) will have changed to kinetic energy.

180. **stored; potential**

Fuels contain chemical energy stored in the bonds between elements. This chemical energy is potential energy.

181. **6.207 kcal**

In this problem, start by writing the number of joules over 1. A calorie equals 4.184 J, so convert joules to calories by multiplying the number of joules by 1 cal/4.184 J. Then convert to kilocalories by multiplying by 1 kcal/1,000 calories: J → cal → kcal:

$$\frac{25,970\ J}{1} \times \frac{1\ cal}{4.184\ J} \times \frac{1\ kcal}{1,000\ cal} = 6.207\ kcal$$

The answer has four significant figures because the given measurement, 25,970 J, has four significant figures. The conversions here are exact numbers, so they have no bearing on the number of significant figures in the answer.

182. **1.3×10^{12} J**

To find the number of joules from a given number of kilocalories, first convert kilocalories to calories by multiplying by 1,000 kcal/1 cal. Then convert calories to joules by multiplying by 4.184 J/1 cal: kcal → cal → J:

$$\frac{3.1 \times 10^8\ kcal}{1} \times \frac{1,000\ cal}{1\ kcal} \times \frac{4.184\ J}{1\ cal} = 1.29704 \times 10^{12}\ J$$

$$\approx 1.3 \times 10^{12}\ J$$

The answer has two significant figures because the given energy measurement has two significant figures. The conversions involve exact numbers, so they have no bearing on the number of significant figures in the answer.

183. **degrees Celsius, degrees Fahrenheit, and kelvins**

Average kinetic energy is the definition of *temperature*. You can measure temperature in kelvins, degrees Celsius (°C), or degrees Fahrenheit (°F).

184. **80.0 K**

Since one degree Celsius is identical to one kelvin, an 80.0 change in one is the same as an 80.0 change in the other.

185. **801°C**

To solve for degrees Celsius from kelvins, use the Kelvin temperature equation:

$$K = °C + 273.15$$
$$°C = K - 273.15$$
$$= 1,074 - 273.15$$
$$= 800.85°C$$

Rounded to the ones place, the answer is 801°C.

186. **–297°F**

To convert from degrees Celsius to degrees Fahrenheit, use the formula $°F = \frac{9}{5}(°C) + 32$.

Be sure to follow the order of operations by doing the multiplication and division before the addition:

$$°F = \frac{9}{5}(°C) + 32$$
$$= \left(\frac{9}{5} \times (-183)\right) + 32$$
$$= -329.4 + 32$$
$$= -297.4°F$$

The answer should have three significant figures because the original measurement (183°C) has three significant figures; the conversions use exact numbers, so they have no bearing on the number of significant figures in the answer. Rounded correctly, the answer is –297°F.

187. **37.0°C**

To go from degrees Fahrenheit to degrees Celsius, use the formula $°C = \frac{5}{9}(°F - 32)$.

Follow the order of operations by completing the subtraction in the parentheses before multiplying by 5 and dividing by 9:

$$°C = \frac{5}{9}(°F - 32)$$
$$= \frac{5}{9}(98.6 - 32)$$
$$= \frac{5}{9}(66.6)$$
$$= 37°C$$

Write the answer with three significant figures, just like the original number. The answer is 37.0°C.

188. **–110°F**

To go from degrees Fahrenheit to degrees Celsius, use the formula $°C = \frac{5}{9}(°F - 32)$.

Plug in –78°C for the $°C$ and solve.

$$°F = \frac{9}{5}(°C) + 32$$

$$= \left(\frac{9}{5} \times (-78)\right) + 32$$

$$= -140.4 + 32$$

$$= -108.4°F$$

Rounded to two significant figures, the answer is –110°F. The answer has two significant figures because the given measurement (–78°C) has two significant figures; the conversions are exact numbers, so they don't have any bearing the on the significant figures in the answer.

---

189. **318 K**

To find the Kelvin temperature from degrees Fahrenheit, first convert from Fahrenheit to Celsius:

$$°C = \frac{5}{9}(°F - 32)$$

$$= \frac{5}{9}(113 - 32)$$

$$= \frac{5}{9}(81)$$

$$= 45°C$$

Then you can convert from degrees Celsius to kelvins by adding 273.15:

$$K = °C + 273.15$$

$$= 45 + 273.15$$

$$= 318.15 \text{ K}$$

Rounded to the ones place, the answer is 318 K.

---

190. **80.°F**

To find the temperature in degrees Fahrenheit, first convert the temperature from kelvins to degrees Celsius by subtracting 273.15 from the Kelvin temperature:

$$°C + 273.15 = K$$

$$°C = K - 273.15$$

$$= 300. - 273.15$$

$$= 26.85°C$$

Next, use the formula $°F = \frac{9}{5}(°C) + 32$ to convert to degrees Fahrenheit:

$$°F = \frac{9}{5}(°C) + 32$$

$$= \left(\frac{9}{5} \times 26.85\right) + 32$$

$$= 48.33 + 32$$

$$= 80.33°F$$

This answer rounds to 80.°F.

191. **11 protons**

The atomic number of an element is equal to the number of protons. Sodium, Na, is atomic number 11, so it has 11 protons.

192. **35 electrons**

The number of protons is equal to the number of electrons in an atom, which is neutral. Bromine, Br, is atomic number 35. It has 35 protons, so it has 35 electrons.

193. **28 electrons**

Nickel, Ni, has 28 protons. Because the number of protons equals the number of electrons in an atom, an atom of nickel contains 28 electrons.

194. **86 protons**

The number of protons in an atom is equal to the element's atomic number. Radon, Rn, is atomic number 86, so it has 86 protons.

195. **21 neutrons**

To find the number of neutrons in an atom, subtract the atomic number (which equals the number of protons) from the atom's mass number (which equals the number of protons + the number of neutrons). Here are the calculations for potassium-40:

# of neutrons = mass number 40 – 19 protons

= 21 neutrons

196. **The atomic number is equal to the number of protons.**

The general definition of atomic number is the number of protons in an atom.

197. **The mass number is equal to the number of neutrons plus the atomic number (the number of protons).**

The majority of the mass of the atom is in the nucleus, where you find the protons and neutrons, collectively called *nucleons*. By adding the number of neutrons to the atomic number (which equals the number of protons), you get the mass number.

198. **29 protons, 29 electrons, and 34 neutrons**

The number of protons is equal to the atomic number; therefore, copper, Cu, has 29 protons. Because the atom is neutral, the number of protons equals the number of electrons.

To get the number of neutrons, subtract the number of protons (29) from the mass number. The mass number, 63, follows the name of the element in *copper-63,* so the number of neutrons is 63 – 29 = 34.

199. **6 protons and 8 neutrons**

If the atom has 6 electrons, then it has 6 protons, because the positive and negative charges have to cancel each other out. The number of neutrons is equal to the mass number minus the number of protons: 14 – 6 = 8.

200. **mass number 91, 40 protons**

The number of protons equals the number of electrons in an atom, so 40 electrons means 40 protons. The element is atomic number 40, zirconium (Zr).

The mass number is equal to the number of protons plus the number of neutrons. The atom contains 51 neutrons, so the mass number is 40p + 51n = 91.

201. **the mass number**

In isotope notation $\left({}_{Z}^{A}X \right)$, the top number A represents the *mass number,* which is the sum of the number of protons and neutrons in the nucleus of the atom.

202. **the number of protons**

In isotope notation $\left({}_{Z}^{A}X \right)$, the bottom number Z represents the *atomic number,* which is the same as the number of protons in the nucleus of the atom.

203. **9 protons, 9 neutrons**

In isotope notation, the bottom number is equal to the number of protons, so ${}_{9}^{18}F$ has nine protons. The number of neutrons is equal to the top number (mass number) minus the bottom number (atomic number): 18 – 9 = 9 neutrons.

204. **11 protons, 14 neutrons**

The bottom number in isotope notation is the atomic number, which is equal to the number of protons, so ${}_{11}^{25}Na$ has 11 protons. The number of neutrons is equal to the mass number (the top number) minus the atomic number (the bottom number): 25 – 11 = 14 neutrons.

205. ${}_{6}^{12}C$

In isotope notation, carbon-12 has the atomic symbol C with a mass number of 12 on top. The bottom number is the atomic number of carbon, 6. The mass number follows the name of the element in the isotope name (carbon-12), and you can find the element's symbol and atomic number on the periodic table.

Answers
201–300

206. $^{37}_{17}Cl$

In isotope notation, chlorine-37 has the atomic symbol Cl with a mass number of 37 on top. The bottom number is the atomic number of chlorine, 17. The mass number follows the name of the element in the isotope name (chlorine-37), and you can find the element's symbol and atomic number on the periodic table.

207. **argon-35**

Ar is the symbol for argon. When you write the isotope name, the top number in isotope notation (the mass number) becomes the number that follows a hyphen and the element's name: argon-35.

208. **An ion has a different number of electrons from the atom.**

An *ion* is an atom that has either gained or lost electrons.

If, instead, the number of protons changes, then the element's identity is changed (transmutation). If the number of neutrons changes, you have an isotope.

209. **isoelectronic**

The prefix for *same* is *iso-*, so two atoms or ions with the same number of electrons are said to be *isoelectronic*.

210. **The number of protons is greater than the number of electrons.**

Protons are positive and electrons are negative, so for an ion to be positive, it needs more protons than electrons. An ion with a positive charge has lost electrons.

211. **The number of protons is less than the number of electrons.**

For an ion to have a negative charge, it must have more electrons than protons. An ion with a negative charge has gained electrons.

212. **20 protons and 18 electrons**

$^{40}_{20}Ca^{2+}$ has a positive charge, so the number of protons must be greater than the number of electrons. The atomic number of Ca is 20 (the bottom number in isotope notation), giving you 20 protons. To find the number of electrons, simply subtract the charge from the number of protons: $20 - 2 = 18$ electrons.

213. **53 protons and 54 electrons**

For an ion to be negative, it needs to have more electrons than protons. Iodine has an atomic number of 53, so it has 53 protons. Because the iodine ion has a –1 charge, the ion must have one more electron than proton, giving you a total of 54 electrons.

Answers
201–300

214. **13 protons and 10 electrons**

For an ion to be positive, the atom must have more protons than electrons. $^{30}_{13}\text{Al}^{3+}$ needs to have three more protons than electrons, because the charge is +3. With an atomic number of 13, the aluminum ion must have three fewer electrons than protons: $13 - 3 = 10$ electrons.

215. **15 protons and 18 electrons**

Sometimes you need to look up the atomic number to determine the number of protons. The atomic number of phosphorus, P, is 15, so phosphorus has 15 protons. You can find the number of electrons by subtracting the charge from the number of protons: $15 - (-3) = 15 + 3 = 18$ electrons.

216. $^{109}_{47}\text{Ag}^{+}$

To write the isotope notation, start with the symbol for silver, Ag. To the left of the symbol and slightly above the base line, write 109, the mass number that follows the hyphen after the element's name. Then write the atomic number, 47 (from the periodic table), below 109. Write a plus sign as the superscript after the Ag to indicate the charge; writing the 1 isn't necessary because it's understood.

217. $^{34}_{16}\text{S}^{2-}$

The symbol for sulfur is S. Write the mass number, 34, to the upper left of the S. Sulfur has an atomic number of 16, so write 16 under the 34. Represent the charge of negative 2 by writing the superscript 2– after the S.

218. **24 protons, 28 neutrons, 18 electrons**

Chromium, Cr, has an atomic number of 24, which is equal to the number of protons. If the mass number is 52, then the atom contains 28 neutrons, because the mass number minus the atomic number (from the periodic table) gives you the number of neutrons: $52 - 24 = 28$ neutrons.

To get the number of electrons, subtract the ion's charge from the number of protons: $24 - 6 = 18$ electrons. This answer makes sense because a +6 charge means that there are six more protons than electrons.

219. **28 protons, 34 neutrons, 25 electrons**

Nickel, Ni, has an atomic number of 28, which is equal to the number of protons. Because the top number in the isotope notation is 62, the mass number is 62. To find the number of neutrons, subtract the atomic number (from the periodic table) from the mass number: $62 - 28 = 34$ neutrons.

To get the number of electrons, subtract the ion's charge from the number of protons: $28 - 3 = 25$ electrons. This answer makes sense because a +3 charge means that there are three more protons than electrons.

220. **Hund's rule**

"Electrons fill subshells singly before doubly" is a shortened restatement of Hund's rule.

221. **Aufbau principle**

The Aufbau principle describes the order in which electrons fill orbitals.

222. **14**

There are seven f orbitals. Each orbital can hold a maximum of two electrons, so a maximum of fourteen electrons are in the f orbitals.

223. **6**

There are three p orbitals. Each orbital can hold a maximum of two electrons, so a maximum of six electrons are in the p orbitals.

224. $1s^22s^22p^2$

Carbon, C, has an atomic number of 6, which means that it has six protons and six electrons. Electrons fill orbitals from lowest energy to highest energy, so the first energy level gets filled before the second energy level. Remember that the s orbital can hold a maximum of two electrons and that the p orbitals can hold a maximum of six electrons.

To check the electron configuration, note that adding the superscripts on $1s^22s^22p^2$ gives you 2 + 2 + 2 = 6 electrons.

225. $1s^22s^22p^63s^2$

Magnesium, Mg, has an atomic number of 12, so it has 12 protons and 12 electrons. Electrons fill orbitals from the first energy level up to the third in magnesium. The first energy level contains two electrons, the second energy level contains eight electrons (two in an s orbital and six in a p orbital), and the third energy level contains the last two electrons, for a total of 12 electrons.

226. $1s^22s^22p^63s^23p^6$

Argon, Ar, has an atomic number of 18, so it has 18 protons and 18 electrons. The electrons fill the first and second energy levels completely (two in each s orbital and six in the p orbital). The remaining electrons fill the 3s and 3p orbitals.

227. $1s^22s^22p^63s^23p^64s^23d^{10}4p^5$

An atom of bromine, Br, has 35 protons and 35 electrons. The first energy level can hold a maximum of two electrons, and the second energy level can hold a maximum of eight electrons (two in the s orbital and six in the p orbital). The third energy level can hold a maximum of 18 electrons (two in an s orbital, six in p orbitals, and 10 in d orbitals), and the fourth energy level contains the remainder of the electrons. Remember that the 4s orbital doesn't require as much energy as the 3d orbital in order to fill with electrons, so the 4s orbital fills first.

228. $1s^22s^22p^63s^23p^64s^23d^{10}4p^65s^24d^2$

An atom of zirconium, Zr, has 40 protons and 40 electrons. The first energy level contains two electrons, the second energy level contains eight electrons (two in the s orbital and six in the p orbital), the third energy level contains 18 electrons, and the fourth energy level contains the remainder of the electrons. Remember that the 5s orbital doesn't require as much energy as the 4d orbital in order to fill with electrons, so the 5s orbital fills first.

229. $1s^22s^22p^63s^23p^64s^23d^{10}4p^65s^24d^{10}5p^66s^24f^{14}5d^{10}6p^67s^25f^6$

Finding the expected electron configuration of plutonium, Pu, is hard because there are so many electrons. When you get down to the bottom of the periodic table, electrons are in every energy level. Just be sure to add them in the correct order based on the energy it takes to fill the orbitals.

Adding the superscripts in the expected electron configuration, $1s^22s^22p^63s^23p^64s^23d^{10}4p^65s^24d^{10}5p^66s^24f^{14}5d^{10}6p^67s^25f^6$, gives you $2 + 2 + 6 + 2 + 6 + 2 + 10 + 6 + 2 + 10 + 6 + 2 + 14 + 10 + 6 + 2 + 6 = 94$ electrons.

230. **the spin quantum number**

The *spin quantum number, s,* describes the direction of the spin of the electron. According to this model, the electron spins either clockwise or counterclockwise.

231. **the principal quantum number**

The *principal quantum number, n,* describes the average distance between the nucleus and the orbital, with the first energy level at a distance of 1.

232. **the magnetic quantum number**

The *magnetic quantum number, m_l,* describes how the orbitals are oriented in three-dimensional space.

233. **the angular momentum quantum number**

The *angular momentum quantum number, l,* describes the shape of the orbital.

234. $-\frac{1}{2}, \frac{1}{2}$

An electron can spin in only two directions: clockwise or counterclockwise. Because direction is relative, each has a numeric value of 1/2 assigned to the electron; one direction is considered positive, and the other is negative.

235.

average atomic mass

The decimal number found in most blocks on the periodic table is the weighted average of the relative abundance of all known isotopes. The more common an isotope is, the more it affects the average atomic mass.

236.

6.940 amu

To find the average atomic mass, take the percent abundance and multiply it by the given mass of the element. Repeat for as many isotopes as are given. Then add the multiplied numbers together:

$$\frac{7.59}{100} \times 6.0151 \text{ amu} \approx 0.45655 \text{ amu}$$

$$\frac{92.41}{100} \times 7.0160 \text{ amu} \approx \underline{+6.48349 \text{ amu}}$$

$$= 6.94004 \approx 6.940 \text{ amu}$$

237.

35.45 amu

You have two isotopes, so take the percent of each isotope, multiply it by the given mass of the isotope, and then add the two numbers together:

$$\frac{75.78}{100} \times 34.96885 \text{ amu} \approx 26.49939 \text{ amu}$$

$$\frac{24.22}{100} \times 36.9659 \text{ amu} \approx \underline{+ 8.95314 \text{ amu}}$$

$$= 35.45253 \text{ amu} \approx 35.45 \text{ amu}$$

238.

24.31 amu

Multiply each atomic mass by the percent abundance of each isotope and add the three numbers together:

$$\frac{78.99}{100} \times 23.985 \text{ amu} \approx 18.9458 \text{ amu}$$

$$\frac{10.00}{100} \times 24.986 \text{ amu} = 2.4986 \text{ amu}$$

$$\frac{11.01}{100} \times 25.983 \text{ amu} \approx \underline{+ 2.8607 \text{ amu}}$$

$$= 24.3051 \text{ amu} \approx 24.31 \text{ amu}$$

239.

39.098 amu

Multiply each atomic mass by the percent abundance of each isotope and add the three numbers together. Be careful in this question — remember to divide the percent that's less than 1 by 100 just like the other percent abundances:

$$\frac{93.258}{100} \times 38.9637 \text{ amu} \approx 36.33677 \text{ amu}$$

$$\frac{0.01170}{100} \times 39.9640 \text{ amu} \approx 0.0046758 \text{ amu}$$

$$\frac{6.7302}{100} \times 40.9618 \text{ amu} \approx \underline{+\ 2.756811 \text{ amu}}$$

$$= 39.0982568 \text{ amu} \approx 39.098 \text{ amu}$$

240. **55.845 amu**

Multiply each atomic mass by the percent abundance of each isotope and add the four numbers together. Don't forget to divide the percent that's less than 1 by 100 just like the other percent abundances.

$$\frac{5.845}{100} \times 53.9396 \text{ amu} \approx \dots3.152770 \text{ amu}$$

$$\frac{91.754}{100} \times 55.9349 \text{ amu} \approx 51.322508 \text{ amu}$$

$$\frac{2.119}{100} \times 56.9354 \text{ amu} \approx \dots1.206461 \text{ amu}$$

$$\frac{0.282}{100} \times 57.9333 \text{ amu} \approx \underline{+\ 0.163372 \text{ amu}}$$

$$= 055.845111 \text{ amu} \approx 55.845 \text{ amu}$$

241. **83.80 amu**

Multiply each atomic mass by the percent abundance of each isotope and add the six numbers together. What makes this problem challenging is the sheer number of numbers and decimal places that you have to keep up with. Having a calculator that can retain your display until you finish entering the numbers is a definite plus here.

$$\frac{0.350}{100} \times 77.9204 \text{ amu} \approx 0.27272 \text{ amu}$$

$$\frac{2.28}{100} \times 79.9164 \text{ amu} \approx 1.82209 \text{ amu}$$

$$\frac{11.58}{100} \times 81.9135 \text{ amu} \approx 9.48558 \text{ amu}$$

$$\frac{11.49}{100} \times 82.9141 \text{ amu} \approx 9.52683 \text{ amu}$$

$$\frac{57.00}{100} \times 83.9115 \text{ amu} \approx 47.82956 \text{ amu}$$

$$\frac{17.30}{100} \times 85.9106 \text{ amu} \approx \underline{+\ 14.86253 \text{ amu}}$$

$$\approx 83.79931 \text{ amu} \approx 83.80 \text{ amu}$$

242. **80.% boron-11**

To solve this problem, you need to rely on your algebra skills. You have two isotopes of boron, and you have to solve for the percent abundance of one of the isotopes. So here's your initial equation:

(% B-11)(mass B-11) + (% B-10)(mass B-10) = average atomic mass

Next, substitute in the given numbers from the problem:

(% B-11)(11.009306 amu) + (% B-10)(10.012397 amu) = 10.81 amu

You have two unknown percentages, and you need to figure out how they're related. The two percentages have to add up to 100%: (% B-11) + (% B-10) = 100%. In decimal form, 100% equals 1, so if you consider the amount of boron-11 to be x, then the amount of boron-10 has to be $1 - x$.

By substituting these values into the initial equation and solving for x, you can find the percent of boron-11:

$$11.009306x + 10.012937(1 - x) = 10.81$$
$$11.009306x + 10.012937 - 10.012937x = 10.81$$
$$0.996369x + 10.012937 = 10.81$$
$$0.996369x = 10.81 - 10.012937$$
$$\frac{0.996369x}{0.996369} = \frac{0.797063}{0.996369}$$
$$x \approx 0.79997$$
$$x \approx 79.997\%$$
$$x \approx 80.\% \text{ boron-11}$$

The subtraction in the fourth line (10.81 – 10.012937) limits the answer to two decimal places (two significant figures), which in turn limits the final answer to two significant figures.

243. nuclear fusion

Nuclear fusion is the process that occurs in the sun to give off energy. During nuclear fusion, two atomic nuclei join (fuse) to form a heavier nucleus.

244. nuclear fission

Nuclear fission is the process by which large atoms are broken into smaller atoms, producing large amounts of energy as well as additional neutrons as byproducts.

245. alpha decay

An *alpha particle* is a helium nucleus (two protons and two neutrons), so when a helium nucleus is ejected from the nucleus of an atom, the nucleus is said to have undergone alpha decay.

246. gamma decay

A $_0^0\gamma$ ray, or *gamma ray,* is evidence of gamma decay. The superscript 0 is the mass number, the subscript 0 is the atomic number, and γ is the Greek letter gamma. A gamma ray is a high-energy form of electromagnetic radiation (light).

247. $_{-1}^{0}e$

Beta decay produces a *beta particle,* which is essentially an electron. The superscript 0 is the mass number, the subscript –1 is the atomic number, and *e* is the symbol for an electron or positron. You can also write beta decay as $_{-1}^{0}\beta$.

248. positron emission

A $_{+1}^{0}e$ particle is a positron. The superscript 0 is the mass number, the subscript +1 is the atomic number, and *e* is the symbol for a positron or electron. When a positron is produced in a nuclear reaction, the atom is said to have undergone *positron emission.*

249. electron capture

An easy way to distinguish between electron capture and beta decay is to determine where the electron is in the equation. In beta decay, the electron is on the product side of the reaction, but with electron capture, the electron is on the reactants side. In this reaction, the beryllium atom has captured the beta particle, making a lithium atom.

For the beryllium atom (Be), electron (e), and lithium atom (Li), the superscripts are the mass numbers, and the subscripts are the atomic numbers. In a nuclear equation, the sum of the superscripts on each side must be equal, and the sum of the subscripts on each side must be equal.

250. alpha decay

The emission of an alpha particle — two protons and two neutrons — from a nucleus is evidence of alpha decay. Remember that an *alpha particle* is the same thing as a helium nucleus.

For the uranium atom (U), thorium atom (Th), and helium (He), the superscripts are the mass numbers and the subscripts are the atomic numbers. In a nuclear equation, the sum of the superscripts on each side must be equal, and the sum of the subscripts on each side must be equal.

251. gamma decay

The emission of a gamma ray $\left(_{0}^{0}\gamma\right)$ is evidence of gamma decay.

252. beta decay

Beta decay occurs when a beta particle, or electron, is emitted from a nucleus. For the phosphorus atom (P), sulfur atom (S), and electron (e), the superscripts are the mass numbers, and the subscripts are the atomic numbers. In a nuclear equation, the sum of the superscripts on each side must be equal, and the sum of the subscripts on each side must be equal.

253. $^{224}_{88}$ Ra

The missing particle in this reaction is the only reactant. In nuclear reactions, the mass numbers on the left side of the arrow have to add up to the sum of the mass numbers on the right side of the arrow. That means the mass number of the missing particle on the left has to be 220 + 4 = 224.

Similarly, the atomic numbers on the left of the arrow have to add up to the sum of the atomic numbers on the right side of the arrow. Therefore, the atomic number of the missing particle on the left has to be 86 + 2 = 88.

Use the atomic number to identify the element symbol to finish the isotope notation. The element with atomic number 88 is radium, Ra.

254. $^{243}_{94}$ Pu

Because the total mass numbers and atomic numbers on the left and right sides of the arrow have to be equal, use a little algebra and solve for the missing piece. Subtract the numbers for helium from the numbers for curium: The mass number of the missing particle is 247 – 4 = 243, and the atomic number of the missing particle is 96 – 2 = 94.

Then look up the identity of the element whose atomic number is 94 to get the symbol, Pu, for plutonium.

255. $^{0}_{-1}\beta$

By subtracting the mass numbers and atomic numbers of the elements on the left side of the arrow from the mass numbers and atomic numbers of the isotopes on the right side of the arrow, you can find the missing piece of this equation. Take the plutonium numbers minus the americium numbers. The mass number of the missing particle is 241 – 241 = 0, and the atomic number of the missing particle is 94 – 95 = –1.

No element has an atomic number with a charge, so this must be something that basically has no mass and a charge of –1, which is a beta particle (electron).

256. $^{0}_{0}\gamma$

When gamma ray emission occurs, the same element with the same mass and atomic numbers shows up on both sides of the equation. No mass was lost, so the emitted particle must be a gamma ray, $^{0}_{0}\gamma$.

257. $^{93}_{42}$ Mo

To find the missing part of this equation, subtract the mass numbers and atomic numbers of the isotopes on the left side of the arrow from the mass numbers and atomic numbers of the isotope on the right side of the arrow. Take the niobium numbers minus the beta particle numbers. The mass number of the missing particle is 93 – 0 = 93, and the atomic number of the missing particle is 41 – (–1) = 42.

Next, use the periodic table to look up atomic number 42 to get the symbol, Mo, for molybdenum.

258. $_1^3\text{H}$

This question involves two isotopes on the right side of the arrow, so add together their mass numbers and then their atomic numbers. The sum of the mass numbers is $4 + 1 = 5$, and the sum of the atomic numbers is $2 + 0 = 2$.

Next, subtract the mass number and the atomic number of the hydrogen isotope on the left side of the arrow: The difference in mass numbers is $5 - 2 = 3$, and the difference in atomic numbers is $2 - 1 = 1$.

The resulting atomic number is 1, which is the atomic number of hydrogen. This form of hydrogen, with a mass number of 3, is a very special isotope known as *tritium*.

259. $_8^{17}\text{O}$

This question involves two isotopes on the left side of the arrow, so add together their mass numbers and then their atomic numbers. The sum of the mass numbers is $4 + 14 = 18$, and the sum of the atomic numbers is $2 + 7 = 9$.

Next, subtract the mass number and the atomic number of the hydrogen isotope on the right side of the arrow. The difference in mass numbers is $18 - 1 = 17$, and the difference in atomic numbers is $9 - 1 = 8$.

The resulting atomic number is 8, which is the atomic number of oxygen, O.

260. 2_0^1n

To solve for the missing piece of this equation, not only do you have to add together the two parts on the left side of the arrow, but you also have to combine the two parts on the right side of the arrow. On the left, the sum of the mass numbers is $235 + 1 = 236$, and the sum of the atomic numbers is $92 + 0 = 92$. On the right, the sum of the mass numbers is $134 + 100 = 134$, and the sum of the atomic numbers is $54 + 38 = 92$.

Next, subtract the sums on the right side of the arrow from the sums on the left. The difference in mass numbers is $236 - 234 = 2$, and the difference in atomic numbers is $92 - 92 = 0$.

The missing part has an atomic mass of 2 and no protons or electrons. A neutron has a mass of 1 and has no charge, so two neutrons must be missing from the right side of this equation.

261. 12.5 g

To find the amount of undecayed atoms that remain after a certain number of half-lives, you can take the original amount and divide it by 2 as many times as there are half-lives. Visually, the process could look like this:

400. g $\div 2$ 200. g $\div 2$ 100. g $\div 2$ 50.0 g $\div 2$ 25.0 g $\div 2$ 12.5 g

Or mathematically, you can write the mass of undecayed atoms as

$$\text{initial mass} \times \left(\frac{1}{2}\right)^n$$

which simplifies to the following:

$$\frac{\text{initial mass}}{2^n}$$

where n is the number of half-lives that have occurred. Here are the calculations for this problem:

$$\frac{400.\text{ g}}{2^5} = 12.5\text{ g}$$

262. 43.8 g

To find the amount of decayed atoms, first calculate the mass of atoms that remain undecayed after a certain number of half-lives. You can take the original amount and divide it by 2 as many times as there are half-lives. Visually, the process could look like this:

50.0 g $\div 2$ 25.0 g $\div 2$ 12.5 g $\div 2$ 6.25 g

Or mathematically, you can write the mass of undecayed atoms as

$$\frac{\text{initial mass}}{2^n}$$

where n is the number of half-lives that have occurred. Enter the numbers and solve:

$$\frac{50.0\text{ g}}{2^3} = 6.25\text{ g}$$

Then subtract the mass of undecayed atoms from the original sample to find out how much of the sample has decayed: 50.0 g – 6.25 g = 43.75 g, which is 43.8 g when you round the answer to the tenths place.

263. 330. g

To find the size of the original sample, simply multiply the mass of the final sample by 2 six times, because it went through six half-lives:

$$5.15\text{ g} \times 2 \times 2 \times 2 \times 2 \times 2 \times 2$$
$$= 5.15\text{ g} \times 2^6$$
$$= 329.6\text{ g}$$

264. 4

The amount of undecayed material is the original amount of material divided by 2 raised to n, a power equal to the number of half-lives:

$$\frac{\text{original amount}}{2^n} = \text{amount remaining}$$

Enter the numbers and start to solve for the number of half-lives:

$$\frac{1.500 \times 10^{20}\text{ atoms}}{2^n} = 9.375 \times 10^{18}\text{ atoms}$$

$$1.500 \times 10^{20}\text{ atoms} = \left(9.375 \times 10^{18}\text{ atoms}\right) \times 2^n$$

$$\frac{1.500 \times 10^{20}\text{ atoms}}{9.375 \times 10^{18}\text{ atoms}} = 2^n$$

$$16 = 2^n$$

Take the log of both sides of the equation and finish solving for n:

$$16 = 2^n$$

$$\log 16 = \log\left(2^n\right)$$

$$\log 16 = n \log 2$$

$$\frac{\log 16}{\log 2} = n$$

$$4 = n$$

Answers 201–300

---

265. $\frac{1}{8}$

First, determine how many half-lives have passed by dividing the total time elapsed by the duration of one half-life:

$$\frac{39 \text{ hours}}{1} \times \frac{1 \text{ half-life}}{13 \text{ hours}} = 3.0 \text{ half-lives}$$

Next, divide 1 by 2 the number of times equal to the number of half-lives:

$$\frac{1}{2^3} = \frac{1}{2 \times 2 \times 2} = \frac{1}{8}$$

---

266. **24.8 g**

First, determine how many half-lives have passed by dividing the total time elapsed by the length of one half-life:

$$\frac{56.14 \text{ days}}{1} \times \frac{1 \text{ half-life}}{8.02 \text{ days}} = 7.00 \text{ half-lives}$$

Next, determine how many grams remain undecayed by dividing 25.0 g by 2 raised to the number of half-lives:

$$\frac{25.0 \text{ g}}{2^7} = 0.195 \text{ g}$$

Finally, subtract the grams of undecayed iodine-131 from the original number of grams: 25.0 g – 0.195 g = 24.8 g decayed.

---

267. **176 g**

First, determine how many half-lives have passed. To do this, divide the total number of years by the length of one half-life:

$$\frac{115.6 \text{ years}}{1} \times \frac{1 \text{ half-life}}{28.9 \text{ years}} = 4.00 \text{ half-lives}$$

After you know how many half-lives have passed, you can get the size of the original sample. Multiply the 11 grams that remain undecayed by 2 multiplied by itself four times (the number of half-lives):

$$11 \text{ g} \times 2 \times 2 \times 2 \times 2$$

$$= 11 \text{ g} \times 2^4$$

$$= 176 \text{ g}$$

268. 1 minute

On this question, you first need to determine how many half-lives have passed. The amount of undecayed material is the original amount of material divided by 2 raised to n, a power equal to the number of half-lives:

$$\frac{\text{original amount}}{2^n} = \text{amount remaining}$$

Plug in the numbers and start to solve for the number of half-lives:

$$\frac{2.56 \times 10^{10} \text{ atoms}}{2^n} = 8.00 \times 10^8 \text{ atoms}$$

$$2.56 \times 10^{10} \text{ atoms} = \left(8.00 \times 10^8 \text{ atoms}\right) \times 2^n$$

$$\frac{2.56 \times 10^{10} \text{ atoms}}{8.00 \times 10^8 \text{ atoms}} = 2^n$$

$$32 = 2^n$$

Take the log of both sides of the equation and solve for n:

$$32 = 2^n$$

$$\log 32 = \log\left(2^n\right)$$

$$\log 32 = n \log 2$$

$$\frac{\log 32}{\log 2} = n$$

$$5 = n$$

Now divide the length of time that has passed, 5 minutes, by the number of half-lives to get the length of one half-life:

$$\frac{5 \text{ min.}}{5 \text{ half-lives}} = 1 \text{ min./half-life}$$

269. $\frac{3}{4}$

First, determine how many half-lives have passed by dividing the total time elapsed by the length of one half-life:

$$\frac{320 \text{ days}}{1} \times \frac{1 \text{ half-life}}{160 \text{ days}} = 2.0 \text{ half-lives}$$

Now you can pick any sample size and think about the decay process. After one half-life, 50% of the sample remains. After a second half-life, 1/2 of the 50%, or 25%, of the sample remains. And 25% is 1/4 of the original amount.

If 1/4 of the sample remains, then 3/4 of the sample has decayed.

270. 110 hours

You first need to determine how many half-lives have passed. The amount of undecayed material is the original amount of material divided by 2 raised to n, a power equal to the number of half-lives:

$$\frac{\text{original amount}}{2^n} = \text{amount remaining}$$

$$\frac{2.5 \text{ kg}}{2^n} = 0.61 \text{ g}$$

The units of mass don't match, so convert the kilograms to grams (1,000 g = 1 kg). Then start to solve for the number of half-lives:

$$\frac{2.5 \cancel{\text{kg}} \times \dfrac{1,000 \text{ g}}{1 \cancel{\text{kg}}}}{2^n} = 0.61 \text{ g}$$

$$2,500 \text{ g} = (0.61 \text{ g}) \times 2^n$$

$$\frac{2,500 \text{ g}}{0.61 \text{ g}} = 2^n$$

$$4,098.36 = 2^n$$

Take the log of both sides of the equation and solve for n:

$$4,098.36 = 2^n$$

$$\log 4,098.36 = \log(2^n)$$

$$\log 4,098.36 = n \log 2$$

$$\frac{\log 4,098.36}{\log 2} = n$$

$$12 = n$$

After you know how many half-lives have passed, you can determine the total amount of time by multiplying the length of one half-life by the number of half-lives:

$$\frac{9.35 \text{ hours}}{1 \cancel{\text{half-life}}} \times 12 \cancel{\text{half-lives}} = 110 \text{ hours}$$

---

271.   **C**

The symbol for carbon is C.

---

272.   **Cl**

The symbol for chlorine is Cl.

---

273.   **Al**

The symbol for aluminum is Al.

---

274.   **Cd**

The symbol for cadmium is Cd.

275. Cu

The symbol for copper is Cu. The symbol originates from the Latin word *cuprum*.

276. As

The symbol for arsenic is As. Ar is the symbol for argon, not arsenic.

277. Na

The symbol for sodium is Na. The symbol originates from the Latin word *natrium*.

278. K

The symbol for potassium is K. The symbol originates from the Latin word *kalium*.

279. Fe

The symbol for iron is Fe. The symbol's origins are in the Latin word *ferrum*, which means iron.

280. Ag

The symbol for silver is Ag. The symbol originates from the Latin word *argentum*, which means silver.

281. nitrogen

The element with the symbol N is nitrogen.

282. sulfur

The element with the symbol S is sulfur.

283. bromine

The element with the symbol Br is bromine.

284. phosphorus

The element with the symbol P is phosphorus. The symbol for potassium is not P but K.

285. manganese

The element with the symbol Mn is manganese. The symbol for magnesium is Mg, not Mn.

Answers
201–300

286. astatine

The element with the symbol At is astatine.

287. radium

The element with the symbol Ra is radium. The symbol for radon is Rn.

288. mercury

The element with the symbol Hg is mercury. The symbol originates from the Latin word *hydrargyrum*, which means liquid silver.

289. tin

The element with the symbol Sn is tin. The symbol comes from the Latin word *stannum*, which means tin. The symbol for silicon is Si.

290. protactinium

The element with the symbol Pa is protactinium.

291. horizontally

A period goes horizontally — from left to right or right to left — on the periodic table.

292. touching the top and bottom of the stair-step line that starts between B and Al and goes down to Te and Po

The *metalloids* are elements that have characteristics of both metals and nonmetals; they form a border between the metals and nonmetals on the periodic table and include Si, Ge, As, Sb, Te, and Po. Elements (except for hydrogen) that are to the left of the stair-step line are metals, and elements to the right of the stair-step line are nonmetals.

293. in the second column of the periodic table

The alkaline earth metals include Be, Mg, Ca, Ba, and Ra, which are in the second column (Group 2 or IIA) of the periodic table.

294. in the middle of the periodic table

The transition metals are in the middle of the periodic table (from Sc to Zn and the rows directly below those ten elements). These groups are usually numbered 3 to 12 or IIIB to IIB.

295. in the rightmost column of the periodic table

The noble gases, He, Ne, Ar, Kr, Xe, and Rn, are in the rightmost column of the periodic table. This column is usually labeled Group 18, 0, or VIIIA.

296. in the bottom two rows of the periodic table

The inner transition metals (elements 58 to 71 and 90 to 103) are located in the bottom two rows of the periodic table, or between La and Hf and Ac and Rf in an expanded view of the periodic table. Elements 58 to 71 comprise the lanthanide series, and elements 90 to 103 comprise the actinide series. In both of these series, the f subshells are being filled.

297. in the leftmost column of the periodic table

Li, Na, K, Rb, Cs, and Fr are all members of the alkali metal family and are located in the leftmost column (Group 1 or IA) of the periodic table.

298. Dmitri Mendeleev

Dmitri Mendeleev arranged all the known elements of his time based on the physical properties of the elements, focusing on atomic mass.

German chemist Lothar Meyer did the same thing independently at the same time as Mendeleev, but Mendeleev receives more credit because he went further in his predictions.

299. Henry Moseley

Henry Moseley, an English physicist, is credited with arranging the modern periodic table with the elements in order of increasing atomic number.

300. above and to the right of the stair-step line that starts between B and Al and goes down to Te and Po

Most of the elements that exist as gases at room temperature (N, O, F, Cl, He, Ne, Ar, Kr, Xe, and Rn) are above and to the right of the stair-step line. The only exception is H, hydrogen, although some versions of the periodic table do place hydrogen above fluorine, in the halogen family.

301. the alkali metals

Potassium, K, is in the leftmost column of the periodic table, the alkali metal family.

302. the transition metals

Silver, Ag, is under copper, Cu, in the transition metals.

303. **the oxygen family**

Selenium, Se, is in the same column as oxygen, O, so selenium is a member of the oxygen family.

304. **the carbon family**

Tin, Sn, is two blocks below carbon on the periodic table, making tin a member of the carbon family.

305. **the halogen family**

Iodine, I, is in the same family as F, Cl, Br, and At, the halogen family.

306. **the alkaline earth metals**

Calcium, Ca, is in the same family as Be, Mg, Sr, Ba, and Ra, making calcium a member of the alkaline earth metals.

307. **the boron family**

Aluminum, Al, is under boron, B, on the periodic table, so aluminum is in the boron family.

308. **the inner transition metals**

The heaviest naturally occurring element is uranium, U. It's in the inner transition metals on the periodic table.

309. **the transition metals**

Mercury, Hg, is the only metal that's liquid at room temperature. Mercury is in the transition metals on the periodic table.

310. **the halogen family**

Bromine, Br, is the only nonmetal that's a liquid at room temperature. Bromine is in the halogen family (Group 17 or VIIA).

311. **the boron family**

Members of the boron family (Group 13 or IIIA) have three valence electrons. Their electron configurations have s^2p^1 in their outermost energy level.

312. **the nitrogen family**

The members of the nitrogen family (Group 15 or VA) have five valence electrons. Their electron configurations have s^2p^3 in their outermost energy level.

313. the alkaline earth metals

The alkaline earth metals (Group 2 or IIA) have two valence electrons in their s orbital.

314. increase; increase

As you go across a period (row), the masses of the elements generally increase — with just a few exceptions. As you go down a period (column), the masses of the elements always increase.

315. decrease; increase

The radii of the elements generally decrease going across a period from left to right, and the radii increase going down a family.

316. ionization energy

Ionization energy is the amount of energy needed to remove an electron from a gaseous atom. The ion that forms has fewer electrons than protons, giving the ion a net positive charge.

317. electron affinity

Affinity means attraction, so if you have a chocolate affinity, you're attracted to chocolate and would gladly add some to your collection. Similarly, if an atom has a high electron affinity, it has an attraction for electrons and would want to add them to its collection.

318. larger than; more

An anion is a negatively charged ion. This means that an anion has more negatively charged particles (electrons) than positively charged particles (protons). The ionic radius of an anion is larger than the radius of the atom because the ion has more electrons (and less of an effective nuclear charge) than the atom.

319. decrease; increases

The atomic radii of the elements decrease in size going from left to right across a period because with each successive element, an additional proton is added to the nucleus. These additional protons pull the electrons in closer to the nucleus, making the distance across the atom smaller.

320. Be < Ca < Ba

Ba, Be, and Ca (barium, beryllium, and calcium) are all in the same family. The atomic radius increases as you go down a family because more energy levels are being added, so Be < Ca < Ba.

321. Cl < S < P

Cl, P, and S (chlorine, phosphorus, and sulfur) are all in the same period on the periodic table. Within a period, the effective nuclear charge increases along with the atomic number. The greater positive charge in the nucleus pulls on the electrons more strongly, leading to a smaller atomic radius.

322. Li < B < C

B, C, and Li (boron, carbon, and lithium) are in the same period on the periodic table. Ionization energy increases going from left to right in a period because the additional protons pull electrons closer to the nucleus, so removing an electron takes more energy.

323. I < Br < Cl

Br, Cl, and I (bromine, chlorine, and iodine) are in the same family. In a family, the ionization energy decreases from top to bottom because the larger the atom is, the less energy it takes to remove an electron. All the extra electrons create shielding between the nucleus and the valence (outermost) electrons.

324. N < O < F

F, O, and N (fluorine, oxygen, and nitrogen) are in the same period on the periodic table. Within a period, the nucleus with the most protons attracts the electrons most strongly. N has the fewest protons of the given elements, so it has the lowest electron affinity; F has the most protons, so it has the highest electron affinity.

325. Te < Se < S

S, Se, and Te (sulfur, selenium, and tellurium) are in the same family on the periodic table. Within a family, the atomic radius increases with increasing atomic number.

The atom with the smallest radius has the highest electron affinity. The atom with the largest atomic radius has a hard time keeping the electrons that it has, so it has low ionization energy. Te has the largest radius and S has the smallest, so Te < Se < S.

326. Sn > Si > Cl

Chlorine (Cl) is a nonmetal, silicon (Si) is a metalloid, and tin (Sn) is a metal. When putting these elements in order from most to least metallic, place Sn before Si and Si before Cl.

327. Mg < Na < K

For any period on the periodic table, the effective nuclear charge increases with increasing atomic number. And for any column on the periodic table, the number of occupied energy levels increases with increasing atomic number.

Mg (magnesium) has a smaller radius than Na (sodium) because Mg has a larger effective nuclear charge. Na has a smaller radius than K (potassium) because Na has one fewer energy level.

328. $S^{2-} > O^{2-} > F^-$

Adding electrons increases the radius. The greater the number of electrons added, the greater the increase in the radius, because the electrons spread out as they repel each other. This is why both O^{2-} and S^{2-} are larger than F^-. S, which is lower on the periodic table than O, has an additional energy level, so it's larger.

329. $F > Li > Cs$

The ionization energy of F (fluorine) is greater than the ionization energy of Li (lithium) because F has the greater effective nuclear charge. Cs (cesium) has a lower ionization energy than Li because the Cs atom has more shielding from the large number of electrons that exist between the nucleus and the valence electrons.

330. $Ba < Bi < N$

Within a period, effective nuclear charge increases with increasing atomic number, and within a column, atomic radius increases with increasing atomic number.

Ba (barium) has a lower ionization energy than Bi (bismuth) because Bi has a greater effective nuclear charge. However, Bi has a lower ionization energy than N (nitrogen) because removing an electron from Bi requires less energy; the valence electrons in Bi are farther away from their nucleus than the valence electrons in N are from their nucleus.

331. **sodium chloride**

A binary ionic compound contains two elements, a metal and a nonmetal. The metal is present as a cation (positive ion), and the nonmetal is present as an anion (negative ion).

The cation name, which matches the original metal name (sometimes with info on the oxidation state), always goes before the anion name. The name of the nonmetal is altered to have an *-ide* suffix (sometimes requiring minor spelling changes). Unlike molecular compounds, ionic compounds do not use prefixes.

In this case, NaCl is composed of Na^+ and Cl^-. Na^+ is the sodium ion, and Cl^- is the chloride ion. Therefore, NaCl is called sodium chloride.

332. **calcium oxide**

CaO is a compound made of calcium ions (Ca^{2+}) and oxide ions (O^{2-}), so CaO is calcium oxide.

333. **aluminum bromide**

To name $AlBr_3$, first name the metal: Al is aluminum. Then name the nonmetal, changing the ending to *-ide*. Br is bromine, which becomes bromide, so $AlBr_3$ is aluminum bromide.

334. potassium sulfide

When naming a binary ionic compound, don't use prefixes; just name the metal and then the nonmetal with the suffix *-ide*. K_2S is composed of potassium ions (K^+) and sulfide ions (S^{2-}), so K_2S is potassium sulfide.

335. aluminum oxide

Al_2O_3 is composed of aluminum ions (Al^{3+}) and oxide ions (O^{2-}), so Al_2O_3 is aluminum oxide.

336. lithium nitride

When naming a binary ionic compound, don't use prefixes; just name the metal and then the nonmetal with the suffix *-ide*. Li_3N is lithium nitride.

337. magnesium iodide

To name MgI_2, first name the metal: Mg is magnesium. Then name the nonmetal, changing the ending to *-ide*. I is iodine, which becomes iodide, so MgI_2 is magnesium iodide.

338. strontium selenide

SrSe is a compound made of strontium and selenium. The nonmetal selenium becomes selenide, so SrSe is strontium selenide.

339. barium fluoride

To name BaF_2, first name the metal: Ba is barium. Then name the nonmetal, changing the ending to *-ide*. F is fluorine, which becomes fluoride, so BaF_2 is barium fluoride.

340. sodium hydride

In this case, hydrogen bonds to Na, sodium. The *-ogen* ending on hydrogen changes to *-ide,* so NaH is sodium hydride.

Note: Hydrogen is an odd nonmetal because it's on the left side of the periodic table. Hydrogen is also unusual because it forms both positive and negative ions.

341. zinc phosphide

Even though zinc (Zn) is a transition metal, it has only one oxidation state. Because only one oxidation state (+2) is possible, you don't need to specify the charge. Phosphorus forms the phosphide ion (P^{3-}), so Zn_3P_2 is zinc phosphide.

342. copper(II) bromide

Cu, copper, is a transition metal that has multiple oxidation states, so you need to use a roman numeral to designate which form of copper is in the compound.

The total charge of ions and cations has to equal zero. Because $CuBr_2$ contains two Br ions, each with a charge of –1, the copper has to have an oxidation number of +2 to cancel out the two negative charges. The first part of the name is copper(II) to take this charge into account. The second part of the name is bromide, as bromine changes to the bromide ion.

343. gold(III) chloride

Au, gold, is a transition metal that has multiple oxidation states, so you need to use a roman numeral to designate which form of gold is in the compound.

The total charge of ions and cations has to equal zero. Because $AuCl_3$ contains three Cl ions, each with a charge of –1, the gold has to have an oxidation number of +3 to cancel out the three negative charges. The first part of the name is gold(III) to take this charge into account. The second part of the name is chloride, as chlorine changes to the chloride ion.

344. cobalt(II) sulfide

Co, cobalt, is a transition metal that has multiple oxidation states, so you need to use a roman numeral to designate which form of cobalt is in the compound.

The total charge of ions and cations has to equal zero. Because the sulfide ion has a –2 charge, the cobalt must have a +2 charge. The first part of the name is cobalt(II) to take this charge into account. The second part of the name is sulfide, as sulfur changes to the sulfide ion. CoS is cobalt(II) sulfide.

345. manganese(III) fluoride

Mn, manganese, is a transition metal that has multiple oxidation states, so you need to use a roman numeral to designate which form of manganese is in the compound.

The total charge of ions and cations has to equal zero. MnF_3 contains three fluoride ions (F^-), each with a charge of –1, so the manganese has to have an oxidation number of +3 to cancel out the three negative charges. The first part of the name is manganese(III) to take this charge into account. Thus, MnF_3 is manganese (III) fluoride.

346. mercury(I) iodide

Hg, mercury, has two oxidation states, Hg_2^{2+} and Hg^{2+}, so you need to use a roman numeral to designate which form of mercury is in the compound.

The total charge of ions and cations has to equal zero. Because Hg_2I_2 contains two iodide ions (I^-), each with a charge of –1, the charge on each mercury ion must be +1. Therefore, Hg_2I_2 is mercury(I) iodide.

Note: Mercury(I) is Hg_2^{2+}, a polyatomic ion consisting of two Hg atoms with a total charge of +2. The charge (+2) divided by the number of atoms (2) gives a +1 charge for each mercury atom. It's improper to split this — or any other polyatomic ion — into smaller pieces.

347. **tin(IV) oxide**

Sn isn't a transition metal, but it does have two common oxidation states, +2 and +4, so you need to use a roman numeral to designate which form of tin is in the compound.

The total charge of ions and cations has to equal zero. Oxide ions have a charge of –2. SnO_2 contains two oxide ions, so the oxygen ions produce a charge of –4. To balance this –4 charge, the tin ion must have a charge of +4. SnO_2 is tin(IV) oxide.

Note: Because tin is a metal, it can lose all four valence electrons to give it a +4 oxidation state. Such high charges rarely occur in stable ions.

348. **sodium hypochlorite**

Ternary ionic compounds (compounds containing three elements) and higher compounds usually contain polyatomic ions. The name of a polyatomic cation simply replaces the name of the cation (metal) in a binary compound, and the name of a polyatomic anion replaces the name of the anion (nonmetal).

Most polyatomic ions have either an *-ate* or an *-ite* suffix. A few, such as hydroxide, have an *-ide* suffix. Ions with *-ate* suffixes have more oxygen than those with an *-ite* suffix.

In this case, the polyatomic ion is ClO^-, the hypochlorite ion. So name the metal ion first (Na is the sodium ion) and then add the name of the polyatomic ion ClO^- to get sodium hypochlorite.

349. **potassium hydroxide**

In this case, the polyatomic ion is OH^-, the hydroxide ion. Name the metal ion and list it first (K is the potassium ion) and then add the name of the polyatomic ion OH^- to get potassium hydroxide.

350. **strontium sulfite**

The polyatomic ion in $SrSO_3$ is SO_3^{2-}, the sulfite ion. Sr^{2+} is the strontium ion, so $SrSO_3$ is strontium sulfite.

351. **calcium carbonate**

The polyatomic ion in $CaCO_3$ is CO_3^{2-}, the carbonate ion. Ca^{2+} is the calcium ion, so $CaCO_3$ is calcium carbonate.

352. **aluminum phosphate**

The polyatomic ion in $AlPO_4$ is PO_4^{3-}, the phosphate ion. Al is aluminum, so $AlPO_4$ is aluminum phosphate.

353. **sodium chlorate**

The polyatomic ion in $NaClO_3$ is ClO_3^-, the chlorate ion. Na is sodium, so $NaClO_3$ is sodium chlorate.

354. **gallium phosphite**

The polyatomic ion in $GaPO_3$ is PO_3^{3-}, the phosphite ion. Ga^{3+} is the gallium ion, so $GaPO_3$ is gallium phosphite.

355. **ammonium chloride**

The polyatomic ion in NH_4Cl is NH_4^+, the ammonium ion. When you combine ammonium with the chloride ion (Cl^-), you get ammonium chloride.

356. **zinc sulfate**

The polyatomic ion in $ZnSO_4$ is SO_4^{2-}, the sulfate ion. Zn^{2+} is the zinc ion, a transition metal that has only one oxidation state (and therefore doesn't require a roman numeral), so $ZnSO_4$ is zinc sulfate.

357. **ammonium oxalate**

$(NH_4)_2C_2O_4$ contains two polyatomic ions, NH_4^+ (ammonium) and $C_2O_4^{2-}$ (oxalate). All you need to do is put both names together: ammonium oxalate.

358. **potassium permanganate**

In $KMnO_4$, the MnO_4^- is the polyatomic ion. K^+ is the potassium ion, and MnO_4^- is permanganate, so $KMnO_4$ is potassium permanganate.

359. **beryllium nitrite**

When a set of parentheses appears in a chemical formula, the contents are often a polyatomic ion. NO_2^- is the nitrite ion, so $Be(NO_2)_2$ is beryllium nitrite.

360. **copper(II) cyanide**

$Cu(CN)_2$ begins with copper, a transition metal that has more than one oxidation state. To determine the oxidation state, look at how many polyatomic ion units are with Cu. The molecule contains two CN^- ions, each with a charge of –1. To balance out this –2 charge, the copper must have a +2 charge.

Name the metal ion using roman numerals to represent its oxidation number, and then name the polyatomic ion (which happens to be inside the parentheses). Cu^{2+} is the copper(II) ion, and CN^- is cyanide, so $Cu(CN)_2$ is copper(II) cyanide. Note that this compound doesn't actually exist, but you can still give it a name.

361. **silver hypobromite**

Even though Ag is a transition metal, it has only the oxidation state of +1, so you don't need to use a roman numeral when naming this compound. Ag^+ is the silver ion, and BrO^- is hypobromite, so AgBrO is silver hypobromite.

362.   **ammonium phosphate**

$(NH_4)_3PO_4$ is made up of two polyatomic ions, so name the cation first and then name the anion. NH_4^+ is the ammonium ion, and PO_4^{3-} is the phosphate ion, so $(NH_4)_3PO_4$ is ammonium phosphate.

363.   **nickel(III) sulfate**

Ni, nickel, is a transition metal with more than one oxidation state, so you have to look at the number of polyatomic ion units and their charge in order to determine which oxidation state to assign to the nickel. The sum of the charges of the ions in the compound has to equal zero, so you can set up an equation to solve for nickel's oxidation state:

$$2Ni + 3SO_4 = 0$$
$$2Ni + 3(-2) = 0$$
$$2Ni + -6 = 0$$
$$2Ni = 6$$
$$Ni = +3$$

Therefore, $Ni_2(SO_4)_3$ is nickel(III) sulfate.

364.   **lead(II) acetate**

Pb, lead, is a metal with more than one oxidation state, so you have to look at the number of polyatomic ion units and their charge in order to determine which oxidation state to assign to the lead. The sum of the charges of the ions in the compound has to equal zero, so you can set up an equation to solve for lead's oxidation state:

$$Pb + 2C_2H_3O_2 = 0$$
$$Pb + 2(-1) = 0$$
$$Pb + -2 = 0$$
$$Pb = +2$$

Therefore, $Pb(C_2H_3O_2)_2$ is lead(II) acetate.

365.   **CsCl**

A binary ionic compound contains two elements, a metal and a nonmetal. The metal is present as a cation (positive ion), and the nonmetal is present as an anion (negative ion).

The cation name always comes before the anion name. If the cation (metal) is able to form different ions, a roman numeral after the name specifies the oxidation state. The total charge of the cations and anions must equal zero.

In this case, the cesium ion is Cs^+, and chloride, Cl^-, is the ionic form of the chlorine atom. The +1 charge of the Cs cancels out the –1 charge of the Cl, so the formula is CsCl.

366. InF$_3$

The indium(III) ion is In^{3+}, and fluoride, F$^-$, is the ion formed by fluorine. The +3 charge of the In needs to be balanced by the negative charge of the F, so you need three –1 fluoride ions to cancel out the +3 charge of the In. The formula is InF$_3$.

367. MgO

The magnesium ion is Mg^{2+}, and oxide, O^{2-}, is the ion formed by oxygen. The +2 charge of the Mg cancels out the –2 charge of the O, so the formula is MgO.

368. BaBr$_2$

The barium ion is Ba^{2+}, and bromide, Br$^-$, is the ionic form of the bromine atom. You need two negatively charged bromide ions to cancel out the +2 charge on the barium ion, so the formula is BaBr$_2$.

369. KI

The potassium ion is K$^+$, and iodide, I$^-$, is the ionic form of the iodine atom. The +1 charge of the K cancels out the –1 charge of the I, so the formula is KI.

370. AlCl$_3$

The aluminum ion is Al^{3+}, and chloride, Cl$^-$, is the ionic form of the chlorine atom. You need three negatively charged chloride ions to cancel out the +3 charge of the aluminum, so the formula is AlCl$_3$.

371. CrF$_3$

Writing formulas of compounds that contain transition metals with variable charges is relatively easy because the roman numeral tells you the charge of the metal ion. Chromium(III) is Cr^{3+}, and fluoride, F$^-$, is the ionic form of the fluorine atom. You need three fluoride ions to balance out the chromium's +3 charge, so the formula is CrF$_3$.

372. FeS

Iron(II) is the Fe^{2+} ion, and sulfide, S^{2-}, is the ionic form of the sulfur atom. The +2 charge cancels out the –2 charge, so the formula is FeS.

373. Cu$_3$N

Copper(I) is the Cu$^+$ ion, and nitride, N^{3-}, is the ionic form of the nitrogen atom. You need three positively charged copper ions to balance out the –3 charge of the nitride ion, so the formula is Cu$_3$N.

374. PbO

Lead(II) is the Pb^{2+} ion, and oxide, O^{2-}, is the ionic form of the oxygen atom. The +2 charge of the Pb cancels out the –2 charge of the O, so the formula is PbO.

375. Ni_2Se

Nickel(I) is the Ni^+ ion, and selenide, Se^{2-}, is the ionic form of the selenium atom. You need two Ni^+ ions to balance out the –2 charge of the selenium, so the formula is Ni_2Se.

376. Ag_2O

The silver ion is Ag^+, and oxide, O^{2-}, is the ionic form of the oxygen atom. You need two Ag^+ ions to balance out the –2 charge of the oxide ion, so the formula is Ag_2O.

377. $SrBr_2$

The strontium ion is Sr^{2+}, and the bromide ion is Br^-. You need two Br^- ions to balance out the +2 charge of the strontium ion, so the formula is $SrBr_2$.

378. Mg_3N_2

The magnesium ion is Mg^{2+}, and nitride, N^{3-}, is the ionic form of the nitrogen atom. The lowest common multiple of the two charges is 6. For the charges to cancel out at +6 and –6, you need three Mg^{2+} ions and two N^{3-} ions. Therefore, the formula for magnesium nitride is Mg_3N_2.

379. LiH

The lithium ion is Li^+. Hydride, H^-, is the anion (negative ion) form that hydrogen takes when bonded to a metal. The +1 and –1 charges cancel out, so the formula is LiH.

380. $ZnCl_2$

Zinc is a transition metal. For the most part, its oxidation number is +2, making the zinc ion Zn^{2+}. Chloride, Cl^-, is the ionic form of the chlorine atom. You need two Cl^- ions to balance out the +2 charge of the zinc ion, so the formula is $ZnCl_2$.

381. CrS

Chromium(II) is Cr^{2+}, and sulfide, S^{2-}, is the ionic form of the sulfur atom. The +2 and –2 charges cancel out, so the formula is CrS.

382. MnSe

Manganese(II) is Mn^{2+}, and selenide, Se^{2-}, is the ionic form of the selenium atom. The +2 and –2 charges cancel out, so the formula is MnSe.

Answers
301–400

383. SnF_4

Tin(IV) is Sn^{4+}, and fluoride, F^-, is the ionic form of the fluorine atom. You need four times as many fluoride ions to balance out the +4 charge on Sn^{4+}, so the formula for tin(IV) fluoride is SnF_4.

384. CuI

The copper(I) ion is Cu^+, and the iodide ion is I^-. You need one –1 iodide ion to balance out the +1 charge of the copper ion, so the formula is CuI.

385. NiP

The nickel(III) ion is Ni^{3+}, and phosphide, P^{3-}, is the ionic form of the phosphorus atom. The +3 charge from the nickel cancels out the –3 charge of the phosphide ion, so the formula is NiP.

386. Al_2Se_3

The aluminum ion is Al^{3+}, and selenide, Se^{2-}, is the ionic form of selenium. The lowest common multiple of the two charges is 6. For the charges to cancel out at +6 and –6, you need two Al^{3+} ions and three Se^{2-} ions. Therefore, the formula for aluminum selenide is Al_2Se_3.

387. SnO_2

Tin(IV) is Sn^{4+}, and oxide, O^{2-}, is the ionic form of the oxygen atom. You need two –2 charged oxide ions to balance out the +4 charge of the tin ion, so the formula is SnO_2.

388. Ca_3P_2

The calcium ion is Ca^{2+}, and the phosphide ion is P^{3-}. The lowest common multiple of the two charges is 6. For the charges to cancel out at +6 and –6, you need three Ca^{2+} ions and two P^{3-} ions. Therefore, the formula for calcium phosphide is Ca_3P_2.

389. Fe_2O_3

The iron(III) ion is Fe^{3+}, and the oxide ion is O^{2-}, is the ionic form of the oxygen atom. The lowest common multiple of the two charges is 6. For the charges to cancel out at +6 and –6, you need two Fe^{3+} ions and three O^{2-} ions. The formula for iron(III) oxide is Fe_2O_3.

390. MnS_2

Manganese(IV) is the Mn^{4+} ion, and sulfide, S^{2-}, is the ionic form of the sulfur atom. You need two –2 sulfide ions to balance out the +4 charge of the manganese(IV) ion, so the formula for manganese(IV) sulfide is MnS_2.

Answers
301–400

391. **RbClO**

In compounds containing polyatomic ions, the name of a polyatomic cation replaces the name of the cation (metal) in a binary compound, or the name of a polyatomic anion replaces the name of the anion (nonmetal).

The cation name always comes before the anion name. If the cation (metal) is able to form different ions, a roman numeral after the name specifies the oxidation state. The total charge of the cations and anions must equal zero.

The rubidium ion is Rb^+, and hypochlorite is ClO^-. The +1 charge cancels out the –1 charge, so the formula for rubidium hypochlorite is RbClO.

392. **BeCO$_3$**

The beryllium ion is Be^{2+}, and carbonate is CO_3^{2-}. The +2 charge cancels out the –2 charge, so the formula for beryllium carbonate is $BeCO_3$.

393. **AlPO$_3$**

The aluminum ion is Al^{3+}, and phosphite is PO_3^{3-}. The +3 charge cancels out the –3 charge, so the formula for aluminum phosphite is $AlPO_3$.

394. **NaHCO$_3$**

The sodium ion is Na^+, and hydrogen carbonate is HCO_3^-. The +1 charge cancels out the –1 charge, so the formula for sodium hydrogen carbonate is $NaHCO_3$. Another name for this compound is sodium bicarbonate.

395. **NaOH**

The sodium ion is Na^+, and hydroxide is OH^-. The +1 charge cancels out the –1 charge, so the formula for sodium hydroxide is NaOH.

396. **KHSO$_4$**

The potassium ion is K^+, and hydrogen sulfate is HSO_4^-. The +1 charge cancels out the –1 charge, so the formula for potassium hydrogen sulfate is $KHSO_4$. Another name for this compound is potassium bisulfate.

397. **LiClO$_4$**

The lithium ion is Li^+, and perchlorate is ClO_4^-. The +1 charge cancels out the –1 charge, so the formula for lithium perchlorate is $LiClO_4$.

398. **BaC$_2$O$_4$**

The barium ion is Ba^{2+}, and oxalate is $C_2O_4^{2-}$. The +2 charge cancels out the –2 charge, so the formula for barium oxalate is BaC_2O_4.

399. CrAsO$_4$

Chromium(III) is Cr^{3+}, and arsenate is AsO_4^{3-}. The +3 charge cancels out the –3 charge, so the formula for chromium(III) arsenate is $CrAsO_4$.

400. AgNO$_3$

The silver ion is Ag^+, and nitrate is NO_3^-. The +1 charge cancels out the –1 charge, so the formula for silver nitrate is $AgNO_3$.

401. PbSO$_3$

Lead(II) is Pb^{2+}, and sulfite is SO_3^{2-}. The +2 charge cancels out the –2 charge, so the formula for lead sulfite is $PbSO_3$.

402. TlBrO$_2$

Thallium(I) is Tl^+, and bromite is BrO_2^-. The +1 charge cancels out the –1 charge, so the formula for thallium(I) bromite is $TlBrO_2$.

403. AuPO$_4$

Gold(III) is Au^{3+}, and phosphate is PO_4^{3-}. The +3 charge cancels out the –3 charge, so the formula for gold(III) phosphate is $AuPO_4$.

404. FeSO$_4$

Iron(II) is Fe^{2+}, and sulfate is SO_4^{2-}. The +2 charge cancels out the –2 charge, so the formula for iron(II) sulfate is $FeSO_4$.

405. CaS$_2$O$_3$

The calcium ion is Ca^{2+}, and thiosulfate is $S_2O_3^{2-}$. The +2 charge cancels out the –2 charge, so the formula for calcium thiosulfate is CaS_2O_3.

406. Na$_2$O$_2$

The sodium ion is Na^+, and peroxide is O_2^{2-}. You need two Na^+ ions to balance out the –2 charge on the peroxide ion, so the formula for sodium peroxide is Na_2O_2.

407. NH$_4$NO$_2$

The ammonium ion is NH_4^+, and nitrite is NO_2^-. The ammonium's +1 charge cancels out the nitrite's –1 charge, so the formula for ammonium nitrite is NH_4NO_2. *Note:* Nitrogen appears twice in the formula because it's in two separate polyatomic ions; combining parts of two separate ions isn't appropriate.

408. $Be(ClO_2)_2$

The beryllium ion is Be^{2+}, and the chlorite ion is ClO_2^-. You need two -1 charged ClO_2^- ions to balance out the $+2$ charge of the beryllium ion, so the formula for beryllium chlorite is $Be(ClO_2)_2$.

409. NaCN

The sodium ion is Na^+, and cyanide is CN^-. The $+1$ charge of the sodium cancels out the -1 charge of the cyanide, so the formula for sodium cyanide is NaCN.

410. $Mg(MnO_4)_2$

The magnesium ion is Mg^{2+}, and the permanganate ion is MnO_4^-. You need two -1 charged permanganate ions to balance out the $+2$ charge of the magnesium ion, so the formula for magnesium permanganate is $Mg(MnO_4)_2$.

411. $(NH_4)_2Cr_2O_7$

The ammonium ion is NH_4^+, and the dichromate ion is $Cr_2O_7^{2-}$. You need two $+1$ charged ammonium ions to balance out the -2 charge of the dichromate ion, so the formula for ammonium dichromate is $(NH_4)_2Cr_2O_7$.

412. $Co(IO_4)_3$

Cobalt(III) is Co^{3+}, and periodate is IO_4^-. You need three -1 charged periodate ions to cancel out the $+3$ charged cobalt ion, so the formula for cobalt(III) periodate is $Co(IO_4)_3$.

413. $Pb(C_2O_4)_2$

The lead(IV) ion is Pb^{4+}, and the oxalate ion is $C_2O_4^{2-}$. You need two -2 charged oxalate ions to balance out the $+4$ charge of the lead(IV) ion, so the formula for lead(IV) oxalate is $Pb(C_2O_4)_2$.

414. $CaHPO_4$

The calcium ion is Ca^{2+}, and the hydrogen phosphate ion is HPO_4^{2-}. The $+2$ charge of the calcium ion cancels out the -2 charge of the hydrogen phosphate ion, so the formula for calcium hydrogen phosphate is $CaHPO_4$.

415. $Fe(C_2H_3O_2)_3$

Iron(III) is Fe^{3+}, and the acetate ion is $C_2H_3O_2^-$. You need three -1 charged acetate ions to cancel out the $+3$ charged iron(III) ion, so the formula for iron(III) acetate is $Fe(C_2H_3O_2)_3$.

Note: You can also write the acetate ion as CH_3COO^-, so $Fe(CH_3COO)_3$ is another way to write the chemical formula for iron(III) acetate.

416. KSCN

The potassium ion is K^+, and the thiocyanate ion is SCN^-. The +1 charge of the potassium ion cancels out the –1 charge of the thiocyanate ion, so the formula for potassium thiocyanate is KSCN.

417. $Cu(HSO_3)_2$

The copper(II) ion is Cu^{2+}, and the hydrogen sulfite ion is HSO_3^-. You need two HSO_3^- ions to balance out the +2 charge of the copper(II) ion, so the formula for copper(II) hydrogen sulfite is $Cu(HSO_3)_2$. This compound is also known as copper(II) bisulfite.

418. HgO_2

The mercury(II) ion is Hg^{2+}, and peroxide is O_2^{2-}. The +2 charge of the Hg^{2+} ion cancels the –2 charge of O_2^{2-}, so the formula for mercury(II) peroxide is HgO_2.

419. AuOCN

Gold(I) is Au^+, and the cyanate ion is OCN^-. The +1 charge cancels out the –1 charge, so the formula for gold(I) cyanate is AuOCN.

420. $Al(H_2PO_4)_3$

The aluminum ion is Al^{3+}, and dihydrogen phosphate is $H_2PO_4^-$. To cancel out the +3 charge of the Al^{3+}, you need three $H_2PO_4^-$ ions, so the formula for aluminum dihydrogen phosphate is $Al(H_2PO_4)_3$.

421. 2

The prefix *di-* represents two atoms.

422. 6

The prefix *hexa-* represents six atoms.

423. 7

The prefix *hepta-* represents seven atoms.

424. 4

The prefix *tetra-* represents four atoms.

425. 9

The prefix *nona-* represents nine atoms.

426. 3

The prefix *tri-* represents three atoms.

427. 1

The prefix *mono-* represents one atom. In more recent naming systems, this prefix is rarely used. *Carbon monoxide* is one of the few cases where you still see *mono-*.

428. 5

The prefix *penta-* represents five atoms.

429. 8

The prefix *octa-* represents eight atoms.

430. 10

The prefix *deca-* represents ten atoms.

431. **carbon monoxide**

CO contains two nonmetals: carbon and oxygen. The molecule has only one atom of carbon, so just use the name of the first element as-is. The second nonmetal, oxygen, gets a prefix (which describes how many oxygen atoms are in the compound) and a suffix of *-ide.* The prefix for one is *mon(o)-,* so CO is carbon monoxide. This is one of the few cases where *mono-* is still in use.

432. **sulfur dibromide**

SBr_2 contains one sulfur atom and two bromine atoms. The molecule has only one sulfur atom, so use the name of the first element as-is. The second element, bromine, gets a suffix of *-ide* and the prefix *di-* (two) to indicate the number of atoms. Therefore, SBr_2 is sulfur dibromide.

433. **iodine chloride**

ICl contains one iodine atom and one chlorine atom. The molecule has only one iodine atom, so use the name of the first element as-is. The second element, chlorine, gets a suffix of *-ide;* the prefix *mono-* (one) is usually omitted in modern naming systems, so you don't need a prefix to indicate the number of chlorine atoms. Therefore, ICl is iodine chloride. The name *iodine monochloride* is also acceptable.

434. sulfur dioxide

SO_2 contains one sulfur atom and two oxygen atoms. The molecule has only one sulfur atom, so use the name of the first element as-is. The second element, oxygen, gets a suffix of -ide and the prefix di- (two) to indicate the number of atoms. Therefore, SO_2 is sulfur dioxide, a compound that contributes to the odor of a freshly struck match.

435. phosphorus pentachloride

PCl_5 contains one phosphorus atom and five chlorine atoms. The molecule has only one phosphorus atom, so phosphorus doesn't need a prefix. The second element, chlorine, gets a suffix of -ide and the prefix penta- (five) to indicate the number of atoms. Therefore, PCl_5 is phosphorus pentachloride.

436. xenon difluoride

XeF_2 contains one xenon atom and two fluorine atoms. The molecule has only one xenon atom, so use the name of the first element as-is. The second element, fluorine, gets a suffix of -ide and the prefix di- (two) to indicate the number of atoms. Therefore, XeF_2 is xenon difluoride.

437. sulfur hexafluoride

SF_6 contains one sulfur atom and six fluorine atoms. The molecule has only one sulfur atom, so use the name of the first element as-is. The second element, fluorine, gets a suffix of -ide and the prefix hexa- (six) to indicate the number of atoms. Therefore, SF_6 is sulfur hexafluoride.

438. carbon tetrabromide

CBr_4 contains one carbon atom and four bromine atoms. The molecule has only one carbon atom, so carbon doesn't need a prefix. The second element, bromine, gets a suffix of -ide and the prefix tetra- (four) to indicate the number of atoms. Therefore, CBr_4 is carbon tetrabromide.

439. boron trichloride

BCl_3 contains one boron atom and three chlorine atoms. The molecule has only one boron atom, so you don't need a prefix on boron. The second element, chlorine, gets a suffix of -ide and the prefix tri- (three) to indicate the number of atoms. Therefore, BCl_3 is boron trichloride.

440. silicon dioxide

SiO_2 contains one silicon atom and two oxygen atoms. The molecule has only one silicon atom, so silicon doesn't need a prefix. The second element, oxygen, gets a suffix of -ide and the prefix di- (two) to indicate the number of atoms. Therefore, SiO_2 is silicon dioxide, which you may recognize as the mineral quartz. This compound is a major component of sand and glass.

441. **arsenic pentachloride**

$AsCl_5$ contains one arsenic atom and five chlorine atoms. The molecule has only one arsenic atom, so use the name of the first element as-is. The second element, chlorine, gets a suffix of *-ide* and the prefix *penta-* (five) to indicate the number of atoms. Therefore, $AsCl_5$ is arsenic pentachloride.

442. **antimony trichloride**

$SbCl_3$ contains one antimony atom and three chlorine atoms. The molecule has only one antimony atom, so antimony doesn't need a prefix. The second element, chlorine, gets a suffix of *-ide* and the prefix *tri-* (three) to indicate the number of atoms. Therefore, $SbCl_3$ is antimony trichloride.

443. **silicon tetraiodide**

SiI_4 contains one silicon atom and four iodine atoms. The molecule has only one silicon atom, so use the name of the first element as-is. The second element, iodine, gets a suffix of *-ide* and the prefix *tetra-* (four) to indicate the number of atoms. Therefore, SiI_4 is silicon tetraiodide.

444. **nitrogen trifluoride**

NF_3 contains one nitrogen atom and three fluorine atoms. The molecule has only one nitrogen atom, so use the name of the first element as-is. The second element, fluorine, gets a suffix of *-ide* and the prefix *tri-* (three) to indicate the number of atoms. Therefore, NF_3 is nitrogen trifluoride.

445. **carbon disulfide**

CS_2 contains one carbon atom and two sulfur atoms. The molecule has only one carbon atom, so you don't need a prefix on the carbon. The second element, sulfur, gets a suffix of *-ide* and the prefix *di-* (two) to indicate the number of atoms. Therefore, CS_2 is carbon disulfide.

446. **chlorine dioxide**

ClO_2 contains one chlorine atom and two oxygen atoms. The molecule has only one chlorine atom, so use the name of the first element as-is. The second element, oxygen, gets a suffix of *-ide* and the prefix *di-* (two) to indicate the number of atoms. Therefore, ClO_2 is chlorine dioxide. This compound is useful as an industrial bleach.

447. **xenon tetroxide**

XeO_4 contains one xenon atom and four oxygen atoms. The molecule has only one xenon atom, so xenon doesn't need a prefix. The second element, oxygen, gets a suffix of *-ide* and the prefix *tetr(a)-* (four) to indicate the number of atoms. Therefore, XeO_4 is xenon tetroxide. This compound is very unstable and will spontaneously detonate.

Answers
401–500

448. **water**

H_2O contains two hydrogen atoms and one oxygen atom. The compound might appear to be dihydrogen oxide, but the current systematic (IUPAC) name is water.

449. **selenium hexafluoride**

SeF_6 contains one selenium atom and six fluorine atoms. The molecule has only one selenium atom, so use the name of the first element as-is. The second element, fluorine, gets a suffix of *-ide* and the prefix *hexa-* (six) to indicate the number of atoms. Therefore, SeF_6 is selenium hexafluoride.

450. **disulfur dichloride**

S_2Cl_2 contains two sulfur atoms and two chlorine atoms. Because you have more than one atom of the first element, it needs a prefix. The prefix for two is *di-*, giving you disulfur. The second element, chlorine, gets a suffix of *-ide* and also takes the prefix *di-*. Therefore, S_2Cl_2 is disulfur dichloride.

451. **dinitrogen trioxide**

N_2O_3 contains two nitrogen atoms and three oxygen atoms. The nitrogen needs the prefix *di-* to represent two atoms, and the oxygen (which becomes *oxide*) needs the prefix *tri-* to indicate three atoms. Therefore, N_2O_3 is dinitrogen trioxide.

452. **tetraphosphorus hexoxide**

P_4O_6 contains four phosphorus atoms and six oxygen atoms. The phosphorus needs the prefix *tetra-* to indicate four atoms, and the oxygen needs the prefix *hex(a)-* (six) and the suffix *-ide*. Therefore, P_4O_6 is tetraphosphorus hexoxide.

453. **diboron tetrachloride**

B_2Cl_4 contains two boron atoms and four chlorine atoms. The boron needs the prefix *di-* (two), and chlorine needs the prefix *tetra-* (four). Chlorine changes to chloride, so B_2Cl_4 is diboron tetrachloride.

454. **bromine trifluoride**

BrF_3 contains one bromine atom and three fluorine atoms. The molecule has only one bromine atom, so use the name of the first element as-is. The second element, fluorine, gets a suffix of *-ide* and the prefix *tri-* (three) to indicate the number of atoms. Therefore, BrF_3 is bromine trifluoride.

455. **disulfur decafluoride**

S_2F_{10} contains two sulfur atoms and ten fluorine atoms. Use the prefix *di-* (two) to represent the number of sulfur atoms and the prefix *deca-* (ten) to represent the number of fluorine atoms. Fluorine changes to fluoride, so S_2F_{10} is disulfur decafluoride.

456. $SiBr_4$

The first element in the name of the compound doesn't have a prefix, and by convention, this means that there's only one atom of that element. The second element in the compound has a prefix to tell you how many atoms of that element are present. In *silicon tetrabromide,* the prefix on the bromide is *tetra-,* which represents four. You have one silicon (Si) atom and four bromine (Br) atoms, so silicon tetrabromide is $SiBr_4$.

457. NI_3

In *nitrogen triiodide,* the absence of a prefix on the nitrogen tells you that the compound has just one nitrogen (N) atom, and the *tri-* prefix on the iodide indicates three iodine (I) atoms. The formula for nitrogen triiodide is NI_3. This compound is very unstable and explodes at the lightest touch.

458. CO_2

You're probably familiar with carbon dioxide and didn't need to think about the formula too much. The absence of a prefix on the carbon and the *di-* (two) prefix on the oxide lead to the formula CO_2. Carbon dioxide is one of the products of respiration and the combustion of most fuels.

459. AsF_5

In *arsenic pentafluoride,* the absence of a prefix on the arsenic means you have just one arsenic (As) atom. The *penta-* prefix on the fluoride tells you that there are five fluorine (F) atoms, so the formula for arsenic pentafluoride is AsF_5.

460. NO

In *nitrogen monoxide,* nitrogen doesn't have a prefix, and the prefix on the oxide is *mon(o)-,* which means one, so this compound has one atom of nitrogen (N) and one atom of oxygen (O). The formula for nitrogen monoxide is NO. This compound is also known as *nitrogen oxide.*

461. SO_3

In *sulfur trioxide,* the sulfur has no prefix, so the formula contains just one atom of sulfur (S). The oxide has a prefix of *tri-,* which means three, so you have three oxygen (O) atoms. This makes sulfur trioxide SO_3.

462. ClF

In *chlorine monofluoride,* chlorine doesn't have a prefix, and the prefix on the fluoride is *mono-,* which means one. So chlorine monofluoride has one atom of chlorine (Cl) and one atom of fluorine (F), making the formula ClF. This compound is also known as *chlorine fluoride.*

Answers
401–500

463. XeF_4

In *xenon tetrafluoride,* xenon doesn't have a prefix, and the prefix on the fluoride is *tetra-,* which means four. So xenon tetrafluoride contains one atom of xenon (Xe) and four atoms of fluorine (F), making the formula XeF_4. This compound was the first binary compound of xenon to be prepared in a laboratory.

464. NO_2

In *nitrogen dioxide,* no prefix on the nitrogen tells you that the formula includes one nitrogen (N) atom, and the *di-* prefix on the oxide indicates two oxygen (O) atoms. The formula for nitrogen dioxide is NO_2. Nitrogen dioxide is one of the contributors to photochemical smog.

465. CCl_4

In *carbon tetrachloride,* no prefix on the carbon means that the formula includes only one carbon (C) atom, and the *tetra-* prefix on the chloride indicates four chlorine (Cl) atoms, so carbon tetrachloride is CCl_4.

466. IF_7

In *iodine heptafluoride,* no prefix on the iodine means that the formula includes only one iodine (I) atom. The *hepta-* prefix represents seven, so there are seven fluorine (F) atoms in the formula. This makes the formula for iodine heptafluoride IF_7. Iodine heptafluoride is one of the few compounds where one atom is surrounded by seven others.

467. PBr_3

In *phosphorus tribromide,* no prefix on the phosphorus tells you that the formula includes just one atom of phosphorus (P), and the *tri-* prefix on the bromide indicates three bromine (Br) atoms. Therefore, the formula for phosphorus tribromide is PBr_3.

468. SeF_4

In *selenium tetrafluoride,* no prefix on the selenium means you have just one atom of selenium (Se), and the prefix on the fluoride is *tetra-,* which means four, so you have four fluorine atoms (F). Therefore, selenium tetrafluoride is SeF_4.

469. ClO_2

In *chlorine dioxide,* the absence of a prefix on the chlorine tells you that there's only one chlorine (Cl) atom, and the *di-* prefix on the oxide tells you that there are two oxygen (O) atoms in the compound. The formula for chlorine dioxide is ClO_2.

470. BF$_3$

In *boron trifluoride,* the absence of a prefix on the boron reflects that there's only one boron (B) atom, and the *tri-* prefix on the fluoride indicates three fluorine (F) atoms. This makes the formula for boron trifluoride BF$_3$.

471. N$_2$O$_5$

Dinitrogen pentoxide has a prefix on both the nitrogen and the oxide. The *di-* represents two, and the *pent(a)-* represents five, so there are two nitrogen (N) atoms and five oxygen (O) atoms in dinitrogen pentoxide. The formula is N$_2$O$_5$.

472. P$_2$O$_3$

Diphosphorus trioxide has a prefix on both the phosphorus and the oxide. The *di-* represents two, and the *tri-* represents three, so there are two phosphorus (P) atoms and three oxygen (O) atoms in diphosphorus trioxide. The formula is P$_2$O$_3$.

Note that the formula P$_2$O$_3$ is the empirical formula. The molecular formula is P$_4$O$_6$, which has the name tetraphosphorus hexoxide (or hexaoxide).

473. N$_2$Cl$_2$

Dinitrogen dichloride has two nitrogen (N) atoms and two chlorine (Cl) atoms because both elements have the prefix *di-,* which means two. Therefore, the formula for dinitrogen dichloride is N$_2$Cl$_2$.

474. P$_4$O$_{10}$

Tetraphosphorus decoxide has the prefix *tetra-,* meaning four, and the prefix *dec(a)-,* meaning ten, so you need four phosphorus (P) atoms and ten oxygen (O) atoms. This makes the formula for tetraphosphorus decoxide P$_4$O$_{10}$.

475. AsF$_3$

In *arsenic trifluoride,* the absence of a prefix on arsenic reflects that there's only one arsenic (As) atom, and the *tri-* prefix on the fluoride indicates three fluorine (F) atoms. This makes the formula for arsenic trifluoride AsF$_3$.

476. N$_2$O

In *dinitrogen oxide,* the prefix *di-* on the nitrogen tells you that there are two nitrogen (N) atoms, and the lack of a prefix on the oxide tells you that there's just one oxygen (O) atom. The formula for dinitrogen oxide is N$_2$O.

This compound is also called *dinitrogen monoxide.* Its common names include *nitrous oxide* and *laughing gas,* which you may recognize as an anesthetic used by dentists.

477. P_4O_3

Tetraphosphorus trioxide has the prefix *tetra-*, meaning four, and the prefix *tri-*, meaning three, so the compound contains four phosphorus (P) atoms and three oxygen (O) atoms. This makes the formula for tetraphosphorus trioxide P_4O_3.

478. XeO_3

In *xenon trioxide,* the absence of a prefix on xenon reflects that there's only one xenon (Xe) atom, and the *tri-* prefix on the oxide indicates three oxygen (O) atoms. This makes the formula for xenon trioxide XeO_3.

479. $SbCl_5$

In *antimony pentachloride,* the absence of a prefix on antimony indicates one antimony (Sb) atom, and the *penta-* prefix on the chloride indicates five chlorine (Cl) atoms, so the formula for antimony pentachloride is $SbCl_5$.

480. Cl_2O_7

The name *dichlorine heptoxide* has the prefix *di-*, meaning two, and the prefix *hept(a)-*, meaning seven, so the formula includes two chlorine (Cl) atoms and seven oxygen (O) atoms. This means that the formula for dichlorine heptoxide is Cl_2O_7.

481. 3

A Lewis dot diagram shows only the valence electrons. Boron has an electron configuration of $1s^2 2s^2 2p^1$. The valence electrons are the ones in the second energy level, $2s^2 2p^1$, so the diagram contains three electrons.

482. 2

Barium has a noble gas electron configuration $[Xe]6s^2$ with two valence electrons. (The electron configuration of a noble gas like xenon is very stable, and any electrons beyond that configuration are an atom's valence electrons.) The Lewis dot diagram shows only valence electrons, so the diagram for barium has two electrons.

483. 7

The noble gas electron configuration of chlorine is $[Ne]3s^2 3p^5$. The Lewis dot diagram shows valence electrons, so the diagram for chlorine has seven electrons.

484. 8

An oxygen atom has eight electrons ($1s^2 2s^2 2p^4$), and the oxygen ion, with a charge of –2, has two more electrons ($1s^2 2s^2 2p^6$). This means that there are eight valence electrons in the oxygen ion ($2s^2 2p^6$), so eight electrons go in the Lewis dot diagram.

485. 0

The sodium ion, Na⁺, is a sodium atom that has lost its one valence electron, going from an electron configuration of $1s^22s^22p^63s^1$ to $1s^22s^22p^6$. The orbitals in the third energy level are still there, but they don't contain any electrons. The Lewis dot diagram, which shows only the valence electrons, has zero electrons for the sodium ion.

486. the nitrogen group

Members of the nitrogen group (Group VA or 15) have five valence electrons because their highest energy level has electrons filling s^2p^3 orbitals.

487. the oxygen group

Members of the oxygen group (Group VIA or 16) have six valence electrons because their highest energy level has electrons filling s^2p^4 orbitals.

488. the boron group

Members of the boron group (Group IIIA or 13) have three valence electrons because their highest energy level has electrons filling s^2p^1 orbitals.

489. the alkaline earth metals

Members of the alkaline earth metals have two valence electrons because their highest energy level has electrons filling s^2 orbitals. Elements with two valence electrons are in Group 2 (or 2A, IIA, or IIB) of the periodic table.

490. the noble gases

The noble gases, except helium, have eight valence electrons because their highest energy level has electrons filling s^2p^6 orbitals (helium, however, has only two valence electrons). The noble gases are in Group 18 (or 8, VIIIA, VIIIB, or 0) of the periodic table.

491. electronegativity

Electronegativity is how strongly an atom attracts a bonding pair of electrons to itself. You can look up electronegativity values in a table.

Tip: In most cases, you need to know only relative electronegativity values. Fluorine has the highest value, so you can assume that the closer an element is to fluorine, the higher its electronegativity. If two elements are the same distance away from fluorine, the one that's closer to the top of the periodic table has the higher value. This tip works for all representative elements except hydrogen.

492. nonpolar covalent

Nonpolar covalent bonds tend to form between elements that have an electronegativity difference between 0.0 and 0.3 (or between 0.0 and 0.4; the cutoff depends on which electronegativity values are used).

Note that in terms of electronegativity, the division between bond types is not a precise division. Additional factors, such as the elements present, neighboring elements, and hybridization, affect the type of bond.

493. **polar covalent**

Polar covalent bonds form between elements that have a moderate electronegativity difference, typically between 0.3 and 1.7. In general, differences less than 0.3 (or 0.4, depending on the electronegativity values used) are considered nonpolar; differences greater than 1.7 are generally considered ionic.

494. **lower; higher**

Electronegativity is the relative attraction that an atom in a compound has for electrons. Metals are more likely to give away electrons than accept them, so metals tend to have lower electronegativity values. Nonmetals are more likely to accept electrons, so nonmetals tend to have higher electronegativity values. Of course, there are exceptions. For example, gold, a metal, has an unusually high electronegativity. The electronegativity of gold (2.4) is higher than that of the nonmetal phosphorus (2.1).

Note: The *representative elements* are the elements in Groups 1, 2, and 13 through 18 on the periodic table. In the Chemical Abstract Services (CAS) version of the periodic table, which uses roman numerals, the representative elements are the A elements (as opposed to the transition elements, which are the B elements). The A elements always have an A in their group number, as in IA and IIA.

495. **polar covalent**

Elements that have a moderate difference in their electronegativity values normally form polar covalent bonds. Nonmetals that aren't adjacent on the periodic table, such as carbon and bromine, usually have moderate differences.

496. **polar covalent**

Elements that have a moderate difference in their electronegativity values normally form polar covalent bonds. Nonmetals that aren't adjacent on the periodic table, such as carbon and fluorine, usually have moderate differences.

497. **polar covalent**

Elements that have a moderate difference in their electronegativity values normally form polar covalent bonds. Nonmetals that aren't adjacent on the periodic table usually have moderate differences.

Elements that are diagonal to each other, such as chlorine and fluorine, are 1.4 times as far away from each other as side-by-side elements. If you go over one and down one, you can think of the elements as though they're nonadjacent; they're twice as far away.

498. **nonpolar covalent**

Elements with similar electronegativity values normally form nonpolar covalent bonds. Similar values occur for a nonmetal bonding with itself or with an adjacent nonmetal on the periodic table. Remember that hydrogen is an exception; its electronegativity places it between boron and carbon. The electronegativity difference between boron and hydrogen is small, so the bond is nonpolar covalent.

499. **nonpolar covalent**

Elements with similar electronegativity values normally form nonpolar covalent bonds. Similar values occur for a nonmetal bonding with itself or with an adjacent nonmetal on the periodic table. Remember that hydrogen is an exception; its electronegativity places it between boron and carbon. The electronegativity difference between carbon and hydrogen is small, so the bond is nonpolar covalent.

500. **polar covalent**

Elements that have a moderate difference in their electronegativity values normally form polar covalent bonds. Nonmetals that aren't adjacent on the periodic table, such as phosphorus and fluorine, usually have moderate differences.

501. **S–Cl < O–S < Si–Cl**

In general, the farther two nonmetals are from each other on the periodic table, the more polar the bond. Remember that hydrogen is an exception in that in terms of electronegativity, it fits between boron and carbon. If two pairs have the same separation, the one closer to the top (or the right) of the periodic table is more polar.

Sulfur and chlorine are adjacent on the periodic table, so the S–Cl bond is slightly polar. Oxygen is higher on the periodic table than chlorine, so its electronegativity is higher. For this reason, even though oxygen is also adjacent to sulfur, the S–O bond is more polar than the S–Cl bond. In this problem, the greatest separation on the periodic table is between silicon and chlorine, so Si–Cl is the most polar bond.

502. **trigonal pyramidal**

This molecule has three bonding pairs of electrons (indicated by the lines between the atoms) and one nonbonding pair of electrons (indicated by the two dots in the region above the central atom). This combination yields a trigonal pyramid shape — a pyramid with a three-sided base. The three black spheres form the triangular base of the pyramid, and the gray sphere — the central atom — represents the apex.

503. **linear**

This molecule has two bonding pairs of electrons (indicated by the lines between the atoms) and no nonbonding pairs of electrons. The bonding pairs orient themselves on opposite sides of the central atom, giving the molecule a linear shape.

504. **tetrahedral**

This molecule has four bonding pairs of electrons (indicated by the four lines between the atoms) and no lone pairs of electrons. The bonding pairs arrange themselves to maximize the distance from each other, so the angle measures are 109.5° at the central atom. A *tetrahedron* is a pyramid with four faces that are equilateral triangles, so this shape is tetrahedral.

505. **trigonal planar**

This molecule has three bonding pairs of electrons (indicated by the lines between the atoms) and no lone pairs of electrons. The three pairs arrange themselves to maximize their distance from each other (120° apart). This arrangement is known as *trigonal planar,* because the three pairs lie in a plane.

506. **linear**

This molecule has one bonding pair (indicated by the line between the atoms) and no nonbonding pairs of electrons. There are only two atoms, so the molecular geometry must be linear.

507. **bent**

This molecule has two bonding areas (indicated by the lines between the atoms) and two nonbonding pairs (indicated by the two pairs of dots). The four total pairs arrange themselves to maximize the distance between pairs. Bonding atoms to two of the pairs results in a bent shape.

508. **linear**

A bond angle of 180° — a straight line — corresponds to the linear shape.

509. **trigonal planar**

A bond angle of 120° is in trigonal planar molecules. In this shape, the three bonding pairs of electrons lie in a plane, maximizing their distance from each other.

510. **tetrahedral**

The tetrahedral molecular shape has a bond angle of approximately 109.5°.

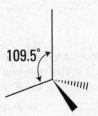

511. **trigonal planar**

Start with the Lewis structure of BCl_3:

$$:\ddot{C}l-B-\ddot{C}l:$$
$$|$$
$$:\ddot{C}l:$$

The lines between the boron and each of the chlorine atoms represent bonding pairs of electrons. There are no other pairs on the boron, and the additional pairs on the non-central atoms aren't important in determining molecular geometry. Because boron has only three pairs, the bonds will be at 120° angles, yielding a trigonal planar structure.

512. **linear**

Bromine, Br_2, is diatomic, so it has only one bonding pair area. Therefore, it's linear in shape.

513. **tetrahedral**

Start by drawing the Lewis structure of CH_4:

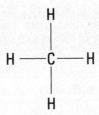

The lines between the carbon and each of the hydrogen atoms represent bonding pairs of electrons. There are no other pairs on the carbon. Because carbon has only four electron pairs, the bonds will be at 109.5° angles, yielding a tetrahedral structure.

514. **bent**

Start by drawing the Lewis structure of H_2O:

$$H-\ddot{O}:$$
$$|$$
$$H$$

The lines between the oxygen and each of the hydrogen atoms represent bonding pairs of electrons. The oxygen also has two lone pairs. Oxygen has four pairs, so the pairs will be at 109.5° angles, yielding a tetrahedral electron-pair structure. But because two of the pairs are lone pairs, they aren't part of the molecular geometry. This leaves a bent geometry. The lone pairs cause the angle between the bonding pairs to be a little less than expected because the electron pairs repel each other, spreading farther apart.

515. **trigonal pyramidal**

Start by drawing the Lewis structure of NH_3:

H——N̈——H
        |
        H

The lines between the nitrogen and each of the hydrogen atoms represent bonding pairs of electrons. The nitrogen also has one lone pair. Nitrogen has four pairs, so they'll be at 109.5° angles, yielding a tetrahedral electron-pair structure. But because one of the pairs is a lone pair, it isn't part of the molecular geometry. This leaves a trigonal pyramidal geometry. The lone pair causes the angle between the bonding pairs to be a little less than expected.

516. **linear**

Start by drawing the Lewis structure of CO_2:

Ö══C══Ö

The lines between the carbon and each of the oxygen atoms represent bonding pairs of electrons; however, when dealing with double bonds, only one of the two bonding pairs counts. Because the carbon has two bonding pair areas, not four, they'll be at 180° angles, yielding a linear geometry. The electron pairs not on the central atom don't affect the molecular geometry.

517. **Its central atom has four electron pairs, two of which are nonbonding.**

Look at the Lewis structure of H_2O:

H——Ö:
        |
        H

The lines between the oxygen and each of the hydrogen atoms represent bonding pairs of electrons. The oxygen also has two lone pairs. Oxygen has four pairs, so the pairs will be at 109.5° angles, yielding a tetrahedral electron-pair structure. But because two of the pairs are lone pairs, they aren't part of the molecular geometry. This leaves a bent geometry. The lone pairs cause the angle between the bonding pairs to be a little less than expected.

518. **Its central atom has three bonding pair areas and one nonbonding pair area.**

Look at the Lewis structure of NF_3:

$$:\ddot{F} - \ddot{N} - \ddot{F}:$$
$$|$$
$$:\ddot{F}:$$

The lines between the nitrogen and each of the fluorine atoms represent bonding pairs of electrons. The nitrogen also has one lone pair. Nitrogen has four pairs, so the pairs will be at 109.5° angles, yielding a tetrahedral electron-pair structure. But because one of the pairs is a lone pair, it isn't part of the molecular geometry. This leaves a trigonal pyramidal geometry. The lone pair causes the angle between the bonding pairs to be a little less than expected.

---

519. **Its central atom has two bonding pairs and no lone pairs of electrons.**

Look at the Lewis structure of CO_2:

$$\ddot{O} = C = \ddot{O}$$

The lines between the carbon and each of the oxygen atoms represent bonding pairs of electrons; however, when dealing with double bonds, only one of the two bonding pairs counts. Because the carbon has two pairs, not four, they'll be at 180° angles, yielding a linear molecular geometry. The pairs not on the central atom don't affect the molecular geometry.

---

520. **It has four bonding pairs and no lone pairs of electrons.**

Look at the Lewis structure of CCl_4:

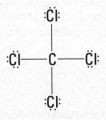

The lines between the carbon and each of the chlorine atoms represent bonding pairs of electrons. There are no other electron pairs on the carbon. Because carbon has only four pairs, they'll be at 109.5° angles, yielding a tetrahedral structure.

---

521. **Its central atom has three bonding pair areas and zero nonbonding pair areas.**

Look at the Lewis structure of BF_3:

$$:\ddot{F} - B - \ddot{F}:$$
$$|$$
$$:\ddot{F}:$$

The lines between the boron and each of the fluorine atoms represent bonding pairs of electrons. There are no other pairs on the boron, and the additional pairs on the non-central atoms aren't important in determining the molecular geometry. Because boron has only three pairs, they'll be at 120° angles, yielding a trigonal planar structure.

522. **octahedral**

This molecule has six bonding pairs of electrons (indicated by the lines between the atoms) and no nonbonding pairs of electrons. The six pairs arrange themselves 90° apart to maximize their separation. The result is an octahedral shape (see the following figure). An *octahedron* has eight faces that are equilateral triangles.

523. *T*-shaped

The drawing shows a molecule with three bonding pairs and two lone pairs of electrons. This combination indicates a *T*-shaped molecule.

524. **seesaw**

This molecule has four bonding pairs of electrons (indicated by the lines between the atoms) and one nonbonding pair of electrons (indicated by the pair of dots). The five electron pairs arrange themselves in a trigonal bipyramidal structure — a double pyramid with a triangular base. The lone pair occupies one of the central positions, leaving a seesaw arrangement.

525. **square pyramidal**

This molecule has five bonding pairs of electrons (indicated by the lines between the atoms) and one nonbonding pair of electrons (indicated by the pair of dots). The six pairs arrange themselves pointing toward the center of an octahedron, with the lone pair occupying one corner. Four of the atoms surrounding the central atom form the base of a square pyramid, and the fifth atom forms the apex, creating a structure known as *square pyramidal:*

526. **trigonal bipyramidal**

The drawing depicts a molecule that has five bonding pairs of electrons (indicated by the lines between the atoms) and no lone pairs of electrons. The five pairs arrange themselves as far apart as possible to minimize repulsion. This results in the only basic structure that has two types of bond angles: 90° and 120°. The three central bonding pairs form a trigonal (three-sided) plane, with the pairs at 120° angles to each other. The upper bonding pair (at 90° to the three in the plane) forms the apex to a *trigonal pyramid* (a pyramid with a three-sided base), and the lower bonding pair (also at 90° to the three in the plane) forms the apex of a second trigonal pyramid.

527. **square planar**

The drawing depicts a molecule with four bonding pairs and two lone pairs of electrons. The bonding pairs form a square around the central atom, with nonbonding pairs above and below the plane.

528. **trigonal bipyramidal**

The trigonal bipyramidal shape has atoms that are 90° apart (the three atoms that are in same the plane are 90° from the atoms above and below the plane), 120° apart (the three atoms in the plane are 120° away from each other), and 180° apart (the atoms above and below the plane are in a line 180° apart).

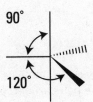

529. **tetrahedral**

A tetrahedral geometry has bond angles of 109.5°. The bond angles are slightly less than 109.5° in the trigonal pyramidal and bent structures because repulsion between electrons causes the nonbonding electron pairs to take up more room in the compound.

530. **seesaw**

Start with the Lewis structure of SF₄:

The diagram shows four bonding pairs of electrons (indicated by the lines) between the sulfur and the fluorine atoms. The sulfur also has a lone pair. You can disregard the lone pairs on the fluorine because they don't determine the structure. The five pairs around the sulfur adopt a trigonal bipyramidal electron-pair geometry (a double pyramid with a triangular base).

The lone pair is one of the trigonal pairs (separated from each other by 120°), leaving atoms at the other two trigonal pairs and at the two pairs at the apexes of the pyramids. This shape is a seesaw molecular geometry.

531. **trigonal bipyramidal**

Look at the Lewis structure of PCl₅:

The diagram shows five bonding pairs of electrons (indicated by the lines) between the phosphorus and the chlorine atoms. The five pairs around the phosphorus adopt a trigonal bipyramidal electron pair geometry (a double pyramid with a triangular base). Three of the pairs form a trigonal planar arrangement (the bonds are 120° apart).

532. **square pyramidal**

Start with the Lewis structure of BrF₅:

The diagram shows five bonding pairs of electrons (indicated by the lines) between the bromine and the fluorine atoms. The bromine also has a lone pair of electrons. The six pairs around the bromine adopt an octahedral electron-pair geometry (all pairs are at 90° angles, forming a double pyramid with a square base). One of the bonding pairs, at 90° to the square plane, forms the apex of a pyramid.

533. octahedral

Look at the Lewis structure of SF_6:

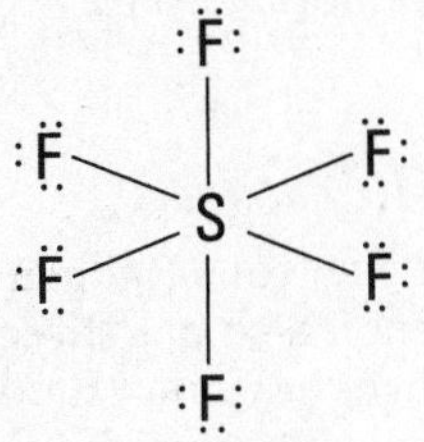

The diagram shows six bonding pairs of electrons (indicated by the lines) between the sulfur and the fluorine atoms. The six pairs around the sulfur adopt an octahedral electron pair geometry (all pairs are at 90° angles, forming a double pyramid with a square base).

534. square planar

Start with the Lewis structure of XeF_4:

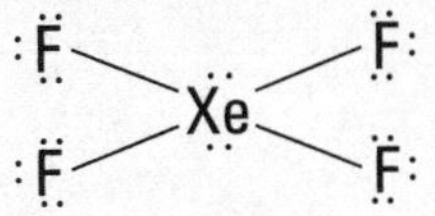

The diagram shows four bonding pairs of electrons (indicated by the lines) between the xenon and the fluorine atoms. The xenon also has two lone pairs of electrons. The six pairs around the xenon adopt an octahedral electron-pair geometry (all pairs are at 90° angles, forming a double pyramid with a square base). The two lone pairs are on opposite sides of the xenon, leaving all the atoms in a plane. This is a square planar geometry.

535. *T*-shaped

Start with the Lewis structure of ClF_3:

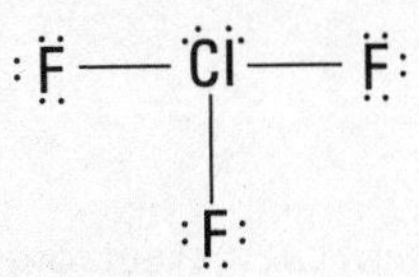

The diagram shows three bonding pairs of electrons (indicated by the lines) between the chlorine and the fluorine atoms. The chlorine also has two lone pairs. You can disregard the lone pairs on the fluorine, because they don't determine the structure. The five pairs around the sulfur adopt a trigonal bipyramidal electron-pair geometry (a double pyramid with a triangular base).

The lone pairs are two of the trigonal pairs (separated by 120°), leaving atoms at the other trigonal pair and at the two pairs at the apexes of the pyramids. This is a *T*-shaped molecular geometry.

536. **trigonal bipyramidal; linear**

Start with the Lewis structure of xenon difluoride, XeF_2:

$$:\ddot{F}\!-\!:\ddot{Xe}\!-\!\ddot{F}:$$

The diagram shows two bonding pairs of electrons (indicated by the lines) between the xenon and the fluorine atoms. The xenon also has three lone pairs. You can disregard the lone pairs on the fluorine, because they don't determine the structure. The five pairs around the xenon adopt a trigonal bipyramidal electron-pair geometry (a double pyramid with a triangular base).

The lone pairs are the trigonal pairs (separated by 120°), leaving atoms at both pairs at the apexes of the pyramids. This is a linear molecular geometry.

537. **the polarity of the bonds, shape of the molecule, and distribution of the electron pairs on the atoms**

To determine the polarity of a molecule, you need to know the polarity of the bonds in the molecule, the shape of the molecule, and the distribution of electron pairs on all the atoms in the molecule.

538. **nonpolar; nonpolar**

Elements with similar electronegativity values normally form nonpolar covalent bonds. Hydrogen is bonding to itself in H_2, so the electronegativity difference is zero. If a molecule has only nonpolar bonds, it's nonpolar.

539. **polar; polar**

Elements that have a moderate difference in their electronegativity values normally form polar covalent bonds. Nonmetals that aren't adjacent on the periodic table, such as hydrogen and chlorine, usually have moderate differences (note that hydrogen is special in that its electronegativity places it between boron and carbon). For diatomic molecules like HCl, if the bond is polar covalent, so is the molecule.

540. **polar; nonpolar**

Elements that have a moderate difference in their electronegativity values normally form polar covalent bonds. Nonmetals that aren't adjacent on the periodic table, such as carbon and chlorine, usually have moderate differences.

In CCl_4, the symmetrical arrangement of the four polar bonds (tetrahedral) results in their polarities canceling, leaving a nonpolar molecule:

541. **polar; polar**

Elements that have a moderate difference in their electronegativity values normally form polar covalent bonds. Nonmetals that aren't adjacent on the periodic table usually have moderate differences (note that hydrogen is special in that its electronegativity places it between boron and the carbon).

In NH_3, the three bonding pairs are on one side of the molecule (the base of the trigonal pyramid), making the base of the pyramid positive and the top negative. This makes the molecule polar.

542. **polar; polar**

Elements that have a moderate difference in their electronegativity values normally form polar covalent bonds. Nonmetals that aren't adjacent on the periodic table usually have moderate differences. (*Note:* Diagonal elements are 1.4 times as far away from each other as side-by-side elements; if you go over one and down one, you can think of the elements as though they're nonadjacent; they're twice as far away.)

The $SeCl_4$ molecule has a seesaw shape, so the polarities of the end atoms (180° apart) cancel. The other two polar bonds pull in the same general direction to make the molecule polar.

543. **polar; nonpolar**

Elements that have a moderate difference in their electronegativity values normally form polar covalent bonds. Nonmetals that aren't adjacent on the period table, such as xenon and oxygen, usually have moderate differences.

In XeO_4, the symmetrical arrangement of the four polar bonds (tetrahedral) results in their polarities canceling, leaving a nonpolar molecule.

544. **polar; polar**

Elements that have a moderate difference in their electronegativity values normally form polar covalent bonds. Nonmetals that aren't adjacent on the periodic table, such as iodine and chlorine, usually have moderate differences.

ICl_3 is a *T*-shaped molecule. The two polar covalent bonds forming the top of the *T* cancel, and the remaining polar covalent bond makes the molecule polar.

545.

OF$_2$, NBr$_3$, PCl$_3$, and IF$_5$

OF$_2$, PCl$_3$, and IF$_5$ are all polar because they contain polar bonds and have uneven distributions of electrons on asymmetric molecules (their respective molecular shapes are bent, trigonal pyramidal, and square pyramidal). NBr$_3$ has nonpolar bonds and an even distribution of electrons all the way around the molecule, but because the molecule is trigonal pyramidal (because it has three bonding pairs of electrons and one nonbonding pair), the molecule is polar.

546.

decomposition

When one reactant forms two or more products, you have a decomposition reaction. In this case, NaCl breaks down into Na and Cl$_2$.

547.

single displacement

In this reaction, the sodium displaces the hydrogen from the HCl. When an element replaces an element with similar charge in a compound, forming a new compound and leaving the replaced element by itself, the reaction is a *single displacement reaction.*

548.

combination

In a c*ombination reaction,* also known as a *synthesis reaction,* two elements react to form a single compound.

549.

combustion

When oxygen in the air reacts with a compound containing carbon (with an ignition source) to form carbon dioxide and water, you have a combustion reaction.

If insufficient oxygen is present, *incomplete combustion* may occur. In that case, CO or even C(s) may form in place of the CO$_2$.

550.

double displacement

A reaction that occurs when two compounds exchange ions is a double displacement reaction. To make the swap easier to see, note that you can write H$_2$O as HOH (or H–OH).

This specific type of double displacement reaction (an acid reacting with a base) is also known as a *neutralization reaction.*

551.

combination

The two elements hydrogen and oxygen react to form a single compound, so this is a combination reaction.

552.   **combustion**

When oxygen in the air reacts with a compound containing carbon (with an ignition source) to form carbon dioxide and water, you have a combustion reaction. Incomplete combustion may also have carbon monoxide as a product.

553.   **combination**

The two elements nitrogen and hydrogen react to form a single compound, so this is a combination reaction.

554.   **decomposition**

When one reactant forms two or more products, the reaction is a decomposition reaction.

555.   **single displacement**

In a single displacement reaction, an element replaces an element of similar charge in a compound, forming a new compound and leaving the replaced element by itself. In this case, potassium displaces silver.

556.   **combination**

$$4Al + 3O_2 \rightarrow 2Al_2O_3$$

When two elements react to form a single compound, you have a combination reaction.

557.   **double displacement**

$$2NaOH + Pb(NO_3)_2 \rightarrow 2NaNO_3 + Pb(OH)_2$$

A reaction that occurs when two compounds exchange ions is a double displacement reaction. In this case, the sodium trades places with the lead (or equivalently, the hydroxide trades places with the nitrate).

558.   **single displacement**

$$Mg + 2HCl \rightarrow MgCl_2 + H_2$$

When an element replaces an element of similar charge in a compound, forming a new compound and leaving the replaced element by itself, the reaction is a single displacement reaction. In this case, magnesium displaces hydrogen.

559. **decomposition**

$$2HgO \rightarrow 2Hg + O_2$$

When one reactant forms two or more products, the reaction is a decomposition reaction.

560. **combustion**

$$2C_4H_{10} + 13O_2 \rightarrow 8CO_2 + 10H_2O$$

When oxygen in the air reacts with a compound containing carbon (with an ignition source) to form carbon dioxide and water, the reaction is a combustion reaction. Incomplete combustion may also have carbon monoxide as a product.

561. **single displacement**

$$Fe + CuCl_2 \rightarrow FeCl_2 + Cu$$

In a single displacement reaction, an element replaces an element of similar charge in a compound, forming a new compound and leaving the replaced element by itself. Here, iron takes the place of copper.

562. **decomposition**

$$2NaHCO_3 \rightarrow Na_2O + H_2O + 2CO_2$$

When one reactant forms two or more products, the reaction is a decomposition reaction.

563. **combustion**

$$C_2H_5OH + 3O_2 \rightarrow 2CO_2 + 3H_2O$$

In a combustion reaction, oxygen in the air reacts with a compound containing carbon (with an ignition source) to form carbon dioxide and water. Incomplete combustion may also have carbon monoxide as a product.

564. **combination**

$$CaO + H_2O \rightarrow Ca(OH)_2$$

When two elements react to form a single compound, it's a combination reaction.

565. combination

$$2CO + O_2 \rightarrow 2CO_2$$

Two elements react to form a single compound, so this is a combination reaction. You can also classify this reaction as a combustion reaction.

566. single displacement

$$Zn + H_2SO_4 \rightarrow ZnSO_4 + H_2$$

When an element replaces an element with similar charge in a compound, forming a new compound and leaving the replaced element by itself, you have a single displacement reaction.

567. decomposition

$$2N_2O \rightarrow 2N_2 + O_2$$

When one reactant forms two or more products, you have a decomposition reaction.

568. double displacement

$$Ca(OH)_2 + H_2SO_4 \rightarrow CaSO_4 + 2H_2O$$

A reaction that occurs when two compounds exchange ions is a double displacement reaction. Here, the calcium trades places with the hydrogen (or equivalently, the hydroxide trades places with the sulfate).

This type of double displacement reaction (an acid plus a base) is also known as a *neutralization reaction*.

569. combination

$$4Li + O_2 \rightarrow 2Li_2O$$

When two elements react to form a single compound, you have a combination reaction.

570. combustion

$$C_3H_8 + 5O_2 \rightarrow 3CO_2 + 4H_2O$$

When oxygen in the air reacts with a compound containing carbon (with an ignition source) to form carbon dioxide and water, you have a combustion reaction. Incomplete combustion may also have carbon monoxide as a product.

571.   **double displacement**

$$AgNO_3 + NaCl \rightarrow NaNO_3 + AgCl$$

A reaction that occurs when two compounds exchange ions of the same charge is a double displacement reaction. In this case, the sodium trades places with the silver (or equivalently, the nitrate trades places with the chloride).

572.   **combination**

$$2Na + H_2 \rightarrow 2NaH$$

When two elements react to form a single compound, the reaction is a combination reaction.

573.   **single displacement**

$$2KBr + Cl_2 \rightarrow 2KCl + Br_2$$

In a single displacement reaction, an element replaces an element of similar charge in a compound, forming a new compound and leaving the replaced element by itself. In this case, chlorine displaces bromine.

574.   **decomposition**

$$Cu(OH)_2 \rightarrow CuO + H_2O$$

When one reactant forms two or more products, you have a decomposition reaction.

575.   **double displacement**

$$Al_2(SO_4)_3 + Ca_3(PO_4)_2 \rightarrow 2AlPO_4 + 3CaSO_4$$

A reaction that occurs when two compounds exchange ions is a double displacement reaction. In this case, the aluminum trades places with the calcium (or equivalently, the sulfate trades places with the phosphate).

576.   1, 2

For every N_2O_4 molecule that decomposes, you need two NO_2 molecules, making the coefficients of the balanced equation 1 and 2:

$$N_2O_4 \rightarrow 2NO_2$$

577.   1, 3, 1, 1

The only element out of balance in this reaction is the chlorine. Two atoms of chlorine are on the reactants side of the equation, and six atoms of chlorine are on the products side of the equation. Multiplying the Cl_2 by 3 makes the coefficients 1, 3, 1, and 1:

$$CS_2 + 3Cl_2 \rightarrow CCl_4 + S_2Cl_2$$

---

578. **1, 1, 1, 2**

In this equation, the Na atoms and the NO_3^- ions are not in balance. Two sodium atoms and two nitrate ions are on the reactants side, and only one sodium atom and one nitrate ion are on the products side. Multiplying the $NaNO_3$ by 2 balances the equation and makes the coefficients 1, 1, 1, and 2:

$$Na_2CO_3 + Cd(NO_3)_2 \rightarrow CdCO_3 + 2NaNO_3$$

---

579. **1, 1, 2**

In this equation, neither the iodine nor the chlorine is balanced. You see two atoms of iodine and two atoms of chlorine on the reactants side but only one atom of iodine and one atom of chlorine on the product side. Multiplying ICl by 2 balances the iodine and the chlorine atoms, making the coefficients 1, 1, and 2:

$$I_2 + Cl_2 \rightarrow 2ICl$$

---

580. **2, 1, 2, 1**

In this equation, neither the chlorine nor the fluorine is balanced. You have one atom of chlorine and two atoms of fluorine on the reactants side but only one atom of fluorine and two atoms of chlorine on the products side. Multiplying the KCl by 2 on the reactants side balances the chlorine atoms, but that gives you two atoms of potassium on the reactants side. Multiplying the KF on the products side then balances the potassium and fluorine atoms, making the coefficients 2, 1, 2, and 1:

$$2KCl + F_2 \rightarrow 2KF + Cl_2$$

---

581. **1, 2, 1, 2**

In this equation, only the carbon atoms are balanced. Multiplying the water on the products side by 2 balances the hydrogen atoms, giving you four on each side; it also increases the number of oxygen atoms on the products side to four. To balance the oxygen atoms, you need to multiply the O_2 on the reactants side by 2, making the coefficients 1, 2, 1, and 2:

$$CH_4 + 2O_2 \rightarrow CO_2 + 2H_2O$$

---

582. **2, 1, 2**

To balance this equation, first balance the oxygen by multiplying CaO by 2. Next, balance the calcium by multiplying the Ca on the reactants side by 2. The coefficients are 2, 1, and 2:

$$2Ca + O_2 \rightarrow 2CaO$$

583.

1, 3, 2

Neither the nitrogen nor the hydrogen is balanced in this equation. First, balance the nitrogen by multiplying the NH_3 by 2, giving you two nitrogen atoms on each side. You now have six hydrogen atoms on the product side. To balance the hydrogen atoms, multiply the H_2 by 3. This final coefficients are 1, 3, and 2:

$$N_2 + 3H_2 \rightarrow 2NH_3$$

584.

2, 1, 2

Initially, only the sulfur is balanced in this equation. You need more oxygen atoms on the product side, so begin by multiplying the SO_3 by 2; this gives you six oxygen atoms and two sulfur atoms on the product side.

To balance the sulfur atoms, multiply the SO_2 by 2 on the reactants side; this gives you two sulfur atoms and four oxygen atoms. When you add the four oxygen atoms from the $2SO_2$ to the two oxygen atoms from the O_2, you see that you have six oxygen atoms on the reactants side, so the oxygen is balanced. The coefficients are 2, 1, and 2:

$$2SO_2 + O_2 \rightarrow 2SO_3$$

585.

3, 2, 1, 6

Nothing is balanced in this equation. Start with the Mg atoms. You have one Mg atom on the reactants side and three on the products side, so multiply the $Mg(NO_3)_2$ by 3; not only does this give you three Mg atoms on both sides, but it also gives you six NO_3 ions on the reactants side, compared to the one NO_3 ion on the products side.

To balance the NO_3 ions, multiply KNO_3 by 6 on the products side; this also gives you six K atoms. Last, multiply the K_3PO_4 by 2 to balance the K and PO_4 ions. The final coefficients 3, 2, 1, and 6:

$$3Mg(NO_3)_2 + 2K_3PO_4 \rightarrow Mg_3(PO_4)_2 + 6KNO_3$$

586.

1, 1, 1

Here's the balanced equation:

$$BaSO_3 \rightarrow BaO + SO_2$$

The coefficients are 1, 1, and 1.

587.

1, 2, 1, 1

Write the unbalanced equation:

$$SiC + Cl_2 \rightarrow SiCl_4 + C$$

Only the chlorine atoms are unbalanced. To balance them, multiply the Cl_2 by 2:

$$SiC + 2Cl_2 \rightarrow SiCl_4 + C$$

The coefficients are 1, 2, 1, and 1.

588. 1, 1, 2, 1

Write the unbalanced equation:

$$Ag_2SO_4 + Cu(ClO_3)_2 \rightarrow AgClO_3 + CuSO_4$$

The silver atoms and the chlorate ions aren't balanced. Multiplying the $AgClO_3$ on the products side by 2 balances the equation:

$$Ag_2SO_4 + Cu(ClO_3)_2 \rightarrow 2AgClO_3 + CuSO_4$$

The coefficients are 1, 1, 2, and 1.

589. 1, 2, 1, 1, 1

Write the unbalanced equation:

$$CaCO_3 + HCl \rightarrow CaCl_2 + H_2O + CO_2$$

Only the hydrogen and chlorine atoms are unbalanced. Multiplying the HCl on the reactants side by 2 balances the equation:

$$CaCO_3 + 2HCl \rightarrow CaCl_2 + H_2O + CO_2$$

The resulting coefficients are 1, 2, 1, 1, and 1.

590. 4, 3, 2

Write the unbalanced equation:

$$B + O_2 \rightarrow B_2O_3$$

To balance the oxygen atoms on both sides of the equation, you look for the lowest common multiple of 2 and 3, which is 6. Multiply the O_2 by 3 and the B_2O_3 by 2 to balance the oxygen atoms, giving you six on each side.

You end up with four boron atoms on the product side, so multiply the B on the reactants side by 4. The coefficients of the balanced reaction are 4, 3, and 2:

$$4B + 3O_2 \rightarrow 2B_2O_3$$

591. 5

Write the unbalanced equation:

$$Na + I_2 \rightarrow NaI$$

To balance the iodine, multiply the NaI by 2. That gives you two sodium atoms on the product side, so multiply Na by 2 on the reactants side:

$$2Na + I_2 \rightarrow 2NaI$$

The coefficients are 2, 1, and 2, and they add up to 5.

592. 7

Write the unbalanced equation, remembering that oxygen gas is diatomic:

$$KClO_3 \rightarrow KCl + O_2$$

To balance this equation, you need to balance the oxygen atoms. The lowest common multiple of 3 and 2 is 6, so multiply the $KClO_3$ by 2 and the O_2 by 3 to get six oxygen atoms on each side.

Now you have two potassium and two chlorine atoms on the reactant side. To balance those atoms, multiply the KCl on the products side by 2, which gives you

$$2KClO_3 \rightarrow 2KCl + 3O_2$$

The coefficients are 2, 2, and 3, and their sum is 7.

593. 9

Write the unbalanced equation, remembering that oxygen gas is diatomic:

$$Fe + O_2 \rightarrow Fe_2O_3$$

First, balance the oxygen atoms. The lowest common multiple of 2 and 3 is 6, so multiply the Fe_2O_3 by 2 (which also gives you four Fe atoms) and the O_2 by 3, giving you six oxygen atoms on each side.

Multiply the Fe on the reactants side by 4 to finish balancing the equation:

$$4Fe + 3O_2 \rightarrow 2Fe_2O_3$$

The coefficients are 4, 3, and 2, and they add up to 9.

594. 12

Write the unbalanced equation, remembering that Ca^{2+} plus N^{3-} gives you Ca_3N_2 (calcium nitride):

$$Ca_3N_2 + H_2O \rightarrow Ca(OH)_2 + NH_3$$

Tip: As you balance this equation, thinking of H_2O as HOH (or H–OH) may be helpful. That way, you can easily see where the OH in $Ca(OH)_2$ and the H in NH_3 come from.

Balance the calcium first by multiplying the $Ca(OH)_2$ on the products side by 3. Then balance the nitrogen by multiplying the NH_3 by 2.

Now you need to balance the hydrogens and hydroxides. You can do this by multiplying the H_2O by 6. Here's the final, balanced equation:

$$Ca_3N_2 + 6H_2O \rightarrow 3Ca(OH)_2 + 2NH_3$$

The coefficients are 1, 6, 3, and 2, and they add up to 12.

595. 2, 25, 16, 18

Write the unbalanced equation:

$$C_8H_{18} + O_2 \rightarrow CO_2 + H_2O$$

Tip: When balancing a combustion reaction with a hydrocarbon, balance the carbon first, the hydrogen second, and the oxygen last.

Balance the carbon and hydrogen by multiplying the carbon dioxide on the products side by 8 and multiplying the water on the products side by 9:

$$C_8H_{18} + O_2 \rightarrow 8CO_2 + 9H_2O$$

Sometimes when balancing the C and H, you end up with an odd number of O atoms on the products side. In this case, $8CO_2 + 9H_2O$ gives you $16 + 9 = 25$ oxygen atoms. To remedy this, multiply the C_8H_{18} by 2. Then continue to balance, doubling your carbon and hydrogen atoms on the products side:

$$2C_8H_{18} + O_2 \rightarrow 16CO_2 + 18H_2O$$

That gives you an even number of oxygen atoms on the products side ($32 + 18 = 50$), so multiply the O_2 by 25. The final coefficients are 2, 25, 16, and 18:

$$2C_8H_{18} + 25O_2 \rightarrow 16CO_2 + 18H_2O$$

---

596.   3

Iron(III) bromide and barium hydroxide are two ionic compounds, so first determine what kind of reaction takes place. You may be able to do that from just the words, or you may need to write the formulas for the reactants first:

$$FeBr_3 + Ba(OH)_2 \rightarrow$$

You should recognize that a double displacement reaction will occur. The anions or cations will change places in the two compounds, and the products will be iron(III) hydroxide and barium bromide. Here's the unbalanced equation:

$$FeBr_3 + Ba(OH)_2 \rightarrow Fe(OH)_3 + BaBr_2$$

The bromine atoms and the hydroxide ions aren't balanced. You can remedy the bromine imbalance by multiplying the $FeBr_3$ by 2 and the $BaBr_2$ by 3, giving you six bromine atoms on each side.

Now you need to balance the Ba^{2+}, Fe^{3+}, and OH^-, which you can do by multiplying the $Ba(OH)_2$ by 3 and the $Fe(OH)_3$ by 2. The balanced equation is

$$2FeBr_3 + 3Ba(OH)_2 \rightarrow 2Fe(OH)_3 + 3BaBr_2$$

The coefficient on the barium hydroxide is 3.

---

597.   1

In this reaction, an element (potassium) reacts with a compound (nickel(II) chloride), and that pattern characterizes this reaction as a single displacement reaction. The positively charged ions trade places, so the potassium replaces the nickel to make potassium chloride. Here's the unbalanced equation:

$$K + NiCl_2 \rightarrow KCl + Ni$$

To balance the chlorine, multiply the potassium chloride on the products side by 2. Then multiply the potassium on the reactants side by 2 to balance the potassium. Here's the balanced equation:

$$2K + NiCl_2 \rightarrow 2KCl + Ni$$

The coefficient on the nickel(II) chloride is 1.

598. 5

When only one reactant shows up in an equation, you can be pretty sure that the reaction is a decomposition reaction. Metallic oxides with high metal charges decompose into metallic oxides with lower metal charges. Lead(IV) oxide decomposes into lead(II) oxide and oxygen gas, so here's the unbalanced equation:

$$PbO_2 \rightarrow PbO + O_2$$

To balance this equation, simply multiply the PbO by 2 to get an even number of oxygen atoms; then multiply the PbO_2 by 2 to balance the lead atoms:

$$2PbO_2 \rightarrow 2PbO + O_2$$

The coefficients are 2, 2, and 1, which add up to 5.

Here's another possible reaction (note that the coefficients still sum to 5):

$$3PbO_2 \rightarrow Pb_3O_4 + O_2$$

599. 3

This reaction involves two compounds, each with a positive and a negative ion, so this must be a double displacement reaction. The silver will change places with the aluminum (or equivalently, the nitrate will change places with the chlorine) to form silver chloride and aluminum nitrate. Here's the unbalanced equation:

$$AgNO_3 + AlCl_3 \rightarrow AgCl + Al(NO_3)_3$$

The chlorine atoms and the nitrate ions aren't balanced. To remedy this situation, multiply the AgCl by 3 to balance the chlorine; then multiply the $AgNO_3$ by 3 to balance the silver and nitrate:

$$3AgNO_3 + AlCl_3 \rightarrow 3AgCl + Al(NO_3)_3$$

The coefficient on the silver nitrate is 3.

600. 6

When only one reactant shows up in an equation, you can be pretty sure you're dealing with a decomposition reaction. With a nonmetallic halide, the products are two elements (the nonmetal and the halogen), so here's the unbalanced equation:

$$NCl_3 \rightarrow N_2 + Cl_2$$

First, multiply the NCl_3 by 2 to even out the nitrogen atoms on each side of the equation. Next, balance the chlorine atoms by multiplying the Cl_2 by 3:

$$2NCl_3 \rightarrow N_2 + 3Cl_2$$

The sum of the coefficients is 2 + 1 + 3 = 6.

601. $2Zn + O_2 \rightarrow 2ZnO$

Here, two elements are reacting with each other to form a compound, making this a combination reaction. Write the unbalanced equation, remembering that oxygen gas is diatomic and that zinc, as a metal, is monatomic:

$$Zn + O_2 \rightarrow ZnO$$

Multiply the ZnO by 2 to balance the oxygen atoms and multiply the Zn by 2 to balance the zinc atoms:

$$2Zn + O_2 \rightarrow 2ZnO$$

602.   9

In this reaction, a compound is reacting with an element — the pattern of a single displacement reaction. The positive hydrogen ion will displace the positive iron ion in iron(III) oxide to make water, leaving Fe as the metal. Here's the unbalanced equation:

$$Fe_2O_3 + H_2 \rightarrow Fe + H_2O$$

First balance the oxygen atoms by multiplying the H_2O by 3. Then multiply the H_2 by 3 to balance the hydrogen atoms. Next, you need to balance the iron atoms, so multiply the Fe on the products side by 2. Here's the balanced equation:

$$Fe_2O_3 + 3H_2 \rightarrow 2Fe + 3H_2O$$

The sum of the coefficients of this reaction is $1 + 3 + 2 + 3 = 9$.

603.   2

Only one reactant is given, so this reaction is a decomposition reaction. Nonmetallic oxides decompose to give you a nonmetal and oxygen gas. Write the unbalanced equation, remembering that elemental gases are diatomic (except for the noble gases, which are monatomic):

$$NO_2 \rightarrow N_2 + O_2$$

To balance the nitrogen, multiply the NO_2 on the reactant side by 2. This gives you four oxygen atoms on the reactant side, so also multiply the O_2 by 2. Here's the balanced equation:

$$2NO_2 \rightarrow N_2 + 2O_2$$

The coefficient on the nitrogen dioxide is 2.

604. $2AgNO_3 + K_2Cr_2O_7 \rightarrow Ag_2Cr_2O_7 + 2KNO_3$

The reactants are two compounds, each with a positive ion and a negative ion, so this must be a double displacement reaction. The silver will change places with the potassium (or equivalently, the nitrate will change places with the dichromate) to form silver dichromate and potassium nitrate. Here's the unbalanced equation:

$$AgNO_3 + K_2Cr_2O_7 \rightarrow Ag_2Cr_2O_7 + KNO_3$$

The potassium atoms and the silver ions aren't balanced. To balance the potassium, multiply the KNO_3 by 2. Then multiply the $AgNO_3$ by 2 to balance the silver and the nitrate:

$$2AgNO_3 + K_2Cr_2O_7 \rightarrow Ag_2Cr_2O_7 + 2KNO_3$$

605.

3

In this case, an element (aluminum) reacts with a compound (copper(II) sulfate), which is the pattern of a single displacement reaction. The ions with the positive charges trade places, so the aluminum replaces the copper to make aluminum sulfate:

$$Al + CuSO_4 \rightarrow Al_2(SO_4)_3 + Cu$$

To balance this equation, you need to multiply the $CuSO_4$ by 3, because you have three SO_4^{2-} ions in aluminum sulfate. Next, multiply the Cu on the products side by 3 to balance the copper atoms. Last, multiply the Al on the reactants side by 2, giving you the balanced equation:

$$2Al + 3CuSO_4 \rightarrow Al_2(SO_4)_3 + 3Cu$$

The coefficient on the copper(II) sulfate is now 3.

606.

+7

To find the oxidation number of an element in a compound, set up an algebraic equation. Because $KMnO_4$ is a neutral compound, the sum of the charges of all the atoms in the compound equals zero:

$$K + Mn + 4O = 0$$

Substitute in the oxidation numbers of the elements other than manganese. Potassium is in Group 1 (IA) on the periodic table, so its expected oxidation state is +1. Oxygen is in Group 16 (VIA), so its expected oxidation state is –2.

$$K + Mn + 4O = 0$$
$$1(+1) + 1(Mn) + 4(-2) = 0$$
$$Mn - 7 = 0$$
$$Mn = 7$$

The oxidation number of manganese is +7.

607.

+5

To find the oxidation number of an element in a compound, set up an algebraic equation. Because $NaBrO_3$ is a neutral compound, the sum of the charges of all the atoms in the compound equals zero:

$$Na + Br + 3O = 0$$

Substitute in the oxidation numbers of the elements other than bromine. Sodium is in Group 1 (IA) on the periodic table, so its expected oxidation state is +1. Oxygen is in Group 16 (VIA), so its expected oxidation state is –2. Enter the values and then solve:

$$Na + Br + 3O = 0$$
$$1(+1) + 1(Br) + 3(-2) = 0$$
$$Br - 5 = 0$$
$$Br = 5$$

The oxidation number of bromine is +5.

608. +6

To find the oxidation number of an element in a compound, set up an algebraic equation. Because $Cr_2O_7^{2-}$ is a polyatomic ion, the sum of the charges of all the atoms in the ion equals the charge on the ion:

$$2Cr + 7O = -2$$

Next, substitute in the oxidation numbers of any elements other than chromium. Oxygen is in Group 16 (VIA), so its most common oxidation state is –2. Then solve:

$$2Cr + 7O = -2$$
$$2Cr + 7(-2) = -2$$
$$2Cr - 14 = -2$$
$$2Cr = 12$$
$$Cr = 6$$

The oxidation number of chromium is +6.

609. +2

To find the oxidation number of an element in a compound, set up an algebraic equation. Because $S_2O_3^{2-}$ is a polyatomic ion, the sum of the charges of all the atoms in the ion equals the charge on the ion:

$$2S + 3O = -2$$

Substitute in the oxidation number of the oxygen, which is more electronegative than sulfur. Oxygen is in Group 16 (VIA) on the periodic table, so its expected oxidation state is –2. Then solve:

$$2S + 3O = -2$$
$$2S + 3(-2) = -2$$
$$2S - 6 = -2$$
$$2S = 4$$
$$S = 2$$

The oxidation number of sulfur is +2.

610. **oxidation**

Oxidation occurs when a substance loses electrons.

611. **reduction**

Reduction occurs when a substance gains electrons.

612. **Fe**

Fe loses two electrons to become Fe^{2+}, so Fe is oxidized.

613. Fe²⁺

Fe^{2+} gains two electrons to become Fe, so it's reduced.

614. Cu

The *reducing agent* is the substance that's been oxidized; that is, it's the one that loses electrons. In this equation, copper loses electrons:

$$Cu \rightarrow Cu^{2+} + 2e^-$$

615. Br₂

The *oxidizing agent* is the substance that's been reduced; that is, it's the one that gains electrons. In this equation, Br_2 gains electrons:

$$Br_2 + 2e^- \rightarrow 2Br^-$$

Notice that Br_2 is also the reducing agent:

$$Br_2 + 12OH^- \rightarrow 2BrO_3^- + 6H_2O + 10e^-$$

616. 3e⁻

To balance this half-reaction, first balance the oxygen atoms:

$$NO_3^- \rightarrow NO$$
$$NO_3^- \rightarrow NO + 2H_2O$$

The water comes from the oxygen and hydrogen present in the acidic environment.

Then balance the hydrogen atoms:

$$4H^+ + NO_3^- \rightarrow NO + 2H_2O$$

Finally, balance the charge, giving you three electrons:

$$3e^- + 4H^+ + NO_3^- \rightarrow NO + 2H_2O$$

617. 10e⁻

To balance this half-reaction, first balance the bromine atoms:

$$Br_2 \rightarrow BrO_3^{1-}$$
$$Br_2 \rightarrow 2BrO_3^-$$

Then balance the oxygen atoms, followed by the hydrogen atoms. Add twice as many OH⁻ as you need to the oxygen-deficient side. Then add half as many H_2O molecules to the opposite side to balance the H atoms:

$$12OH^- + Br_2 \rightarrow 2BrO_3^-$$
$$12OH^- + Br_2 \rightarrow 2BrO_3^- + 6H_2O$$

Finally, balance the charge, giving you 10 electrons:

$$12OH^- + Br_2 \rightarrow 2BrO_3^- + 6H_2O + 10e^-$$

618. 8

The two half-reactions for this equation are as follows:

$$Zn \rightarrow Zn^{2+}$$
$$H_3AsO_4 \rightarrow AsH_3$$

You need to balance the first half-reaction only for charge. To do this, add two electrons to the products side:

$$Zn \rightarrow Zn^{2+} + 2e^-$$

The second half-reaction needs oxygen atoms on the products side, so add water to the products side and then add hydrogen ions to the reactants side:

$$H_3AsO_4 \rightarrow AsH_3 + 4H_2O$$
$$8H^+ + H_3AsO_4 \rightarrow AsH_3 + 4H_2O$$

Next, add eight electrons to the reactants side to balance the charge:

$$8e^- + 8H^+ + H_3AsO_4 \rightarrow AsH_3 + 4H_2O$$

The number of electrons on each side of the half-reactions needs to match, so finish balancing the equation by multiplying the zinc half-reaction $\left(Zn \rightarrow Zn^{2+} + 2e^-\right)$ by 4 and then adding the two half-reactions back together:

$$4Zn \rightarrow 4Zn^{2+} + \cancel{8e^-}$$
$$\underline{8e^- + 8H^+ + H_3AsO_4 \rightarrow AsH_3 + 4H_2O}$$
$$4Zn + 8H^+ + H_3AsO_4 \rightarrow 4Zn^{2+} + AsH_3 + 4H_2O$$

The equation is now balanced, with eight hydrogen ions on the reactants side.

619. 6

The two half-reactions for this equation are as follows:

$$Bi^{3+} \rightarrow Bi$$
$$SnO_2{}^{2-} \rightarrow SnO_3{}^{2-}$$

You need to balance the first half-reaction only for charge. To do this, add three electrons to the reactant side:

$$3e^- + Bi^{3+} \rightarrow Bi$$

For the second half-reaction, you need to balance the oxygen atoms. To do this in the presence of a base (OH^-), note how many oxygen atoms you need to add to the oxygen-deficient side and add twice as many hydroxide ions; then balance out the hydrogen atoms by adding half as many water molecules to the opposite side. In this case, you add two hydroxide ions to the reactants and one water molecule to the products side, for a net gain of one oxygen atom:

$$2OH^- + SnO_2{}^{2-} \rightarrow SnO_3{}^{2-} + H_2O$$

Next, balance the charge by adding two electrons to the products side:

$$2OH^- + SnO_2{}^{2-} \rightarrow SnO_3{}^{2-} + H_2O + 2e^-$$

When you add the two half-reactions together, the number of electrons needs to cancel. So multiply the first half-reaction $\left(3e^- + Bi^{3+} \rightarrow Bi\right)$ by 2 and multiply the second half-reaction $\left(2OH^- + SnO_2{}^{2-} \rightarrow SnO_3{}^{2-} + H_2O + 2e^-\right)$ by 3. Here's what you get when you add the half-reactions:

$$6e^- + 2Bi^{3+} \rightarrow 2Bi$$

$$\underline{6OH^- + 3SnO_2{}^{2-} \rightarrow 3SnO_3{}^{2-} + 3H_2O + 6e^-}$$

$$2Bi^{3+} + 6OH^- + 3SnO_2{}^{2-} \rightarrow 2Bi + 3SnO_3{}^{2-} + 3H_2O$$

Both the atoms and charge are in balance. The question asks for the coefficient on the hydroxide ion in the balanced equation, so the answer is 6.

620.	**1, 5, 3, 3, 3, 3**

First, separate the substances that would dissolve in water and eliminate the spectator ions:

$$KIO_3 + KI + H_2SO_4 \rightarrow K_2SO_4 + I_2 + H_2O$$

$$K^+ + IO_3{}^- + K^+ + I^- + H^+ + SO_4{}^{2-} \rightarrow K^+ + SO_4{}^{2-} + I_2 + H_2O$$

Then write your two half-reactions:

$$IO_3{}^- \rightarrow I_2$$

$$I^- \rightarrow I_2$$

Next, balance the iodine atoms, then the oxygen atoms, and then the hydrogen atoms in the first half-reaction. Note that in an acidic solution, excess oxygen combines with hydrogen atoms from the acidic environment to produce water. Here's the first half-reaction:

$$2IO_3{}^- \rightarrow I_2$$

$$2IO_3{}^- \rightarrow I_2 + 6H_2O$$

$$12H^+ + 2IO_3{}^- \rightarrow I_2 + 6H_2O$$

For the second half-reaction, you just balance the iodine atoms:

$$2I^- \rightarrow I_2$$

Now balance the charge in each half-reaction by adding electrons where needed:

$$10e^- + 12H^+ + 2IO_3{}^- \rightarrow I_2 + 6H_2O$$

$$2I^- \rightarrow I_2 + 2e^-$$

Next, balance the charge — that is, make the number of electrons in each half-reaction equal. In this case, the lowest common multiple of 10 and 2 is 10, so you multiply the second half-reaction $\left(2I^- \rightarrow I_2 + 2e^-\right)$ by 5 to get 10 electrons. Add the half-reactions and cancel the electrons:

$$10e^- + 12H^+ + 2IO_3{}^- \rightarrow I_2 + 6H_2O$$

$$\underline{10I^- \rightarrow 5I_2 + 10e^-}$$

$$12H^+ + 2IO_3{}^- + 10I^- \rightarrow 6I_2 + 6H_2O$$

Now substitute this equation into the original molecular equation, reinserting the spectator ions (K^+ and $SO_4{}^{2-}$):

$$KIO_3 + KI + H_2SO_4 \rightarrow K_2SO_4 + I_2 + H_2O$$

$$2KIO_3 + 10KI + H_2SO_4 \rightarrow K_2SO_4 + 6I_2 + 6H_2O$$

You have 12 hydrogen ions on the reactants side, so multiply the H_2SO_4 by 6; then multiply K_2SO_4 by 6 to balance the potassium and sulfate ions with the ones on the reactants side of the equation:

$$2KIO_3 + 10KI + 6H_2SO_4 \rightarrow 6K_2SO_4 + 6I_2 + 6H_2O$$

You can divide everything by 2, so here's the final equation:

$$KIO_3 + 5KI + 3H_2SO_4 \rightarrow 3K_2SO_4 + 3I_2 + 3H_2O$$

The resulting coefficients are 1, 5, 3, 3, 3, and 3.

621. 3

At first, this equation may appear to be balanced. In fact, the atoms are balanced, but the charge isn't. The two half-reactions for this equation are as follows:

$$MnO_4^- \rightarrow MnO_2$$

$$ClO_2^- \rightarrow ClO_4^-$$

You need to balance the oxygen atoms in each half-reaction. To do this in a basic solution, think of how many oxygen atoms you need on the oxygen-deficient side and add twice as many hydroxide ions; then balance out the hydrogen atoms by adding half as many water molecules to the opposite side. In this case, you add four hydroxide ions to the oxygen-deficient side and two water molecules to the opposite side, for a net gain of two oxygen atoms:

$$2H_2O + MnO_4^- \rightarrow MnO_2 + 4OH^-$$

$$4OH^- + ClO_2^- \rightarrow ClO_4^- + 2H_2O$$

Next, balance the charge in each half-reaction by adding electrons. In this case, add three electrons to the reactants in the first half-reaction:

$$3e^- + 2H_2O + MnO_4^- \rightarrow MnO_2 + 4OH^-$$

Then add four electrons to the products in the second half-reaction:

$$4OH^- + ClO_2^- \rightarrow ClO_4^- + 2H_2O + 4e^-$$

The number of electrons in each half-reaction must be equal. The lowest common multiple of 3 and 4 is 12, so multiply the first half-reaction $\left(3e^- + 2H_2O + MnO_4^- \rightarrow MnO_2 + 4OH^-\right)$ by 4 and the second half-reaction $\left(4OH^- + ClO_2^- \rightarrow ClO_4^- + 2H_2O + 4e^-\right)$ by 3. Then add the half-reactions together and simplify:

$$\cancel{12e^-} + 8H_2O + 4MnO_4^- \rightarrow 4MnO_2 + 16OH^-$$
$$12OH^- + 3ClO_2^- \rightarrow 3ClO_4^- + 6H_2O + \cancel{12e^-}$$
$$\overline{8H_2O + 4MnO_4^- + 12OH^- + 3ClO_2^- \rightarrow 4MnO_2 + 16OH^- + 3ClO_4^- + 6H_2O}$$

Last, remove the terms that appear on both sides of the equation. Six water molecules cancel out, reducing the eight water molecules on the reactants side to two. Twelve hydroxide ions cancel out, reducing the number of hydroxide ions on the products side from sixteen to four. Here's the reaction:

$$2H_2O + 4MnO_4^- + 3ClO_2^- \rightarrow 4MnO_2 + 4OH^- + 3ClO_4^-$$

Double-check to make sure all the atoms and the charge are in balance. They're balanced, so the coefficient of ClO_4^- is 3.

622. **74.56 g/mol**

Find the molar masses of the elements in KCl by looking up their atomic masses on the periodic table. The molar mass of potassium is 39.102 g/mol, and the molar mass of chlorine is 35.453 g/mol. Adding the masses gives you 39.102 + 35.453 = 74.555 g/mol, which rounds to 74.56 g/mol.

623. **56.08 g/mol**

Find the molar masses of the elements in CaO by looking up their atomic masses on the periodic table. The molar mass of calcium is 40.08 g/mol, and the molar mass of oxygen is 15.9994 g/mol. The sum is 40.08 + 15.9994 = 56.0794 g/mol, which rounds to 56.08 g/mol.

624. **83.98 g/mol**

Find the molar masses of the elements in AlF_3 by looking up their atomic masses on the periodic table. Based on the formula, you need to multiply the molar mass of fluorine (18.9984 g/mol) by 3, the number of fluorine atoms. You can use the molar mass of aluminum (26.9815 g/mol) directly.

Add the masses of the elements together and then round to the hundredths place. The sum is $26.9815 + (18.9984 \times 3) = 83.9767$ g/mol, which rounds to 83.98 g/mol.

625. **69.62 g/mol**

Find the molar masses of the elements in B_2O_3 by looking up their atomic masses on the periodic table. The formula includes two boron atoms and three oxygen atoms, so multiply the molar mass of boron (10.811 g/mol) by 2 and multiply the molar mass of oxygen (15.9994 g/mol) by 3.

Add the masses of the elements and round to the hundredths place. The sum is $(10.811 \times 2) + (15.9994 \times 3) = 69.6202$ g/mol, which rounds to 69.62 g/mol.

626. **331.63 g/mol**

Find the molar masses of the elements in CBr_4 by looking up their atomic masses on the periodic table. Carbon has a molar mass of 12.01115 g/mol, and bromine has a molar mass of 79.904 g/mol. The formula includes four bromine atoms, so multiply the molar mass of bromine by 4.

Add the masses of the elements and round to the hundredths place. The sum is $12.01115 + (79.904 \times 4) = 331.62715$ g/mol, which rounds to 331.63 g/mol.

627. **53.49 g/mol**

Find the molar masses of the elements in NH_4Cl by looking up their atomic masses on the periodic table. Nitrogen has a molar mass of 14.0067 g/mol, hydrogen has a molar mass of 1.00797 g/mol, and chlorine has a molar mass of 35.453 g/mol. The formula includes four hydrogen atoms, so multiply the molar mass of hydrogen by 4.

Add the masses of the elements and round to the hundredths place. The sum is $14.0067 + (1.00797 \times 4) + 35.453 = 53.49158$ g/mol, which rounds to 53.49 g/mol.

628. **261.35 g/mol**

Find the molar masses of the elements in $Ba(NO_3)_2$ by looking up their atomic masses on the periodic table. Based on the formula for barium nitrate, you need to multiply the molar mass of nitrogen (14.0067 g/mol) by 2, the number of nitrogen atoms present; you also need to multiply the molar mass of oxygen (15.9994 g/mol) by 6, the number of oxygen atoms present. You can use the molar mass of barium (137.34 g/mol) directly.

Add the masses of the elements and round to the hundredths place. The sum is $137.34 + (14.0067 \times 2) + (15.9994 \times 3 \times 2) = 261.3498$ g/mol, which rounds to 261.35 g/mol.

629. **132.14 g/mol**

Find the molar masses of the elements in $(NH_4)_2SO_4$ by looking up their atomic masses on the periodic table. Ammonium sulfate contains two ammonium ions for every sulfate ion, giving you two nitrogen atoms, eight hydrogen atoms, one sulfur atom, and four oxygen atoms. Multiply the molar mass of each element by the proper number of atoms, add the results, and round to the hundredths place. The sum is $(14.0067 \times 2) + (1.00797 \times 4 \times 2) + 32.064 + (15.9994 \times 4) = 132.13876$ g/mol, which rounds to 132.14 g/mol.

630. **247.80 g/mol**

First, write the formula for silver sulfide. Silver is Ag^+ and sulfide is S^{2-}, so the formula is Ag_2S. Next, find the molar masses of the elements by looking up their atomic masses on the periodic table.

Based on the formula, you need to multiply the molar mass of silver (107.8682 g/mol) by 2, the number of silver atoms. You can use the molar mass of sulfur (32.064 g/mol) directly. Add the masses together and round to the hundredths place. The sum is $(107.8682 \times 2) + 32.064 = 247.8004$ g/mol, which rounds to 247.80 g/mol.

631. **105.99 g/mol**

First, write the formula for sodium carbonate. The sodium ion is Na^+ and carbonate is CO_3^{2-}, so the formula is Na_2CO_3. Next, find the molar masses of the elements by looking up their atomic masses on the periodic table.

Based on the formula, you need to multiply the molar mass of sodium (28.9898 g/mol) by 2, the number of sodium atoms, and multiply the molar mass of oxygen (15.9994 g/mol) by 3, the number of oxygen atoms. You can use the molar mass of carbon (12.0115 g/mol) directly. Add the masses together and round to the hundredths place. The sum is $(22.9898 \times 2) + 12.0115 + (15.9994 \times 3) = 105.9893$ g/mol, which rounds to 105.99 g/mol.

632. **318.02 g/mol**

Find the molar masses of the elements in $Al_2(C_2O_4)_3$ by looking up their atomic masses on the periodic table. In aluminum oxalate, you have two aluminum ions and three oxalate ions. Each oxalate ion $(C_2O_4^{2-})$ contains two carbon atoms and four oxygen atoms. When you multiply the atoms in the oxalate ion by 3, you get a total of six carbon atoms (2×3) and twelve oxygen atoms (4×3) in the formula. The sum of the molar masses is $(26.9815 \times 2) + (12.0115 \times 2 \times 3) + (15.9994 \times 4 \times 3) = 318.0248$ g/mol, which rounds to 318.02 g/mol.

633. **204.40 g/mol**

Find the molar masses of the elements in $Zn(NH_3)_4Cl_2$ by looking up their atomic masses on the periodic table. This compound, which contains a complexion, includes four ammonia molecules combined with a zinc ion and two chlorine ions. Based on the formula, you need to multiply the molar mass of nitrogen (14.0067 g/mol) by 4, the number of nitrogen atoms; multiply the molar mass of hydrogen (1.00979 g/mol) by 12, the number of hydrogen atoms; multiply the molar mass of chlorine (35.453 g/mol) by 2, the number of chlorine atoms; and use the molar mass of zinc (65.37 g/mol) directly. The sum of the molar masses is $65.37 + (14.0067 \times 4) + (1.00797 \times 3 \times 4) + (35.453 \times 2) = 204.39844$ g/mol, which rounds to 204.40 g/mol.

634. **249.68 g/mol**

This formula is for copper(II) sulfate pentahydrate. To calculate its molar mass, think of the molecule as copper sulfate plus five water molecules. So not only do you have one copper atom, one sulfur atom, and four oxygen atoms, but you also have ten hydrogen atoms and five more oxygen atoms to account for when calculating the molar mass.

Find the molar masses of the elements in $CuSO_4 \cdot 5H_2O$ by looking up their atomic masses on the periodic table, multiply the molar mass of each element by the proper number of atoms, and add the results. The sum is $[63.546 + 32.064 + (15.9994 \times 4)] + [(5 \times 1.00797 \times 2) + (5 \times 15.9994)] = 249.6843$ g/mol, which rounds to 249.68 g/mol.

635. **287.04 g/mol**

The formula for manganese(II) nitrate hexahydrate begins with the manganese(II) (Mn^{2+}) ion, followed by the nitrate ion (NO_3^-). You need two nitrate ions to balance the +2 charge on the manganese. After the $Mn(NO_3)_2$, you need to account for the hexahydrate, which is six water molecules. The formula becomes $Mn(NO_3)_2 \cdot 6H_2O$.

The formula shows one manganese atom, two nitrogen atoms, six oxygen atoms, twelve hydrogen atoms, and six more oxygen atoms. Find the molar masses of the elements in $Mn(NO_2)_3 \cdot 6H_2O$ by looking up their atomic masses on the periodic table, multiply the molar mass of each element by the proper number of atoms, and add the results. The sum is $54.9380 + (14.0067 \times 2) + (15.9994 \times 3 \times 2) + (6 \times 1.00797 \times 2) + (6 \times 15.9994) = 287.0384$ g/mol, which rounds to 287.04 g/mol.

636. **501.61 g/mol**

The formula for iron(II) phosphate octahydrate begins with the iron(II) ion (Fe^{2+}), followed by the phosphate ion (PO_4^{3-}), giving you the formula $Fe_3(PO_4)_2$. Add the octahydrate (eight water molecules) to the $Fe_3(PO_4)_2$, and you get $Fe_3(PO_4)_2 \cdot 8H_2O$.

The formula shows three iron atoms, two phosphorus atoms, eight oxygen atoms, sixteen hydrogen atoms, and eight more oxygen atoms. Find the molar masses of the elements in $Fe_3(PO_4)_2 \cdot 8H_2O$ by looking up their atomic masses on the periodic table, multiply the molar mass of each element by the proper number of atoms, and add the results. The sum is $[(55.847 \times 3) + (30.9738 \times 2) + (15.9994 \times 4 \times 2)] + [(8 \times 1.00797 \times 2) + (8 \times 15.9994)] = 501.60652$ g/mol, which rounds to 501.61 g/mol.

637. **22.34% Na**

Divide the mass of 1 mol Na by the mass of 1 mol NaBr. The mass of NaBr is the sum of the masses of the atoms present. Then multiply by 100 to get the percent.

$$\% \text{ Na} = \frac{\text{mass of Na}}{\text{mass of NaBr}} \times 100$$

$$= \frac{22.9898 \text{ g}}{(22.9898 + 79.904) \text{ g}} \times 100$$

$$= 22.34\% \text{ Na}$$

638. **73.21% Sr**

Divide the mass of 1 mol Sr by the mass of 1 mol SrS. The mass of SrS is the sum of the masses of the atoms present. Then multiply by 100 to get the percent.

$$\% \text{ Sr} = \frac{\text{mass of Sr}}{\text{mass of SrS}} \times 100$$

$$= \frac{87.62 \text{ g}}{(87.62 + 32.064) \text{ g}} \times 100$$

$$= 73.21\% \text{ Sr}$$

639. **28.93% Cl**

Divide the mass of 1 mol Cl by the mass of 1 mol $KClO_3$. In finding the mass of $KClO_3$, multiply the mass of the oxygen by 3, because three oxygen atoms are present in the compound. Then multiply by 100 to get the percent.

$$\% \text{ Cl} = \frac{\text{mass of Cl}}{\text{mass of KClO}_3} \times 100$$

$$= \frac{35.453 \text{ g}}{(39.102 + 35.453 + (15.9994 \times 3)) \text{ g}} \times 100$$

$$= 28.93\% \text{ Cl}$$

640. **49.96% O**

Divide the mass of 4 mol O by the mass of 1 mol CaC_2O_4. Remember to multiply the molar mass of oxygen by 4, because four oxygen atoms are present in the compound. The mass of CaC_2O_4 is the sum of the masses of the atoms present, written in grams.

After dividing the masses, multiply by 100 to get the percent:

$$\% \ O = \frac{\text{mass of 4O}}{\text{mass of } CaC_2O_4} \times 100$$

$$= \frac{(15.9994 \times 4) \ g}{(40.08 + (12.01115 \times 2) + (15.9994 \times 4)) \ g} \times 100$$

$$= 49.96\% \ O$$

641. **40.56% S**

Divide the mass of 2 mol S by the mass of 1 mol $Na_2S_2O_3$. Remember to multiply the molar mass of sulfur by 2, because two sulfur atoms are present in the compound. The mass of $Na_2S_2O_3$ is the sum of the atomic masses of the atoms present, written in grams.

After dividing the masses, multiply by 100 to get the percent:

$$\% \ S = \frac{\text{mass of 2S}}{\text{mass of } Na_2S_2O_3} \times 100$$

$$= \frac{(32.064 \times 2) \ g}{((22.9898 \times 2) + (32.064 \times 2) + (15.9994 \times 3)) \ g} \times 100$$

$$= 40.56\% \ S$$

642. **35.00% N**

Divide the mass of 2 mol N by the mass of 1 mol NH_4NO_3. Make sure you include both the nitrogen from the ammonium and the nitrogen from the nitrate in the mass of the nitrogen. The mass of NH_4NO_3 is the sum of the atomic masses of the atoms present, written in grams.

After dividing the masses, multiply by 100 to get the percent:

$$\% \ N = \frac{\text{mass of 2N}}{\text{mass of } NH_4NO_3} \times 100$$

$$= \frac{(14.0067 \times 2) \ g}{(14.0067 + (1.00797 \times 4) + 14.0067 + (15.9994 \times 3)) \ g} \times 100$$

$$= 35.00\% \ N$$

643. **10.21% Li**

The first step in finding the mass percent of lithium is to write the formula of lithium hydrogen carbonate. The lithium ion is Li^+ and the hydrogen carbonate ion is HCO_3^-, so the formula is $LiHCO_3$.

Next, divide the mass of 1 mol Li by the mass of 1 mol $LiHCO_3$. The mass of $LiHCO_3$ is the sum of the atomic masses of the atoms present, written in grams. Then multiply by 100 to get the percent:

$$\% \text{ Li} = \frac{\text{mass of Li}}{\text{mass of LiHCO}_3} \times 100$$

$$= \frac{6.939 \text{ g}}{\left(6.939 + 1.00797 + 12.01115 + (15.9994 \times 3)\right) \text{ g}} \times 100$$

$$= 10.21\% \text{ Li}$$

644. **87.06% Ag**

First, write the formula for silver sulfide. The silver ion is Ag^+ and the sulfide ion is S^{2-}, so you need two positive silver ions to balance out the –2 charge of the sulfide ion; the formula is Ag_2S.

Next, divide the mass of 2 mol Ag by the mass of 1 mol Ag_2S. The mass of Ag_2S is the sum of the atomic masses of the atoms present, written in grams. Multiply the result by 100 to get the percent silver:

$$\% \text{ Ag} = \frac{\text{mass of 2Ag}}{\text{mass of Ag}_2\text{S}} \times 100$$

$$= \frac{(107.868 \times 2) \text{ g}}{\left((107.868 \times 2) + 32.064\right) \text{ g}} \times 100$$

$$= 87.06\% \text{ Ag}$$

645. **34.59% Al, 61.53% O, and 3.88% H**

To find the mass percent of each element in aluminum hydroxide, first write the formula for the compound. The aluminum ion is Al^{3+} and the hydroxide ion is OH^-, so you need three hydroxide ions to balance the +3 charge of the aluminum; therefore, the formula is $Al(OH)_3$.

Next, find the mass percent of each element by taking the molar mass of each element (based on the number of atoms in the formula), dividing by the mass of the compound, and then multiplying by 100 to get the percent:

$$\% \text{ Al} = \frac{\text{mass of Al}}{\text{mass of Al(OH)}_3} \times 100$$

$$= \frac{26.9815 \text{ g}}{\left(26.9815 + (15.9994 \times 3) + (1.00797 \times 3)\right) \text{ g}} \times 100 = 34.59\% \text{ Al}$$

$$\% \text{ O} = \frac{\text{mass of 3O}}{\text{mass of Al(OH)}_3} \times 100$$

$$= \frac{(15.9994 \times 3) \text{ g}}{\left(26.9815 + (15.9994 \times 3) + (1.00797 \times 3)\right) \text{ g}} \times 100 = 61.53\% \text{ O}$$

$$\% \text{ H} = \frac{\text{mass of 3H}}{\text{mass of Al(OH)}_3} \times 100$$

$$= \frac{(1.00797 \times 3) \text{ g}}{\left(26.9815 + (15.9994 \times 3) + (1.00797 \times 3)\right) \text{ g}} \times 100 = 3.88\% \text{ H}$$

Answers
601–700

646.   **15.75% Zn, 61.13% I, and 23.12% O**

To find the mass percent of each element in zinc iodate, first write the formula for the compound. The formula comes from the Zn^{2+} ion and the IO_3^- ion. You need two negatively charged iodate ions to balance the +2 charge of the zinc ion, so the formula is $Zn(IO_3)_2$.

To find the mass percent of each element, take the molar mass of each element (based on the number of atoms in the formula), divide by the molar mass of the compound, and then multiply by 100 to get the percent.

$$\% \text{ Zn} = \frac{\text{mass of Zn}}{\text{mass of Zn}(IO_3)_2} \times 100$$

$$= \frac{65.37 \text{ g}}{\left(65.37 + (126.9044 \times 2) + (15.9994 \times 3 \times 2)\right) \text{ g}} \times 100 = 15.75\% \text{ Zn}$$

$$\% \text{ I} = \frac{\text{mass of 2I}}{\text{mass of Zn}(IO_3)_2} \times 100$$

$$= \frac{(126.9044 \times 2) \text{ g}}{\left(65.37 + (126.9044 \times 2) + (15.9994 \times 3 \times 2)\right) \text{ g}} \times 100 = 61.13\% \text{ I}$$

$$\% \text{ O} = \frac{\text{mass of 6O}}{\text{mass of Zn}(IO_3)_2} \times 100$$

$$= \frac{(15.9994 \times 3 \times 2) \text{ g}}{\left(65.37 + (126.9044 \times 2) + (15.9994 \times 3 \times 2)\right) \text{ g}} \times 100 = 23.12\% \text{ O}$$

647.   **N_2O_5, NO_2, N_2O_3, NO, N_2O**

In this problem, calculating the actual mass percent for each compound isn't necessary because the compounds contain only nitrogen and oxygen (although you may still choose to calculate the values). Because only approximate values are necessary, you can simplify the calculations by rounding the molar masses to whole numbers, with 14 g/mol for nitrogen and 16 g/mol for oxygen.

For each compound, set up a ratio of the mass of nitrogen to the mass of the compound; then compare the ratios with a common numerator:

$$\text{NO: } \frac{14}{14+16} = \frac{14}{30} = \frac{28}{60}$$

$$\text{NO}_2 \text{: } \frac{14}{14+(16 \times 2)} = \frac{14}{46} = \frac{28}{92}$$

$$\text{N}_2\text{O: } \frac{14 \times 2}{(14 \times 2)+16} = \frac{28}{44}$$

$$\text{N}_2\text{O}_3 \text{: } \frac{14 \times 2}{(14 \times 2)+(16 \times 3)} = \frac{28}{76}$$

$$\text{N}_2\text{O}_5 \text{: } \frac{14 \times 2}{(14 \times 2)+(16 \times 5)} = \frac{28}{108}$$

Arranging the ratios from smallest to largest gives you $\frac{28}{108} < \frac{28}{92} < \frac{28}{76} < \frac{28}{60} < \frac{28}{44}$, so the correct order is N_2O_5, NO_2, N_2O_3, NO, N_2O.

648. C_2H_6, CH_4, CaC_2O_4, K_2CO_3, CCl_4

Find the mass percent of carbon by taking the molar mass of carbon in each compound (based on the number of atoms in the formula), dividing that mass by the mass of the compound, and then multiplying the result by 100 to get the percent:

$$CH_4: \frac{12.01115}{12.01115+(1.00797\times4)}\times100=74.87\%\ C$$

$$CCl_4: \frac{12.01115}{12.01115+(35.453\times4)}\times100=7.81\%\ C$$

$$C_2H_6: \frac{12.01115\times2}{(12.01115\times2)+(1.00797\times6)}\times100=79.89\%\ C$$

$$K_2CO_3: \frac{12.01115}{(39.102\times2)+12.01115+(15.9994\times3)}\times100=8.69\%\ C$$

$$CaC_2O_4: \frac{12.01115\times2}{40.08+(12.01115\times2)+(15.9994\times4)}\times100=18.75\%\ C$$

Arranging the percentages from largest to smallest gives you 79.89% C > 74.87% C > 18.75% C > 8.69% C > 7.81% C, so the correct order is C_2H_6, CH_4, CaC_2O_4, K_2CO_3, CCl_4.

649. K_2SO_4, K_2SO_3, K_2S, $KSCN$, $K_2S_2O_3$

To find the mass percent of sulfur, take the molar mass of sulfur in each compound (based on the number of atoms in the formula), divide by the mass of the compound, and then multiply by 100 to get the percent:

$$K_2S: \frac{32.064}{(39.102\times2)+32.064}\times100=29.08\%\ S$$

$$K_2SO_3: \frac{32.064}{(39.102\times2)+32.064+(15.9994\times3)}\times100=20.26\%\ S$$

$$K_2SO_4: \frac{32.064}{(39.102\times2)+32.064+(15.9994\times4)}\times100=18.40\%\ S$$

$$K_2S_2O_3: \frac{32.064\times2}{(39.102\times2)+(32.064\times2)+(15.9994\times3)}\times100=33.69\%\ S$$

$$KSCN: \frac{32.064}{39.102+32.064+12.01115+14.0067}\times100=32.99\%\ S$$

Arranging the percentages from smallest to largest gives you 18.40% S < 20.26% S < 29.08% S < 32.99% S < 33.69% S, so the correct order is K_2SO_4, K_2SO_3, K_2S, $KSCN$, $K_2S_2O_3$.

650. 36.08% H_2O

To find the mass percent of water in the hydrate, take the mass of 1 mol of water (H_2O), divide it by the mass 1 mol of the hydrated crystal ($CuSO_4\cdot5H_2O$), and multiply by 100 to get the percent:

$$\frac{(5\times1.00797\times2)+(5\times15.9994)}{63.546+32.064+(15.9994\times4)+(5\times1.00797\times2)+(5\times15.9994)}\times100=36.08\%\ H_2O$$

Answers 601–700

651.

67.34% O

When given the name of the compound, first write the formula. Sodium phosphate contains the sodium ion Na^+ and the phosphate ion PO_4^{3-}. To balance out the –3 charge on the phosphate ion, you need three sodium ions, so the first part of the formula is Na_3PO_4.

The *dodecahydrate* part of the name represents the 12 water molecules in the hydrated crystal. Remember to insert the dot between the sodium phosphate part of the formula and the water part ($Na_3PO_4 \cdot 12H_2O$), but don't treat it as a multiplication sign when calculating the molar mass of the hydrated crystal. Think of it more like sodium phosphate plus 12 water molecules.

Take the mass of the oxygen (accounting for the four oxygen atoms in the phosphate and the twelve oxygen atoms in the water), divide it by the mass of 1 mol of the compound, and multiply by 100 to get the percent:

$$\frac{(15.9994 \times 4) + (12 \times 15.9994)}{(22.9898 \times 3) + 30.9738 + (15.9994 \times 4) + (12 \times 1.00797 \times 2) + (12 \times 15.9994)} \times 100$$

$$= 67.34\% \text{ O}$$

652.

$C_2H_3O_2$

The *empirical formula* is the simplest whole-number ratio of atoms in a compound. Out of the formulas listed, only $C_2H_3O_2$ can't be simplified. In each of the other compounds, the number of atoms in the formula is divisible by 2.

653.

CH_2

To find the empirical formula of a compound, you need to have counting units (or groupings) of each element, not the mass. If the question gives you the mass percent of each element, suppose that you have a 100.00-g sample such that 85.62% = 85.62 g. The first step for each element is to change the percent to grams. The second step is to convert grams to moles by dividing by the molar mass:

$$\frac{85.62 \text{ g C}}{1} \times \frac{1 \text{ mol C}}{12.01115 \text{ g C}} = 7.128 \text{ mol C}$$

$$\frac{14.38 \text{ g H}}{1} \times \frac{1 \text{ mol H}}{1.00797 \text{ g H}} = 14.27 \text{ mol H}$$

Next, compare the number of moles of each element: Divide the moles of each element by the smaller number of moles to get a ratio. Here, divide by 7.128 mol:

$$\frac{7.128 \text{ mol C}}{7.128 \text{ mol}} = 1.000 \text{ C}$$

$$\frac{14.27 \text{ mol H}}{7.128 \text{ mol}} = 2.000 \text{ H}$$

You have two hydrogen atoms for every atom of carbon, so the empirical formula is CH_2.

654. $CHCl_3$

To find the empirical formula of a compound, you need to have counting units (or groupings) of each element, not the mass. This question gives you the mass in grams, so convert the grams to moles by dividing by the molar mass:

$$\frac{10.06 \, g\,C}{1} \times \frac{1 \, mol \, C}{12.01115 \, g\,C} = 0.8376 \, mol \, C$$

$$\frac{0.84 \, g\,H}{1} \times \frac{1 \, mol \, H}{1.00797 \, g\,H} = 0.83 \, mol \, H$$

$$\frac{89.09 \, g\,Cl}{1} \times \frac{1 \, mol \, Cl}{35.453 \, g\,Cl} = 2.513 \, mol \, Cl$$

Next, compare the number of moles of each element: Divide the moles of each element by the smallest number of moles to get a ratio. Here, divide by 0.83 mol:

$$\frac{0.8376 \, mol \, C}{0.83 \, mol} = 1.0 \, C$$

$$\frac{0.83 \, mol \, H}{0.83 \, mol} = 1.0 \, H$$

$$\frac{2.513 \, mol \, Cl}{0.83 \, mol} = 3.0 \, Cl$$

You have three chlorine atoms for every carbon and hydrogen atom, so the empirical formula for the compound is $CHCl_3$.

655. $FeCrO_4$

To find the empirical formula of a compound, you need to have counting units (or groupings) of each element, not the mass. This question gives you the mass in grams, so convert the grams to moles by dividing by the molar masses:

$$\frac{1.4066 \, g\,Fe}{1} \times \frac{1 \, mol \, Fe}{55.847 \, g\,Fe} = 0.025187 \, mol \, Fe$$

$$\frac{1.3096 \, g\,Cr}{1} \times \frac{1 \, mol \, Cr}{51.996 \, g\,Cr} = 0.025187 \, mol \, Cr$$

$$\frac{1.6118 \, g\,O}{1} \times \frac{1 \, mol \, O}{15.9994 \, g\,O} = 0.10074 \, mol \, O$$

Next, compare the number of moles of each element: Divide the moles of each element by the smallest number of moles to get a ratio. Here, divide by 0.025187 mol:

$$\frac{0.025187 \, mol \, Fe}{0.025187 \, mol} = 1.0000 \, Fe$$

$$\frac{0.025187 \, mol \, Cr}{0.025187 \, mol} = 1.0000 \, Cr$$

$$\frac{0.10074 \, mol \, O}{0.025187 \, mol} = 3.9997 \, O$$

You have four oxygen atoms for every atom of iron and chromium, so the empirical formula for the compound is $FeCrO_4$.

656. $CdSO_4$

First, determine the mass of the anhydrous salt. If the 100.00-g sample includes 18.73 g of water, then the anhydrous salt has a mass of 100.00 g – 18.73 g = 81.27 g.

To find the number of grams of oxygen, subtract the mass of the cadmium and sulfur from the mass of the anhydrous salt: 81.27 g – (43.82 g + 12.50 g) = 24.95 g O. Use these numbers to find the empirical formula. First, change the masses to moles by dividing by the molar mass of each element:

$$\frac{43.82 \text{ g Cd}}{1} \times \frac{1 \text{ mol Cd}}{112.40 \text{ g Cd}} = 0.3899 \text{ mol Cd}$$

$$\frac{12.50 \text{ g S}}{1} \times \frac{1 \text{ mol S}}{32.064 \text{ g S}} = 0.3898 \text{ mol S}$$

$$\frac{24.95 \text{ g O}}{1} \times \frac{1 \text{ mol O}}{15.9994 \text{ g O}} = 1.559 \text{ mol O}$$

Then compare the number of moles of each element: Divide the moles of each element by the smallest number of moles to get a ratio. Here, divide by 0.3898 mol:

$$\frac{0.3899 \text{ mol Cd}}{0.3898 \text{ mol}} = 1.000 \text{ Cd}$$

$$\frac{0.3898 \text{ mol S}}{0.3898 \text{ mol}} = 1.000 \text{ S}$$

$$\frac{1.559 \text{ mol O}}{0.3898 \text{ mol}} = 3.999 \text{ O}$$

You have four oxygen atoms for every atom of cadmium and sulfur, so the anhydrous salt is $CdSO_4$.

657. II only

The empirical formula is the lowest whole-number ratio of the each kind of atom in the compound. CH_4 and C_3H_8 can't be reduced, but you can reduce C_2H_2 to CH.

658. II and III

The empirical formula is the lowest whole-number ratio of the each kind of atom in the compound. CO_2 can't be reduced, but you can reduce C_6H_{12} to CH_2 and reduce $C_6H_{12}O_6$ to CH_2O.

659. C_6H_6

To find the molecular formula given the empirical formula and the molar mass of the molecular formula, find the ratio of the molecular formula to the empirical formula. Divide the molar mass of the molecular formula by the molar mass of the empirical formula:

$$\frac{78.11 \text{ g/mol}}{(12.0115 + 1.00797) \text{ g/mol}} = 6$$

Use that number to scale the empirical formula: $6 \times CH = C_6H_6$.

660. $C_6H_{14}O_2N_4$

To find the empirical formula of a compound, you need to have counting units (or groupings) of each element, not the mass. If the question gives you the mass percent of each element as this question does, suppose you have a 100.00-g sample, such that 41.37% = 41.37 g. The first step for each element is to change the percent to grams. The second step for each element is to convert grams to moles by dividing by the molar mass:

$$\frac{41.37\ g\,C}{1} \times \frac{1\ mol\ C}{12.01115\ g\,C} = 3.444\ mol\ C$$

$$\frac{8.10\ g\,H}{1} \times \frac{1\ mol\ H}{1.00797\ g\,H} = 8.04\ mol\ H$$

$$\frac{18.39\ g\,O}{1} \times \frac{1\ mol\ O}{15.9994\ g\,O} = 1.149\ mol\ O$$

$$\frac{32.16\ g\,N}{1} \times \frac{1\ mol\ N}{14.0067\ g\,N} = 2.296\ mol\ N$$

Next, compare the number of moles of each element. Divide the moles of each element by the smallest number of moles to get a ratio. Here, divide by 1.149 mol:

$$\frac{3.444\ mol\ C}{1.149\ mol} = 2.997\ C$$

$$\frac{8.04\ mol\ H}{1.149\ mol} = 7.00\ H$$

$$\frac{1.149\ mol\ O}{1.149\ mol} = 1.000\ O$$

$$\frac{2.296\ mol\ N}{1.149\ mol} = 1.998\ N$$

The ratios tell you that the empirical formula of the compound is $C_3H_7ON_2$. To find the molecular formula, take the given molar mass and divide it by the molecular mass of the empirical formula:

$$\frac{174.20\ g/mol}{\left((12.01115 \times 3) + (1.00797 \times 7) + 15.9994 + (14.0067 \times 2)\right) g/mol} = 1.997$$

The molecular formula is twice as big as the empirical formula, so the molecular formula for the compound is $C_6H_{14}O_2N_4$.

661. $C_9H_9O_5NI_2$ and $C_9H_9O_5NI_2$

To find the empirical formula of a compound, you need to have counting units (or groupings) of each element, not the mass. When a question gives you the mass percent of each element, suppose you have a 100.00-g sample, such that 23.25% = 23.25 g. The first step for each element is the change the percent to grams. The second step for each element is to convert grams to moles by dividing by the molar mass. To find the amount of iodine, subtract the masses of the other elements from 100.0 g before converting to moles.

$$\frac{23.25 \text{ g C}}{1} \times \frac{1 \text{ mol C}}{12.01115 \text{ g C}} = 1.936 \text{ mol C}$$

$$\frac{1.95 \text{ g H}}{1} \times \frac{1 \text{ mol H}}{1.00797 \text{ g H}} = 1.93 \text{ mol H}$$

$$\frac{17.20 \text{ g O}}{1} \times \frac{1 \text{ mol O}}{15.9994 \text{ g O}} = 1.075 \text{ mol O}$$

$$\frac{3.01 \text{ g N}}{1} \times \frac{1 \text{ mol N}}{14.0067 \text{ g N}} = 0.215 \text{ mol N}$$

$$\frac{(100.00 - 23.25 - 1.95 - 17.20 - 3.01) \text{ g I}}{1} \times \frac{1 \text{ mol I}}{126.9044 \text{ g I}} = 0.4302 \text{ mol I}$$

Next, compare the number of moles of each element. Divide the moles of each element by the smallest number of moles to get a ratio. Here, divide by 0.215 mol:

$$\frac{1.936 \text{ mol C}}{0.215 \text{ mol}} = 9.00 \text{ C}$$

$$\frac{1.93 \text{ mol H}}{0.215 \text{ mol}} = 8.98 \text{ H}$$

$$\frac{1.075 \text{ mol O}}{0.215 \text{ mol}} = 5.00 \text{ O}$$

$$\frac{0.215 \text{ mol N}}{0.215 \text{ mol}} = 1.00 \text{ N}$$

$$\frac{0.4302 \text{ mol I}}{0.215 \text{ mol}} = 2.00 \text{ I}$$

Therefore, the empirical formula of the compound is $C_9H_9O_5NI_2$. To find the molecular formula, take the given molar mass and divide it by the molar mass of the empirical formula:

$$\frac{465 \text{ g/mol}}{((12.01115 \times 9) + (1.00797 \times 9) + (15.9994 \times 5) + 14.0067 + (126.9044 \times 2)) \text{ g/mol}} = 1.000$$

This means that the molecular formula is the same as the empirical formula, $C_9H_9O_5NI_2$.

---

662.   **0.171 mol NaCl**

To find the number of moles when given the number of grams, divide the given mass by the molar mass of the compound and then round to the correct number of significant figures:

$$\frac{10.0 \text{ g NaCl}}{1} \times \frac{1 \text{ mol NaCl}}{(22.9898 + 35.453) \text{ g NaCl}} = 0.171 \text{ mol NaCl}$$

---

663.   **0.361 mol BH$_3$**

To find the number of moles when given the number of grams, divide the given mass by the molar mass of the compound and then round to the correct number of significant figures:

$$\frac{5.00 \text{ g BH}_3}{1} \times \frac{1 \text{ mol BH}_3}{(10.811 + (1.00797 \times 3)) \text{ g BH}_3} = 0.361 \text{ mol BH}_3$$

664. 2.59 mol Na$_2$CO$_3$

To find the number of moles when given the number of grams, divide the given mass by the molar mass of the compound and round to the correct number of significant figures:

$$\frac{275 \text{ g Na}_2\text{CO}_3}{1} \times \frac{1 \text{ mol Na}_2\text{CO}_3}{\left((22.9898 \times 2) + 12.01115 + (15.9994 \times 3)\right) \text{ g Na}_2\text{CO}_3}$$

$$= 2.59 \text{ mol Na}_2\text{CO}_3$$

665. 11.4 mol NH$_4$OH

To find the number of moles when given the number of grams, divide the given mass by the molar mass of the compound and round to the correct number of significant figures:

$$\frac{400. \text{ g NH}_4\text{OH}}{1} \times \frac{1 \text{ mol NH}_4\text{OH}}{\left(14.0067 + (1.00797 \times 4) + 15.9994 + 1.00797\right) \text{ g NH}_4\text{OH}}$$

$$= 11.4 \text{ mol NH}_4\text{OH}$$

666. 0.63 mol KMnO$_4$

To find the number of moles when given the number of grams, divide the given mass by the molar mass of the compound and round to the correct number of significant figures:

$$\frac{99 \text{ g KMnO}_4}{1} \times \frac{1 \text{ mol KMnO}_4}{\left(39.102 + 54.9380 + (15.9994 \times 4)\right) \text{ g KMnO}_4} = 0.63 \text{ mol KMnO}_4$$

667. 3.65 mol BeS

First, write the formula for beryllium sulfide. The beryllium ion is Be^{2+}, and the sulfide ion is S^{2-}. The +2 charge on the beryllium ion cancels out the –2 charge on the sulfide ion, so the formula is BeS.

To find the number of moles when given the number of grams, divide the given mass by the molar mass of the compound and round to the correct number of significant figures:

$$\frac{150. \text{ g BeS}}{1} \times \frac{1 \text{ mol BeS}}{\left(9.0122 + 32.064\right) \text{ g BeS}} = 3.65 \text{ mol BeS}$$

668. 0.245 mol Al$_2$O$_3$

First, write the formula for aluminum oxide. The aluminum ion is Al^{3+}, and the oxide ion is O^{2-}. You need two +3 charged aluminum ions to balance out the three –2 charged oxygen ions, so the formula is Al$_2$O$_3$.

To find the number of moles when given the number of grams, divide the given mass by the molar mass of the compound and round to the correct number of significant figures:

$$\frac{25.0 \text{ g Al}_2\text{O}_3}{1} \times \frac{1 \text{ mol Al}_2\text{O}_3}{\left(\left(26.9815 \times 2\right) + \left(15.9994 \times 3\right)\right) \text{g Al}_2\text{O}_3} = 0.245 \text{ mol Al}_2\text{O}_3$$

669. 1.17 mol CCl$_4$

First, write the formula for carbon tetrachloride. The prefix *tetra-* indicates that there are four chlorine atoms in the formula, and the lack of a prefix on the carbon means that there's only one carbon atom, so the formula is CCl$_4$.

Next, to find the number of moles when given the number of grams, divide the given mass by the molar mass of the compound and round to the correct number of significant figures:

$$\frac{180. \text{ g CCl}_4}{1} \times \frac{1 \text{ mol CCl}_4}{\left(12.01115 + \left(35.453 \times 4\right)\right) \text{g CCl}_4} = 1.17 \text{ mol CCl}_4$$

670. 1.76 mol Ca$_3$P$_2$

First, write the formula for calcium phosphide. The calcium ion is Ca^{2+}, and the phosphide ion is P^{3-}. You need three +2 charged calcium ions to cancel out the two –3 charged phosphorus ions, so the formula is Ca$_3$P$_2$.

Next, to find the number of moles when given the number of grams, divide the given mass by the molar mass of the compound and round to the correct number of significant figures:

$$\frac{320. \text{ g Ca}_3\text{P}_2}{1} \times \frac{1 \text{ mol Ca}_3\text{P}_2}{\left(\left(40.08 \times 3\right) + \left(30.9738 \times 2\right)\right) \text{g Ca}_3\text{P}_2} = 1.76 \text{ mol Ca}_3\text{P}_2$$

671. 0.00648 mol Mg(MnO$_4$)$_2$

First, write the formula for magnesium permanganate. The magnesium ion is Mg^{2+}, and the permanganate ion is MnO$_4^-$. You need two –1 charged permanganate ions to cancel out the +2 charged magnesium ion, so the formula is Mg(MnO$_4$)$_2$.

Next, to find the number of moles when given the number of grams, divide the given mass by the molar mass of the compound and round to the correct number of significant figures:

$$\frac{1.70 \text{ g Mg(MnO}_4)_2}{1} \times \frac{1 \text{ mol Mg(MnO}_4)_2}{\left(24.305 + \left(54.9380 \times 2\right) + \left(15.9994 \times 8\right)\right) \text{g Mg(MnO}_4)_2}$$
$$= 0.00648 \text{ mol Mg(MnO}_4)_2$$

672. 377 g FeO

To find the number of grams when given the number of moles, multiply the given number of moles by the molar mass of the compound and round to the correct number of significant figures:

$$\frac{5.25 \text{ mol FeO}}{1} \times \frac{\left(55.847 + 15.9994\right) \text{g FeO}}{1 \text{ mol FeO}} = 377 \text{ g FeO}$$

673. **101 g S₂Cl₂**

To find the number of grams when given the number of moles, multiply the given number of moles by the molar mass of the compound and round to the correct number of significant figures:

$$\frac{0.750 \text{ mol } S_2Cl_2}{1} \times \frac{\left((32.064\times2)+(35.453\times2)\right)g\ S_2Cl_2}{1\text{ mol } S_2Cl_2} = 101 \text{ g } S_2Cl_2$$

674. **16,600 g KI**

To find the number of grams when given the number of moles, multiply the given number of moles by the molar mass of the compound and round to the correct number of significant figures:

$$\frac{100. \text{ mol KI}}{1} \times \frac{(39.102+126.9044)g\ KI}{1\text{ mol KI}} = 16,600 \text{ g KI}$$

675. **4,230 g CaCO₃**

To find the number of grams when given the number of moles, multiply the given number of moles by the molar mass of the compound and round to the correct number of significant figures:

$$\frac{42.3 \text{ mol } CaCO_3}{1} \times \frac{(40.08+12.01115+(15.9994\times3))g\ CaCO_3}{1\text{ mol } CaCO_3} = 4,230 \text{ g CaCO}_3$$

676. **1,700 g Al(NO₃)₃**

To find the mass of given substance when given the number of moles, multiply the given number of moles by the molar mass of the compound and round to the correct number of significant figures:

$$\frac{7.9 \text{ mol } Al(NO_3)_3}{1} \times \frac{(26.9815+(14.0067\times3)+(15.9994\times3\times3))g\ Al(NO_3)_3}{1\text{ mol } Al(NO_3)_3}$$
$$= 1,700 \text{ g Al}(NO_3)_3$$

677. **50. g Pb(NO₃)₂**

First, write the formula for lead(II) nitrate. The lead(II) ion is Pb^{2+}, and the nitrate ion is NO_3^-. You need two –1 charged nitrate ions to cancel out the +2 charged lead ion, so the formula is $Pb(NO_3)_2$.

Next, to find the number of grams when given the number of moles, multiply the given number of moles by the molar mass of the compound and round to the correct number of significant figures:

$$\frac{0.15 \text{ mol } Pb(NO_3)_2}{1} \times \frac{(207.19+(14.0067\times2)+(15.9994\times3\times2))g\ Pb(NO_3)_2}{1\text{ mol } Pb(NO_3)_2}$$
$$= 50. \text{ g Pb}(NO_3)_2$$

678. **3,400 g P$_4$O$_{10}$**

First, write the formula for tetraphosphorus decoxide. The prefix *tetra-* indicates that there are four phosphorus atoms in the formula, and the prefix *dec(a)-* on the oxide means that there are ten oxygen atoms, so the formula is P$_4$O$_{10}$.

Next, to find the number of grams when given the number of moles, multiply the given number of moles by the molar mass of the compound and round to the correct number of significant figures:

$$\frac{12 \, \text{mol} \, P_4O_{10}}{1} \times \frac{\left((30.9728 \times 4) + (15.9994 \times 10)\right) g \, P_4O_{10}}{1 \, \text{mol} \, P_4O_{10}} = 3,400 \, g \, P_4O_{10}$$

679. **36 g Cu$_2$O**

First, write the formula for copper(I) oxide. The copper(I) ion is Cu$^+$, and the oxide ion is O^{2-}. You need two +1 charged copper ions to cancel out the –2 charged oxygen ion, so the formula is Cu$_2$O.

Next, to find the number of grams when given the number of moles, multiply the given number of moles by the molar mass of the compound and round to the correct number of significant figures.

$$\frac{0.25 \, \text{mol} \, Cu_2O}{1} \times \frac{\left((63.546 \times 2) + 15.9994\right) g \, Cu_2O}{1 \, \text{mol} \, Cu_2O} = 36 \, g \, Cu_2O$$

680. **24.6 kg Zn(OH)$_2$**

First, write the formula for zinc hydroxide. The zinc ion is Zn^{2+}, and the hydroxide ion is OH$^-$. You need two –1 charged hydroxide ions to cancel out the +2 charged zinc ion, so the formula is Zn(OH)$_2$.

Next, to find the number of kilograms when given the number of moles, multiply the given number of moles by the molar mass of the compound, convert to kilograms, and round to the correct number of significant figures:

$$\frac{248 \, \text{mol} \, Zn(OH)_2}{1} \times \frac{\left(65.37 + (15.9994 \times 2) + (1.00797 \times 2)\right) g \, Zn(OH)_2}{1 \, \text{mol} \, Zn(OH)_2}$$

$$\times \frac{1 \, kg}{1,000 \, g} = 24.6 \, kg \, Zn(OH)_2$$

681. **48.5 kg (NH$_4$)$_2$SO$_4$**

First, write the formula for ammonium sulfate. The ammonium ion is NH$_4^+$, and the sulfate ion is SO$_4^{2-}$. You need two +1 charged ammonium ions to cancel out the –2 charged sulfate ion, so the formula is (NH$_4$)$_2$SO$_4$.

Next, to find the number of kilograms when given the number of moles, multiply the given number of moles by the molar mass of the compound, convert to kilograms, and round to the correct number of significant figures:

$$\frac{367 \, \text{mol}(NH_4)_2 SO_4}{1}$$

$$\times \frac{\left((14.0067 \times 2) + (1.00797 \times 4 \times 2) + 32.064 + (15.9994 \times 4)\right) g \ (NH_4)_2 SO_4}{1 \, \text{mol} \ (NH_4)_2 SO_4}$$

$$\times \frac{1 \, \text{kg}}{1,000 \, g} = 48.5 \, \text{kg} \ (NH_4)_2 SO_4$$

---

682. 3.76×10^{22} **atoms of He**

To find the number of atoms in a given the mass of a substance, first divide the given mass by the molar mass to get moles. Then multiply the result by Avogadro's number of atoms in a mole. Finally, round the answer to the correct number of significant figures.

$$\frac{0.250 \, g \, He}{1} \times \frac{1 \, \text{mol He}}{4.0026 \, g \, He} \times \frac{6.022 \times 10^{23} \, \text{atoms He}}{1 \, \text{mol He}} = 3.76 \times 10^{22} \, \text{atoms of He}$$

---

683. 6.7×10^{23} **atoms of Al**

To find the number of atoms in a given the mass of a substance, first divide the given mass by the molar mass to get moles. Then multiply the result by Avogadro's number of atoms in a mole. Finally, round the answer to the correct number of significant figures.

$$\frac{30. \, g \, Al}{1} \times \frac{1 \, \text{mol Al}}{26.9815 \, g \, Al} \times \frac{6.022 \times 10^{23} \, \text{atoms Al}}{1 \, \text{mol Al}} = 6.7 \times 10^{23} \, \text{atoms of Al}$$

---

684. 2.74×10^{24} **molecules of CO_2**

Before starting the math on this question, be sure to write the formula for carbon dioxide: CO_2.

To find the number of molecules in a given the mass of a substance, divide the given mass by the molar mass to get moles. Then multiply the result by Avogadro's number of molecules in a mole. Finally, round the answer to the correct number of significant figures.

$$\frac{200. \, g \, CO_2}{1} \times \frac{1 \, \text{mol } CO_2}{(12.01115 + (15.9994 \times 2)) \, g \, CO_2} \times \frac{6.022 \times 10^{23} \, \text{molecules } CO_2}{1 \, \text{mol } CO_2}$$

$$= 2.74 \times 10^{24} \, \text{molecules of } CO_2$$

---

685. 3.3×10^{20} **molecules of N_2O_4**

First, write the formula for dinitrogen trioxide. The *di-* prefix on the nitrogen indicates that there are two nitrogen atoms in the formula, and the prefix *tetr(a)-* on the oxide tells you that there are four atoms of oxygen, so the formula is N_2O_4.

To find the number of molecules in a given the mass of a substance, first divide the given mass by the molar mass to get moles. Then multiply the result by Avogadro's number of molecules in a mole. Finally, round answer to the correct number of significant figures.

$$\frac{0.050\ g\ N_2O_4}{1} \times \frac{1\ mol\ N_2O_4}{\left((14.0067 \times 2) + (15.9994 \times 4)\right) g\ N_2O_4} \times \frac{6.022 \times 10^{23}\ molecules\ N_2O_4}{1\ mol\ N_2O_4}$$

$$= 3.3 \times 10^{20}\ molecules\ of\ N_2O_4$$

686. **4.70×10^{24} atoms of S**

In this question, you need to take into account that there are eight atoms of sulfur in a molecule of S_8, which will add one step to the end of the problem. First divide the given mass of S_8 by the molar mass of S_8 to get the moles of S_8. Then multiply the result by Avogadro's number of molecules in a mole. Finally, use the ratio of atoms of S in a molecule of S_8 to get your answer, rounding to the correct number of significant figures.

$$\frac{250.\ g\ S_8}{1} \times \frac{1\ mol\ S_8}{(32.064 \times 8)\ g\ S_8} \times \frac{6.022 \times 10^{23}\ molecules\ S_8}{1\ mol\ S_8} \times \frac{8\ atoms\ S}{1\ molecule\ S_8}$$

$$= 4.70 \times 10^{24}\ atoms\ of\ S$$

687. **2.5×10^5 g Br_2**

To find the mass in grams of a given number of molecules of a substance, first divide the given number of molecules by Avogadro's number of molecules in a mole. Then multiply the result by the molar mass of the substance. Finally, round the answer to the correct number of significant figures.

$$\frac{9.5 \times 10^{26}\ molecules\ Br_2}{1} \times \frac{1\ mol\ Br_2}{6.022 \times 10^{23}\ molecules\ Br_2} \times \frac{(79.904 \times 2)\ g\ Br_2}{1\ mol\ Br_2}$$

$$= 2.5 \times 10^5\ g\ Br_2$$

688. **8.095×10^{-1} g Ni**

To find the mass in grams of a given number of atoms of a substance, first divide the given number of atoms by Avogadro's number of atoms in a mole. Then multiply the result by the molar mass of the substance. Finally, round your answer to the correct number of significant figures.

$$\frac{8.306 \times 10^{21}\ atoms\ Ni}{1} \times \frac{1\ mol\ Ni}{6.022 \times 10^{23}\ atoms\ Ni} \times \frac{58.69\ g\ Ni}{1\ mol\ Ni} = 8.095 \times 10^{-1}\ g\ Ni$$

689. **8.98×10^1 g $C_6H_{12}O_6$**

To find the mass in grams of a given number of molecules of a substance, first divide the given number of molecules by Avogadro's number of molecules in a mole. Then multiply the result by the molar mass of the substance. Finally, round your answer to the correct number of significant figures.

$$\frac{3.00 \times 10^{23}\ molecules\ C_6H_{12}O_6}{1} \times \frac{1\ mol\ C_6H_{12}O_6}{6.022 \times 10^{23}\ molecules\ C_6H_{12}O_6}$$

$$\times \frac{\left((12.01115 \times 6) + (1.00797 \times 12) + (15.9994 \times 6)\right)\ g\ C_6H_{12}O_6}{1\ mol\ C_6H_{12}O_6} = 89.8\ g\ C_6H_{12}O_6$$

690. 9.0×10^{-15} g Kr

Krypton is a noble gas, so it's made up of single atoms, not diatomic molecules. To find the mass in grams of a given number of atoms of a substance, first divide the given number of atoms by Avogadro's number of atoms in a mole. Then multiply the result by the molar mass of the substance. Finally, round your answer to the correct number of significant figures.

$$\frac{65,000,000 \text{ atoms Kr}}{1} \times \frac{1 \text{ mol Kr}}{6.022 \times 10^{23} \text{ atoms Kr}} \times \frac{83.80 \text{ g Kr}}{1 \text{ mol Kr}} = 9.0 \times 10^{-15} \text{ g Kr}$$

691. 1.7×10^1 g AuF_3

Before starting the math on this question, write the formula for gold(III) fluoride. The gold(III) ion is Au^{3+}, and the fluoride ion is F^-, so the formula is AuF_3.

To find the number of grams in a given number of formula units of a substance, first divide the given number of formula units by Avogadro's number of formula units in a mole. Then multiply the result by the molar mass of the substance. Finally, round the answer to the correct number of significant figures.

$$\frac{4.0 \times 10^{22} \text{ formula units } AuF_3}{1} \times \frac{1 \text{ mol } AuF_3}{6.022 \times 10^{23} \text{ formula units } AuF_3}$$

$$\times \frac{\left(196.967 + \left(18.9984 \times 3\right)\right) \text{ g } AuF_3}{1 \text{ mol } AuF_3} = 17 \text{ g } AuF_3$$

692. 10.0 mol NH_3

To find the moles of ammonia produced, use the mole ratio from the balanced equation. For every mole of nitrogen gas consumed, 2 mol of ammonia are made.

$$\frac{5.00 \text{ mol } N_2}{1} \times \frac{2 \text{ mol } NH_3}{1 \text{ mol } N_2} = 10.0 \text{ mol } NH_3$$

693. 8.00 mol NH_3

To find the moles of ammonia produced, use the mole ratio from the balanced equation. For every 3 mol of hydrogen gas consumed, 2 mol of ammonia are made.

$$\frac{12.0 \text{ mol } H_2}{1} \times \frac{2 \text{ mol } NH_3}{3 \text{ mol } H_2} = 8.00 \text{ mol } NH_3$$

694. 12.0 mol N_2

To find the moles of nitrogen gas that would be necessary to make 24.0 mol of ammonia, use the mole ratio from the balanced equation. For every 2 mol of ammonia produced, 1 mol of nitrogen gas is consumed.

$$\frac{24.0 \text{ mol } NH_3}{1} \times \frac{1 \text{ mol } N_2}{2 \text{ mol } NH_3} = 12.0 \text{ mol } N_2$$

695. 54.0 mol H$_2$

To find the moles of hydrogen gas that would be necessary to make 36.0 mol of ammonia, use the mole ratio from the balanced equation. For every 2 mole of ammonia produced, 3 mol of hydrogen gas are consumed:

$$\frac{36.0 \text{ mol NH}_3}{1} \times \frac{3 \text{ mol H}_2}{2 \text{ mol NH}_3} = 54.0 \text{ mol H}_2$$

696. 9.00 mol CO$_2$ and 12.0 mol H$_2$O

To find the moles of each product produced, you need to complete two separate calculations using the mole ratio from the balanced equation. To find the moles of CO$_2$, note that there are 3 mol of CO$_2$ produced for every mole of C$_3$H$_8$ combusted:

$$\frac{3.00 \text{ mol C}_3\text{H}_8}{1} \times \frac{3 \text{ mol CO}_2}{1 \text{ mol C}_3\text{H}_8} = 9.00 \text{ mol CO}_2$$

To find the moles of H$_2$O, note that there are 4 mol of H$_2$O produced for every mole of C$_3$H$_8$ combusted:

$$\frac{3.00 \text{ mol C}_3\text{H}_8}{1} \times \frac{4 \text{ mol H}_2\text{O}}{1 \text{ mol C}_3\text{H}_8} = 12.0 \text{ mol H}_2\text{O}$$

697. 3.0×10^0 g KCl

To find the mass of a substance when given the mass of a different substance in the same reaction, complete the following conversions:

g given → mol given → mol desired substance → g desired substance

To change from grams to moles of the given substance, divide by the molar mass. To get from moles of the given substance to moles of the desired substance, use the coefficients from the balanced reaction to form a mole ratio. And to get the mass of the desired substance, multiply the moles of the desired substance by the molar mass of the desired substance. *Tip:* To avoid rounding errors, complete all the calculations on your calculator and round the final answer to the correct number of significant figures.

$$\frac{5.0 \text{ g KClO}_3}{1} \times \frac{1 \text{ mol KClO}_3}{\left(39.102 + 35.453 + (15.9994 \times 3)\right) \text{ g KClO}_3} \times \frac{2 \text{ mol KCl}}{2 \text{ mol KClO}_3}$$

$$\times \frac{\left(39.102 + 35.453\right) \text{ g KCl}}{1 \text{ mol KCl}} = 3.0 \text{ g KCl}$$

698. 2.0 g O$_2$

To find the mass of a substance when given the mass of a different substance in the same reaction, complete the following conversions:

g given → mol given → mol desired substance → g desired substance

To change from grams of the given substance to moles, divide by the molar mass. To get from moles of the given substance to moles of the desired substance, use the coefficients from the balanced reaction to form a mole ratio. To get the mass of the desired substance, multiply the moles of the desired substance by the molar mass of the desired substance.

$$\frac{5.0 \text{ g KClO}_3}{1} \times \frac{1 \text{ mol KClO}_3}{\left(39.102 + 35.453 + (15.9994 \times 3)\right) \text{ g KClO}_3} \times \frac{3 \text{ mol O}_2}{2 \text{ mol KClO}_3}$$

$$\times \frac{(15.9994 \times 2) \text{ g O}_2}{1 \text{ mol O}_2} = 2.0 \text{ g O}_2$$

---

699. **27.0 g CaCO$_3$**

To find the mass of a substance when given the mass of a different substance in the same reaction, complete the following conversions:

g given → mol given → mol desired substance → g desired substance

To change from grams of the given substance to moles, divide by the molar mass. To get from moles of the given substance to moles of the desired substance, use the coefficients from the balanced reaction to form a mole ratio. To get the mass of the desired substance, multiply the moles of the desired substance by the molar mass of the desired substance.

$$\frac{20.0 \text{ g Ca(OH)}_2}{1} \times \frac{1 \text{ mol Ca(OH)}_2}{\left(40.08 + (15.9994 \times 2) + (1.00797 \times 2)\right) \text{ g Ca(OH)}_2}$$

$$\times \frac{1 \text{ mol CaCO}_3}{1 \text{ mol Ca(OH)}_2} \times \frac{\left(40.08 + 12.01115 + (15.9994 \times 3)\right) \text{ g CaCO}_3}{1 \text{ mol CaCO}_3} = 27.0 \text{ g CaCO}_3$$

---

700. **117 g NaCl**

To find the mass of a substance when given the mass of a different substance in the same reaction, complete the following conversions:

g given → mol given → mol desired substance → g desired substance

To change from grams of the given substance to moles, divide by the molar mass. To get from moles of the given substance to moles of the desired substance, use the coefficients from the balanced reaction to form a mole ratio. To get the mass of the desired substance, multiply the moles of the desired substance by the molar mass of the desired substance.

$$\frac{80.0 \text{ g NaOH}}{1} \times \frac{1 \text{ mol NaOH}}{(22.9898 + 15.9994 + 1.00797) \text{ g NaOH}} \times \frac{1 \text{ mol NaCl}}{1 \text{ mol NaOH}}$$

$$\times \frac{(22.9898 + 35.453) \text{ g NaCl}}{1 \text{ mol NaCl}} = 117 \text{ g NaCl}$$

---

701. **4.6 g Ag$_3$PO$_4$**

First, you need a balanced chemical equation from the formulas. Potassium phosphate consists of three K^+ and one PO_4^{3-}. Silver nitrate consists of one Ag^+ and one NO_3^-.

Potassium nitrate consists of one K^+ and one NO_3^-. Silver phosphate consists of three Ag^+ and one PO_4^{3-}. Putting all these together gives you the following reaction:

$$K_3PO_4 + AgNO_3 \rightarrow KNO_3 + Ag_3PO_4$$

To balance this reaction, the potassium and silver atoms need to be evened out, so multiply the silver nitrate and the potassium nitrate by 3:

$$K_3PO_4 + 3AgNO_3 \rightarrow 3KNO_3 + Ag_3PO_4$$

To find the mass of silver phosphate, first divide the given mass of silver nitrate by its molar mass. To get from moles of silver nitrate to the moles of silver phosphate, use the coefficients from the balanced reaction to form a mole ratio. To get the mass of silver phosphate, multiply the moles of the silver phosphate by its molar mass.

$$\frac{5.6 \text{ g AgNO}_3}{1} \times \frac{1 \text{ mol AgNO}_3}{(107.968 + 14.0067 + (15.9994 \times 3)) \text{ g AgNO}_3} \times \frac{1 \text{ mol Ag}_3\text{PO}_4}{3 \text{ mol AgNO}_3}$$

$$\times \frac{((107.9688 \times 3) + 30.9738 + (15.9994 \times 4)) \text{ g Ag}_3\text{PO}_4}{1 \text{ mol Ag}_3\text{PO}_4} = 4.6 \text{ g Ag}_3\text{PO}_4$$

---

702.   **8.20 g NaOH**

First, you need a balanced chemical equation from the formulas. Sodium hydroxide consists of one Na^+ and one OH^-. Copper(II) sulfate consists of one Cu^{2+} and one SO_4^{2-}. Copper(II) hydroxide consists of one Cu^{2+} and two OH^-. Sodium sulfate consists of two Na^+ and one SO_4^{2-}. Putting all these together gives you the following equation:

$$NaOH + CuSO_4 \rightarrow Cu(OH)_2 + Na_2SO_4$$

To balance this equation, you need two sodium ions and two hydroxide ions on the reactants side, so multiply the NaOH by 2:

$$2NaOH + CuSO_4 \rightarrow Cu(OH)_2 + Na_2SO_4$$

Now that you have the formula, you can find the number of grams of NaOH that would be necessary to make 10.0 g of $Cu(OH)_2$. Start by dividing the given mass of $Cu(OH)_2$ by its molar mass. Then use the mole ratio from the balanced equation to get the number of moles of NaOH. Last, convert the moles of NaOH to grams of NaOH by multiplying by the molar mass of NaOH.

$$\frac{10.0 \text{ g Cu(OH)}_2}{1} \times \frac{1 \text{ mol Cu(OH)}_2}{(63.546 + (15.9994 \times 2) + (1.00797 \times 2)) \text{ g Cu(OH)}_2}$$

$$\times \frac{2 \text{ mol NaOH}}{1 \text{ mol Cu(OH)}_2} \times \frac{(22.9898 + 15.9994 + 1.00797) \text{ g NaOH}}{1 \text{ mol NaOH}} = 8.20 \text{ g NaOH}$$

---

703.   **66.8 g Zn**

This equation is incomplete, so the first thing you need to do is determine what type of reaction it is. A single element is reacting with a compound, so this is a single displacement reaction: The Zn will displace the H in the HCl. Remember that hydrogen gas is diatomic. Zinc as an ion is Zn^{2+}, so it will take two negatively charged chlorine ions to make the compound neutral:

$$Zn + HCl \rightarrow ZnCl_2 + H_2$$

To balance the equation, multiply the HCl on the reactants side by 2:

$$Zn + 2HCl \rightarrow ZnCl_2 + H_2$$

To find the number of grams of Zn from the HCl, first take the number of grams of HCl and divide by HCl's molar mass. Then use the mole ratio from the balanced equation to find the number of moles of Zn. Next, convert the moles of Zn to grams of Zn by multiplying by the molar mass of Zn.

$$\frac{74.5 \text{ g HCl}}{1} \times \frac{1 \text{ mol HCl}}{(1.00797 + 35.453) \text{ g HCl}} \times \frac{1 \text{ mol Zn}}{2 \text{ mol HCl}} \times \frac{65.37 \text{ g Zn}}{1 \text{ mol Zn}} = 66.8 \text{ g Zn}$$

704. **430 g KNO_3**

The identities of the products are given in the question, so complete the equation with the formulas. You need three +2 charged calcium ions to balance out two –3 charged phosphate ions, and you need one positively charged potassium ion to balance out the one negatively charged nitrate ion:

$$Ca(NO_3)_2 + K_3PO_4 \rightarrow Ca_3(PO_4)_2 + KNO_3$$

This equation is unbalanced, so start with balancing the calcium ions by multiplying the $Ca(NO_3)_2$ by 3:

$$3Ca(NO_3)_2 + K_3PO_4 \rightarrow Ca_3(PO_4)_2 + KNO_3$$

Next, balance the nitrate ions by multiplying the KNO_3 by 6:

$$3Ca(NO_3)_2 + K_3PO_4 \rightarrow Ca_3(PO_4)_2 + 6KNO_3$$

Now you can finish balancing the potassium by multiplying the K_3PO_4 by 2:

$$3Ca(NO_3)_2 + 2K_3PO_4 \rightarrow Ca_3(PO_4)_2 + 6KNO_3$$

To find the number of grams of potassium nitrate (KNO_3) that would be produced along with 220 g of calcium phosphate ($Ca_3(PO_4)_2$), start by dividing 220 g of $Ca_3(PO_4)_2$ by its molar mass. Next, use the mole ratio from the balanced equation to find the moles of KNO_3 needed. Then multiply the number of moles of KNO_3 by the molar mass of KNO_3.

$$\frac{220 \text{ g Ca}_3(PO_4)_2}{1} \times \frac{1 \text{ mol Ca}_3(PO_4)_2}{((40.08 \times 3) + (30.9738 \times 2) + (15.9994 \times 4 \times 2)) \text{ g Ca}_3(PO_4)_2}$$

$$\times \frac{6 \text{ mol KNO}_3}{1 \text{ mol Ca}_3(PO_4)_2} \times \frac{(39.102 + 14.0067 + (15.9994 \times 3)) \text{ g KNO}_3}{1 \text{ mol KNO}_3} = 430 \text{ g KNO}_3$$

705.   **0.15 g $BaCl_2$ and 0.21 g $K_2Cr_2O_7$**

To work this problem, the first thing you need is a complete, balanced equation. Determine the formulas from the names, determine the type of reaction so you can predict the products, and then balance the equation.

Barium chloride comes from the combination of barium ions, Ba^{2+}, and chloride ions, Cl^-. Potassium dichromate comes from the combination of potassium ions, K^+, and dichromate ions, $Cr_2O_7^{2-}$. This gives you

$$BaCl_2 + K_2Cr_2O_7 \rightarrow$$

The reactants are a pair of ionic compounds that can switch partners, so this is a double displacement reaction. Therefore, the products are barium dichromate and potassium chloride:

$$BaCl_2 + K_2Cr_2O_7 \rightarrow BaCr_2O_7 + KCl$$

Balancing this equation requires making the number of chlorine and potassium atoms equal on each side. Simply multiply the KCl on the products side of the reaction by 2:

$$BaCl_2 + K_2Cr_2O_7 \rightarrow BaCr_2O_7 + 2KCl$$

Now calculate the number of grams of $BaCl_2$ and $K_2Cr_2O_7$ needed. You need to do the following conversions:

g given $\rightarrow$ mol given $\rightarrow$ mol desired substance $\rightarrow$ g desired substance

To change from grams to moles of barium dichromate, divide by the molar mass. To get from moles of barium dichromate to moles of the reactants, use the coefficients from the balanced reaction to form a mole ratio. To get the mass of the desired substance, multiply the moles of the desired substance by the molar mass of the desired substance.

$$\frac{0.25 \text{ g } BaCr_2O_7}{1} \times \frac{1 \text{ mol } BaCr_2O_7}{(137.34 + (51.996 \times 2) + (15.9994 \times 7)) \text{ g } BaCr_2O_7} \times \frac{1 \text{ mol } BaCl_2}{1 \text{ mol } BaCr_2O_7}$$

$$\times \frac{(137.34 + (35.453 \times 2)) \text{ g } BaCl_2}{1 \text{ mol } BaCl_2} = 0.15 \text{ g } BaCl_2$$

$$\frac{0.25 \text{ g } BaCr_2O_7}{1} \times \frac{1 \text{ mol } BaCr_2O_7}{(137.34 + (51.996 \times 2) + (15.9994 \times 7)) \text{ g } BaCr_2O_7} \times \frac{1 \text{ mol } K_2Cr_2O_7}{1 \text{ mol } BaCr_2O_7}$$

$$\times \frac{((39.102 \times 2) + (51.996 \times 2) \times (15.9994 \times 7)) \text{ g } K_2Cr_2O_7}{1 \text{ mol } K_2Cr_2O_7} = 0.21 \text{ g } K_2Cr_2O_7$$

706.   **167 g $AlBr_3$**

First, complete and balance the equation. There's only one product, so this is a synthesis (combination) reaction. Aluminum as an ion has a charge of +3, and the bromine ion has a charge of –1, so you need three bromine ions to balance out the +3 charge of the aluminum ion. The formula is $AlBr_3$. Here's the unbalanced equation:

$$Al + Br_2 \rightarrow AlBr_3$$

To balance this equation, you need the lowest common multiple of 2 and 3, which is 6, to balance the bromine atoms. Multiply the Br_2 on the reactants side by 3, and multiply the $AlBr_3$ on the product side by 2:

$$Al + 3Br_2 \rightarrow 2AlBr_3$$

Then you can balance the aluminum atoms by multiplying the Al on the left by 2:

$$2Al + 3Br_2 \rightarrow 2AlBr_3$$

Now find the maximum amount of aluminum bromide that can form if all of each reactant is used up. You need to do the following conversions:

g given → mol given → mol desired substance → g desired substance

To change from grams of a given substance to moles of that substance, divide by the molar mass. To get from moles of the given substance to moles of the desired substance, use the coefficients from the balanced reaction to form a mole ratio. To get the mass of the desired substance, multiply the moles of the desired substance by the molar mass of the desired substance.

$$\frac{50.0 \text{ g Al}}{1} \times \frac{1 \text{ mol Al}}{26.9815 \text{ g Al}} \times \frac{2 \text{ mol AlBr}_3}{2 \text{ mol Al}} \times \frac{(26.9815 + (79.904 \times 3)) \text{ g AlBr}_3}{1 \text{ mol AlBr}_3} = 494 \text{ g AlBr}_3$$

$$\frac{150. \text{ g Br}_2}{1} \times \frac{1 \text{ mol Br}_2}{(79.904 \times 2) \text{ g Br}_2} \times \frac{2 \text{ mol AlBr}_3}{3 \text{ mol Br}_2} \times \frac{(26.9815 + (79.904 \times 3)) \text{ g AlBr}_3}{1 \text{ mol AlBr}_3}$$

$$= 167 \text{ g AlBr}_3$$

If 50.0 g Al can make 494 g $AlBr_3$ but 150. g Br_2 can make only 167 g $AlBr_3$, then the maximum amount that can be made with the reactants provided is the smaller number, 167 g.

707. 2.01×10^{25} molecules of O_2

First, balance the equation. To balance the oxygen atoms, you need the lowest common multiple of 2 and 5, which is 10. So multiply the P_2O_5 by 2 and multiply the O_2 by 5:

$$P + 5O_2 \rightarrow 2P_2O_5$$

Next, balance the atoms of phosphorus by multiplying the P on the reactants side by 4:

$$4P + 5O_2 \rightarrow 2P_2O_5$$

To find how many molecules of O_2 you need, start with the given number of P_2O_5 and divide by Avogadro's number of molecules in a mole to get moles of P_2O_5. Next, use the mole ratio to get moles of O_2. Then multiply by Avogadro's number of molecules in a mole to get the molecules of O_2.

$$\frac{8.02 \times 10^{24} \text{ molecules P}_2O_5}{1} \times \frac{1 \text{ mol P}_2O_5}{6.022 \times 10^{23} \text{ molecules P}_2O_5} \times \frac{5 \text{ mol O}_2}{2 \text{ mol P}_2O_5}$$

$$\times \frac{6.022 \times 10^{23} \text{ molecules O}_2}{1 \text{ mol O}_2} = 2.01 \times 10^{25} \text{ molecules O}_2$$

708. 8.45×10^{21} molecules of H_2O

This equation is already balanced, so you can start the math part of the problem. To find the number of molecules of H_2O needed, start with the given number of HNO_3 molecules and divide by Avogadro's number of molecules in a mole to get moles of HNO_3. Then use the mole ratio to get moles of H_2O. Multiply by Avogadro's number of molecules in a mole to get the molecules of H_2O.

$$\frac{1.69 \times 10^{22} \text{ molecules } HNO_3}{1} \times \frac{1 \text{ mol } HNO_3}{6.022 \times 10^{23} \text{ molecules } HNO_3} \times \frac{2 \text{ mol } H_2O}{4 \text{ mol } HNO_3}$$

$$\times \frac{6.022 \times 10^{23} \text{ molecules } H_2O}{1 \text{ mol } H_2O} = 8.45 \times 10^{21} \text{ molecules } H_2O$$

709. 3.16×10^{21} molecules of $BaSO_3$

First, write and balance the equation for this reaction. Barium sulfite contains barium ions, Ba^{2+}, and sulfite ions, SO_3^{2-}, which gives you a formula of $BaSO_3$. Barium sulfide contains barium ions, Ba^{2+}, and sulfide ions, S^{2-}, which gives you a formula of BaS. Sulfur dioxide contains one sulfur atom and two oxygen atoms, so the formula is SO_2. Here's the reaction:

$$BaSO_3 \rightarrow BaS + SO_2$$

This just happens to be balanced, so on to the calculations! To find the number of molecules of $BaSO_3$ needed, start with the given number of SO_2 molecules and divide by Avogadro's number of molecules in a mole to get moles of SO_2. Then use the mole ratio to get moles of $BaSO_3$. Multiply by Avogadro's number of molecules in a mole to get the molecules of $BaSO_3$.

$$\frac{3.16 \times 10^{21} \text{ molecules } SO_2}{1} \times \frac{1 \text{ mol } SO_2}{6.022 \times 10^{23} \text{ molecules } SO_2} \times \frac{1 \text{ mol } BaSO_3}{1 \text{ mol } SO_2}$$

$$\times \frac{6.022 \times 10^{23} \text{ molecules } BaSO_3}{1 \text{ mol } BaSO_3} = 3.16 \times 10^{21} \text{ molecules } BaSO_3$$

710. 2.64×10^{24} molecules of I_2

First, write and balance the equation for this reaction. Iodine and chlorine gases are diatomic, so the reactants are I_2 and Cl_2. Iodine monochloride has one iodine atom and one chlorine atom, for the formula ICl. Here's the equation:

$$I_2(g) + Cl_2(g) \rightarrow ICl(g)$$

This equation isn't balanced, but to balance it, you only need to put a 2 in front of the ICl:

$$I_2(g) + Cl_2(g) \rightarrow 2ICl(g)$$

To find the number of molecules of iodine needed, start with the given number of ICl molecules and divide by Avogadro's number of molecules in a mole to get moles of ICl.

Then use the mole ratio to get moles of iodine. Multiply by Avogadro's number of molecules in a mole to get the molecules of iodine.

$$\frac{5.28 \times 10^{24} \text{ molecules ICl}}{1} \times \frac{1 \text{ mol ICl}}{6.022 \times 10^{23} \text{ molecules ICl}} \times \frac{1 \text{ mol I}_2}{2 \text{ mol ICl}}$$

$$\times \frac{6.022 \times 10^{23} \text{ molecules I}_2}{1 \text{ mol I}_2} = 2.64 \times 10^{24} \text{ molecules I}_2$$

711.　　**5.65×10^{23} molecules C_8H_{18}**

This combustion reaction is not balanced. Start by balancing the carbon:

$$C_8H_{18} + O_2 \rightarrow 8CO_2 + H_2O$$

Then balance the hydrogen:

$$C_8H_{18} + O_2 \rightarrow 8CO_2 + 9H_2O$$

Then balance the oxygen:

$$C_8H_{18} + \frac{25}{2}O_2 \rightarrow 8CO_2 + 9H_2O$$

Fractions aren't appropriate for balanced equations, so multiply everything by 2:

$$2C_8H_{18} + 25O_2 \rightarrow 16CO_2 + 18H_2O$$

Now calculate the number of molecules of octane needed. Here's the plan for the conversions:

g given → mol given → mol desired substance → g desired substance

To change from molecules to moles of a given substance, divide by Avogadro's number of molecules in a mole. To get from moles of the given substance to moles of the desired substance, use the coefficients from the balanced reaction to form a mole ratio. To get the molecules of the desired substance, multiply the moles of the desired substance by Avogadro's number of molecules in a mole of the desired substance.

$$\frac{4.52 \times 10^{24} \text{ molecules CO}_2}{1} \times \frac{1 \text{ mol CO}_2}{6.022 \times 10^{23} \text{ molecules CO}_2} \times \frac{2 \text{ mol C}_8H_{18}}{16 \text{ mol CO}_2}$$

$$\times \frac{6.022 \times 10^{23} \text{ molecules C}_8H_{18}}{1 \text{ mol C}_8H_{18}} = 5.65 \times 10^{23} \text{ molecules C}_8H_{18}$$

$$\frac{4.52 \times 10^{24} \text{ molecules H}_2O}{1} \times \frac{1 \text{ mol H}_2O}{6.022 \times 10^{23} \text{ molecules H}_2O} \times \frac{2 \text{ mol C}_8H_{18}}{18 \text{ mol H}_2O}$$

$$\times \frac{6.022 \times 10^{23} \text{ molecules C}_8H_{18}}{1 \text{ mol C}_8H_{18}} = 5.02 \times 10^{23} \text{ molecules C}_8H_{18}$$

Compare the two numbers. You need at least 5.65×10^{23} molecules of C_8H_{18}, because 5.02×10^{23} molecules C_8H_{18} would produce less than the required amount of carbon dioxide.

712.

84.39%

To find the percent yield, divide the actual yield (the amount produced in the experiment) by the theoretical yield (the maximum amount possible based on calculations) and multiply by 100:

$$\% \text{ yield} = \frac{\text{actual yield}}{\text{theoretical yield}} \times 100$$

$$= \frac{1.146 \text{ g}}{1.358 \text{ g}} \times 100$$

$$= 84.39\%$$

713.

104%

To find the percent yield, divide the actual yield (the amount produced in the experiment) by the theoretical yield (the maximum amount possible based on calculations) and multiply by 100:

$$\% \text{ yield} = \frac{\text{actual yield}}{\text{theoretical yield}} \times 100$$

$$= \frac{5.15 \text{ g}}{4.95 \text{ g}} \times 100$$

$$= 104\%$$

The student in this lab hasn't violated the law of conservation of mass by creating matter. Most likely, there was an error in finding the mass of product synthesized, or some contaminants got in the sample.

714.

67.4%

To solve for the percent yield, you need both the actual yield and the theoretical yield. The actual yield and the mass of a reactant are given, so you need to calculate the theoretical yield. To do this, convert the grams of NH_3 to moles by dividing by the molar mass of NH_3; then use the mole ratio to get to moles of NH_4Br and multiply the moles of NH_4Br by the molar mass:

$$\frac{316 \text{ g } NH_3}{1} \times \frac{1 \text{ mol } NH_3}{(14.0067 + (1.00797 \times 3)) \text{ g } NH_3} \times \frac{1 \text{ mol } NH_4Br}{1 \text{ mol } NH_3}$$

$$\times \frac{(14.0067 + (1.00797 \times 4) + 79.904) \text{ g } NH_4Br}{1 \text{ mol } NH_4Br}$$

$$= 1,817 \text{ g } NH_4Br$$

To find the percent yield, divide the actual yield by the theoretical yield and multiply by 100:

$$\frac{1,225 \text{ g}}{1,817 \text{ g}} \times 100 = 67.4\% \ NH_4Br$$

715. **87.1%**

To solve for percent yield, you need both the actual yield and the theoretical yield. The problem gives only the actual yield and the mass of a reactant, so you need to calculate the theoretical yield. To do this, convert the grams of NH_3 to moles by dividing by the molar mass of NH_3; then use the mole ratio to get to moles of NH_4Br and multiply the moles of NH_4Br by the molar mass:

$$\frac{623 \text{ g NH}_3}{1} \times \frac{1 \text{ mol NH}_3}{(14.0067 + (1.00797 \times 3)) \text{ g NH}_3} \times \frac{6 \text{ mol NH}_4Br}{8 \text{ mol NH}_3}$$

$$\times \frac{(14.0067 + (1.00797 \times 4) + 79.904) \text{ g NH}_4Br}{1 \text{ mol NH}_4Br}$$

$$= 2{,}687 \text{ g NH}_4Br$$

To find the percent yield, divide the actual yield by the theoretical yield and multiply by 100:

$$\frac{2{,}341 \text{ g}}{2{,}687 \text{ g}} \times 100 = 87.1\% \text{ NH}_4Br$$

716. **93.2% BaSO₄**

First, having a balanced reaction would be good. Both reactants and one of the products are given in the question, so here's your initial equation:

$$BaCl_2 + Na_2SO_4 \rightarrow BaSO_4 + ?$$

This reaction appears to be a double displacement, so the other product is sodium chloride:

$$BaCl_2 + Na_2SO_4 \rightarrow BaSO_4 + NaCl$$

To balance this equation, just multiply the NaCl by 2:

$$BaCl_2 + Na_2SO_4 \rightarrow BaSO_4 + 2NaCl$$

To find the percent yield of the reaction, you first need to calculate the theoretical yield. To do this, convert the grams of $BaCl_2$ to moles by dividing by the molar mass of $BaCl_2$; then use the mole ratio to get to moles of $BaSO_4$ and multiply the moles of $BaSO_4$ by the molar mass:

$$\frac{0.527 \text{ g BaCl}_2}{1} \times \frac{1 \text{ mol BaCl}_2}{(137.34 + (35.453 \times 2)) \text{ g BaCl}_2} \times \frac{1 \text{ mol BaSO}_4}{1 \text{ mol BaCl}_2}$$

$$\times \frac{(137.34 + 32.064 + (15.9994 \times 4)) \text{ g BaSO}_4}{1 \text{ mol BaSO}_4}$$

$$= 0.591 \text{ g BaSO}_4$$

The actual yield divided by the theoretical yield times 100 gives you the percent yield:

$$\frac{0.551 \text{ g}}{0.591 \text{ g}} \times 100 = 93.2\% \text{ BaSO}_4$$

717. 3.60×10^3 g SO_3

The equation is already balanced, so the first step is to determine which reactant is the limiting reactant. Divide the moles of each substance by the coefficient of that substance from the balanced chemical equation. The lowest value from this division is the limiting reactant. The limiting reactant is the key to all subsequent calculations.

$$\frac{45.0 \text{ mol SO}_2}{2 \text{ mol}} = 22.5 \text{ (limiting)}$$

$$\frac{25.0 \text{ mol O}_2}{1 \text{ mol}} = 25.0 \text{ (in excess)}$$

The limiting reactant controls the amount of product formed, so start with 45.0 mol of SO_2 and perform a normal stoichiometry calculation to find the maximum amount of SO_3 that can be produced:

$$\frac{45.0 \text{ mol SO}_2}{1} \times \frac{2 \text{ mol SO}_3}{2 \text{ mol SO}_2} \times \frac{(32.064 + (15.9994 \times 3)) \text{ g SO}_3}{1 \text{ mol SO}_3} = 3.60 \times 10^3 \text{ g SO}_3$$

718. 99.5 g NaCl

The chlorine is out of balance in the equation: You have three chlorine atoms on the reactants side and two chlorine atoms on the products side. An easy way to try to balance the odd number of chlorine atoms is to multiply each appearance of chlorine that appears singly in the formula (the $NaClO_2$, the ClO_2, and the $NaCl$) by 2, which gives you

$$Cl_2 + 2NaClO_2 \rightarrow 2ClO_2 + 2NaCl$$

Now the equation is balanced. To find the maximum amount of a substance that can be produced, you need to figure out which reactant will run out first. First, change the grams of each reactant to moles (so you have counting units to work with) by dividing by molar mass:

$$\frac{106 \text{ g Cl}_2}{1} \times \frac{1 \text{ mol Cl}_2}{(35.453 \times 2) \text{ g Cl}_2} = 1.50 \text{ mol Cl}_2$$

$$\frac{154 \text{ g NaClO}_2}{1} \times \frac{1 \text{ mol NaClO}_2}{(22.9898 + 35.453 + (15.9994 \times 2)) \text{ g NaClO}_2} = 1.70 \text{ mol NaClO}_2$$

To determine the limiting reactant, divide the moles of each substance by the coefficient of that substance from the balanced chemical equation. The lowest value from this division is the limiting reactant. The limiting reactant is the key to all subsequent calculations.

$$\frac{1.50 \text{ mol Cl}_2}{1 \text{ mol}} = 1.50 \text{ (in excess)}$$

$$\frac{1.70 \text{ mol NaClO}_2}{2 \text{ mol}} = 0.851 \text{ (limiting)}$$

Now knowing the limiting reactant, you can find the mass of the product from the given mass. The limiting reactant controls the amount of product formed, so start with 154 g of $NaClO_2$ and perform a normal stoichiometry calculation to find the grams of NaCl produced:

$$\frac{154\ \text{g NaClO}_2}{1} \times \frac{1\ \text{mol NaClO}_2}{\left(22.9898 + 35.453 + \left(15.9994 \times 2\right)\right)\text{g NaClO}_2} \times \frac{2\ \text{mol NaCl}}{2\ \text{mol NaClO}_2}$$

$$\times \frac{\left(22.9898 + 35.453\right)\text{g NaCl}}{1\ \text{mol NaCl}} = 99.5\ \text{g NaCl}$$

---

719.   **188.9 g Al$_2$O$_3$, and the Al is limiting**

First, you need a balanced equation. Aluminum is a metal that you write as Al; oxygen gas is diatomic, so it's written O$_2$; and aluminum oxide is given as Al$_2$O$_3$. With this info, you can write the unbalanced equation:

$$Al + O_2 \rightarrow Al_2O_3$$

To balance the oxygen atoms, you need the lowest common multiple of 2 and 3, which is 6. Multiply the O$_2$ by 3 and the Al$_2$O$_3$ by 2:

$$Al + 3O_2 \rightarrow 2Al_2O_3$$

Then balance the aluminum by multiplying the Al on the reactants side by 4:

$$4Al + 3O_2 \rightarrow 2Al_2O_3$$

To find the limiting reactant, change the grams of each reactant to moles (so you have counting units to work with) by dividing by molar mass:

$$\frac{100.0\ \text{g Al}}{1} \times \frac{1\ \text{mol Al}}{26.9815\ \text{g Al}} = 3.706\ \text{mol Al}$$

$$\frac{100.0\ \text{g O}_2}{1} \times \frac{1\ \text{mol O}_2}{\left(15.9994 \times 2\right)\text{g O}_2} = 3.125\ \text{mol O}_2$$

To determine the limiting reactant, divide the moles of each substance by the coefficient of that substance from the balanced chemical equation. The lowest value from this division is the limiting reactant. The limiting reactant is the key to all subsequent calculations.

$$\frac{3.706\ \text{mol Al}}{4\ \text{mol}} = 0.9266\ \left(\text{limiting}\right)$$

$$\frac{3.125\ \text{mol O}_2}{3\ \text{mol}} = 1.042\ \left(\text{in excess}\right)$$

The Al is limiting because you have a smaller relative amount of it available to react. Now find the mass of Al$_2$O$_3$. Take the original mass of aluminum and perform a normal stoichiometry calculation to find the grams of Al$_2$O$_3$ produced:

$$\frac{100.0\ \text{g Al}}{1} \times \frac{1\ \text{mol Al}}{26.9815\ \text{g Al}} \times \frac{2\ \text{mol Al}_2\text{O}_3}{4\ \text{mol Al}} \times \frac{\left(\left(26.9815 \times 2\right) + \left(15.9994 \times 3\right)\right)\text{g Al}_2\text{O}_3}{1\ \text{mol Al}_2\text{O}_3}$$

$$= 188.9\ \text{g Al}_2\text{O}_3$$

---

720.   **9.34 g NO, and HNO$_3$ is in excess**

The equation is balanced, so the first step is to determine which reactant is limiting. First, change the grams of each reactant to moles (so you have counting units to work with) by dividing by molar mass:

$$\frac{155 \text{ g KI}}{1} \times \frac{1 \text{ mol KI}}{(39.102 + 126.9044) \text{ g KI}} = 0.934 \text{ mol KI}$$

$$\frac{175 \text{ g HNO}_3}{1} \times \frac{1 \text{ mol HNO}_3}{(1.00797 + 14.0067 + (15.9994 \times 3)) \text{ g HNO}_3} = 2.78 \text{ mol HNO}_3$$

The mole ratio from the balanced equation shows that it takes 6 mol of KI for every 8 mol of HNO_3.

To determine the limiting reactant, divide the moles of each substance by the coefficient of that substance from the balanced chemical equation. The lowest value from this division is the limiting reactant. The limiting reactant is the key to all subsequent calculations.

$$\frac{0.934 \text{ mol KI}}{6 \text{ mol}} = 0.156 \text{ (limiting)}$$

$$\frac{2.78 \text{ mol HNO}_3}{8 \text{ mol}} = 0.348 \text{ (in excess)}$$

KI is the limiting reactant because you have a smaller amount of it available compared to the amount of HNO_3.

Now knowing the limiting reactant, you can calculate the number of grams of NO. The limiting reactant controls the amount of product formed, so begin with 155 g of KI and perform a normal stoichiometry calculation:

$$\frac{155 \text{ g KI}}{1} \times \frac{1 \text{ mol KI}}{(39.102 + 126.9044) \text{ g KI}} \times \frac{2 \text{ mol NO}}{6 \text{ mol KI}} \times \frac{(14.0067 + 15.9994) \text{ g NO}}{1 \text{ mol NO}} = 9.34 \text{ g NO}$$

721. **O_2 is the limiting reactant, 56.3 g of water is produced, and 43.7 g of H_2 is in excess**

First, you need an equation for the reaction. Hydrogen and oxygen both exist as diatomic gases, and with sufficient energy, they react to form water:

$$H_2 + O_2 \rightarrow H_2O$$

Balance the oxygen atoms by multiplying the water by 2, and then balance the hydrogen atoms by multiplying the H_2 by 2:

$$2H_2 + O_2 \rightarrow 2H_2O$$

Now you can determine which reactant is limiting and which is in excess. First, change the grams of each reactant to moles (so you have counting units to work with) by dividing by molar mass:

$$\frac{50.0 \text{ g H}_2}{1} \times \frac{1 \text{ mol H}_2}{(1.00797 \times 2) \text{ g H}_2} = 24.8 \text{ mol H}_2$$

$$\frac{50.0 \text{ g O}_2}{1} \times \frac{1 \text{ mol O}_2}{(15.9994 \times 2) \text{ g O}_2} = 1.56 \text{ mol O}_2$$

To determine the limiting reactant, divide the moles of each substance by the coefficient of that substance from the balanced chemical equation. The lowest value from this division is the limiting reactant. The limiting reactant is the key to all subsequent calculations.

$$\frac{24.8 \text{ mol H}_2}{2 \text{ mol}} = 12.4 \text{ (in excess)}$$

$$\frac{1.56 \text{ mol O}_2}{1 \text{ mol}} = 1.56 \text{ (limiting)}$$

The O_2 is the limiting reactant because a smaller relative amount of it is available. Now knowing the limiting reactant, you can find the mass of the water produced from the given mass of the oxygen. The limiting reactant controls the amount of product formed, so begin with 50.0 g of O_2 and perform a normal stoichiometry calculation:

$$\frac{50.0\,\text{g}\,O_2}{1} \times \frac{1\,\text{mol}\,O_2}{(15.9994 \times 2)\,\text{g}\,O_2} \times \frac{2\,\text{mol}\,H_2O}{1\,\text{mol}\,O_2} \times \frac{((1.00797 \times 2)+15.9994)\,\text{g}\,H_2O}{1\,\text{mol}\,H_2O}$$

$$= 56.3\,\text{g}\,H_2O$$

Using the law of conservation of mass, the mass of the hydrogen used plus the mass of the water used should equal the mass of the water produced (x g H_2 + 50.0 g O_2 = 56.3 g H_2O), so the mass of the hydrogen used is 6.3 grams. The original mass of hydrogen available was 50.0 g, so subtract the hydrogen used from 50.0 g to find the H_2 in excess:

$$50.0\,\text{g}\,H_2 - 6.3\,\text{g}\,H_2 = 43.7\,\text{g}\,H_2$$

722. **573 K**

To find the Kelvin temperature when given degrees Celsius, add 273 to the Celsius temperature:

$$K = {}^\circ C + 273$$
$$= 300 + 273$$
$$= 573\,K$$

723. **423 K**

To find the Kelvin temperature when given degrees Celsius, add 273 to the Celsius temperature:

$$K = {}^\circ C + 273$$
$$= 150 + 273$$
$$= 423\,K$$

724. **73 K**

To find the Kelvin temperature when given degrees Celsius, add 273 to the Celsius temperature:

$$K = {}^\circ C + 273$$
$$= -200 + 273$$
$$= 73\,K$$

Answers
701–800

725. **195 K**

To find the Kelvin temperature when given degrees Celsius, add 273 to the Celsius temperature:

$$K = {}^\circ C + 273$$
$$= -78 + 273$$
$$= 195\,K$$

726. 310 K

To find the Kelvin temperature when given degrees Celsius, add 273 to the Celsius temperature:

$$K = °C + 273$$
$$= 37 + 273$$
$$= 310 \text{ K}$$

727. –173°C

To find the temperature in degrees Celsius when given kelvins, subtract 273 from the Kelvin temperature:

$$°C = K - 273$$
$$= 100 - 273$$
$$= -173°C$$

728. 27°C

To find the temperature in degrees Celsius when given kelvins, subtract 273 from the Kelvin temperature:

$$°C = K - 273$$
$$= 300 - 273$$
$$= 27°C$$

729. –273°C

To find the temperature in degrees Celsius when given kelvins, subtract 273 from the Kelvin temperature:

$$°C = K - 273$$
$$= 0 - 273$$
$$= -273°C$$

730. 40°C and 104°F

To find the temperature in degrees Celsius when given kelvins, subtract 273 from the Kelvin temperature:

$$°C = K - 273$$
$$= 313 - 273$$
$$= 40°C$$

To find the temperature in degrees Fahrenheit, multiply the temperature in degrees Celsius by 9/5 and then add 32:

$$°F = \frac{9}{5}(40°C) + 32$$
$$= 72 + 32$$
$$= 104°F$$

731. −40.00°C and −40.00°F

To find the temperature in degrees Celsius when given kelvins, subtract 273.15 from the Kelvin temperature. Use 273.15 (not 273) for this conversion to match the number of decimal places in the given temperature (233.15 K):

$$°C = K - 273.15$$
$$= 233.15 - 273.15$$
$$= -40.00°C$$

To find the temperature in degrees Fahrenheit, multiply the temperature in degrees Celsius by 9/5 and then add 32.00:

$$°F = \frac{9}{5}(-40.00°C) + 32.00$$
$$= -72.00 + 32.00$$
$$= -40.00°F$$

Note that answers of −40°C and −40°F have the wrong number of significant figures. You need two places after the decimal point to correspond to the original temperature.

732. $\overline{BC}$

The lower segment that has no slope (from B to C) represents *melting,* turning from a solid into a liquid. All added heat energy goes toward the phase change, so no heat is available to change the temperature. The curve begins with a solid, at Point A, and heat raises the temperature of the solid to Point B, where the substance begins to melt. Melting is complete at Point C, and heating of the liquid begins.

733. $\overline{BC}$

Condensation is the change from a gas to a liquid, and this occurs when the substance cools from Point B to C. All the heat energy removed comes from the phase change, so there's no change in temperature; therefore, the slope in this region is zero.

Cooling is the removal of heat energy and may or may not involve a temperature change. The two line segments representing phase changes, $\overline{BC}$ and $\overline{DE}$, have different lengths because the amount of energy released when a substance condenses is greater than the amount released when the same amount of material freezes.

734. $\overline{CD}$

Areas on the graph that have a slope are areas where the substance is being heated but not changing phase. The liquid phase exists between Points C and D. The curve begins with a solid, at Point A, and heating warms the solid to Point B, where the substance begins to melt. Melting is complete at Point C, and heating of the liquid begins. Heating of the liquid continues until the substance reaches the boiling point (Point D), where vaporization begins.

735. $\overline{DE}$

Boiling is the change from a liquid to a gas. This occurs with no change in temperature (a line with no slope in the graph). The substance is a liquid at Point D and a gas at

Point E. The curve begins with a solid, at Point A, and heating warms the solid to Point B, where the substance begins to melt. Melting is complete at Point C, and heating of the liquid begins. Heating of the liquid continues until the substance reaches the boiling point (Point D), where vaporization begins. Vaporization continues until all the liquid is converted to gas (Point E). The gas begins to heat at Point E and continues to heat until the end of the experiment at Point F.

736. $\overline{AB}$

The solid is being heated at the bottom left-hand corner of the graph in the segment going from A to B. The curve begins with a solid, at Point A, and heating warms the solid to Point B, where the substance begins to melt.

737. $\overline{DE}$

Freezing is the change from a liquid to a solid. This process takes place from Point D to Point E. The curve begins with a gas, at Point A, and removing heat cools the gas to Point B, where it begins to condense. Condensation is complete at Point C, and cooling of the liquid begins. Cooling of the liquid continues until it reaches the freezing point (Point D), where freezing begins. Freezing continues until all the liquid is converted to a solid (Point E). The substance cools from Point E to the end of the experiment (Point F).

738. $\overline{EF}$

A gas is being heated starting at Point E and going to Point F and beyond. The curve begins with a solid, at Point A, and heating warms the solid to Point B, where the substance begins to melt. Melting is complete at Point C, and heating of the liquid begins. Heating of the liquid continues until the substance reaches the boiling point (Point D), where vaporization begins. Vaporization continues until all the liquid is converted to gas (Point E). The gas heats from Point E to the end of the experiment (Point F).

739. 16,700 J

To calculate the amount of energy needed to melt a substance, you can use the following formula:

$$Q = m\Delta H_{fus}$$

where Q is the quantity of heat energy, m is the mass, and ΔH_{fus} is the enthalpy of fusion. The *enthalpy of fusion*, sometimes called the *heat of fusion*, is the energy necessary to fuse (melt) a quantity of solid.

Enter the numbers in the formula and solve for Q:

$$Q = 50.0 \text{ g} \times \frac{334 \text{ J}}{\text{g}} = 16,700 \text{ J}$$

You need 16,700 J of heat energy to melt the ice.

740. –2,500 calories

The temperature at the melting point and the freezing point are the same, so you can use the following formula:

$$Q = m\Delta H_{sol}$$

where Q is the quantity of heat energy, m is the mass, and ΔH_{sol} is the enthalpy of solidification. The *enthalpy of solidification,* sometimes called the *heat of solidification,* is the energy that has to be removed to freeze (solidify) a quantity of liquid; at a substance's melting point, the heats of fusion and of solidification (freezing) are numerically the same; however, the signs are opposite. This means $\Delta H_{sol} = -\Delta H_{fus}$. Therefore, if the heat of fusion is 25cal/g, then the heat of solidification is –25 cal/g.

Enter the numbers in the formula and solve for Q:

$$Q = 100.0 \text{ g} \times \frac{-25 \text{ calories}}{\text{g}} = -2{,}500 \text{ calories}$$

When the ethanol freezes, it releases –2,500 calories of heat energy.

---

741. **27,000 calories**

To calculate the amount of energy needed to evaporate a substance, use the following formula:

$$Q = m\Delta H_{vap}$$

where Q is the quantity of heat energy, m is the mass, and ΔH_{vap} is the enthalpy of vaporization. The *enthalpy of vaporization,* sometimes called the *heat of vaporization,* is the energy necessary to vaporize a quantity of liquid.

Enter the numbers in the formula and solve for Q:

$$Q = 50.0 \text{ g} \times \frac{540 \text{ calories}}{\text{g}} = 27{,}000 \text{ calories}$$

You need 27,000 calories to evaporate the water.

---

742. **–4,020 J**

To calculate the amount of energy released when a substance is condensed, use the following formula:

$$Q = m\Delta H_{cond}$$

where Q is the quantity of heat energy, m is the mass, and ΔH_{cond} is the enthalpy of condensation. The *enthalpy of condensation,* sometimes called the *heat of condensation,* is the energy removed to condense a quantity of gas; at a substance's boiling point, the heats of vaporization and of condensation are numerically the same; however, the signs are opposite. This means $\Delta H_{vap} = -\Delta H_{cond}$. Therefore, if the heat of vaporization is 201 J/g, then the heat of condensation is –201 J/g.

Enter the numbers in the formula and solve for Q:

$$Q = 20.0 \text{ g} \times \frac{-201 \text{ J}}{\text{g}} = -4{,}020 \text{ J}$$

When the nitrogen condenses, it releases –4,020 J of heat energy.

---

743. **50. calories**

If the substance doesn't go through any phase changes, you can use the following formula to calculate the amount of energy needed to raise the substance's temperature:

$$Q = mC_p\Delta T$$

In this equation, Q is the quantity of heat energy, m is the mass, C_p is the specific heat of the material, and ΔT is the change in temperature ($T_{final} - T_{initial}$). The *specific heat* is the quantity of heat energy necessary to increase the temperature of 1 g of a substance by 1°C.

Enter the numbers in the formula and solve for Q:

$$Q = 10.0\ \cancel{g} \times \frac{0.500\ \text{cal}}{\cancel{g}\ \cancel{°C}} \times (112 - 102)\cancel{°C} = 50.\ \text{calories}$$

You need 50. calories to raise the temperature of the steam to 112°C.

---

744.   **19,000 J**

If the substance doesn't go through any phase changes, you can use the following formula to calculate the amount of energy needed to raise the substance's temperature:

$$Q = mC_p\Delta T$$

In this equation, Q is the quantity of heat energy, m is the mass, C_p is the specific heat of the material, and ΔT is the change in temperature ($T_{final} - T_{initial}$). The *specific heat* is the quantity of heat energy necessary to increase the temperature of 1 g of a substance by 1°C.

Enter the numbers in the formula and solve for Q:

$$Q = 75.0\ \cancel{g} \times \frac{4.18\ \text{J}}{\cancel{g}\ \cancel{°C}} \times (80. - 20.)\cancel{°C} = 18,810\ \text{J} \approx 19,000\ \text{J}$$

Round the intermediate answer (18,810 J) to 19,000 J so that the answer has the same number of significant figures as the temperatures given in the problem. (**Remember:** In multiplication problems, the answer should have the same number of significant figures as the measurement with the fewest significant figures.) You need 19,000 J to raise the temperature of the water to 80.°C.

---

745.   **2,130 calories**

You need heat to raise the temperature of the ice to the melting point, 0.0°C, and additional heat to melt the ice. The first step, raising the temperature, requires the formula $Q = mC_p\Delta T$, where Q is the heat energy, m is the mass, C_p is the specific heat of the material, and ΔT is the change in temperature ($T_{final} - T_{initial}$). The second step, melting the ice, requires the formula $Q = m\Delta H_{fus}$, where Q is the quantity of heat energy, m is the mass, and ΔH_{fus} is the enthalpy (heat) of fusion. Combining the formulas gives you the following:

$$Q = mC_p\Delta T + m\Delta H_{fus}$$
$$= (25.00\ \text{g})(0.500\ \text{cal/g°C})(0.0 - (-10.0))°C + (25.00\ \text{g})(80.00\ \text{cal/g})$$
$$= 125\ \text{cal} + 2,000\ \text{cal}$$
$$= 2,125\ \text{cal} \approx 2,130\ \text{cal}$$

---

746.   **-5.48×10^4 J**

You first need to remove enough energy to lower the temperature to the freezing point of water, 0.0°C; then remove enough energy from the water for it to freeze. The first step, bringing the water to its freezing point, removes $Q = mC_p\Delta T$ energy, where Q is the energy, m is the mass, C_p is the specific heat of the material, and ΔT is the change in temperature $(T_{final} - T_{initial})$. The second step, freezing the water, removes $Q = m\Delta H_{sol}$ energy, where Q is the quantity of heat energy, m is the mass, and ΔH_{sol} is the enthalpy (heat) of solidification, which is the opposite of the enthalpy of fusion: $\Delta H_{sol} = -\Delta H_{fus}$.

This combination of formulas gives you the following:

$$Q = mC_p\Delta T + m\Delta H_{sol}$$
$$= (125 \text{ g})(4.18 \text{ J/g}^\circ\text{C})(0.0 - 25.0)^\circ\text{C} + (125 \text{ g})(-334.0 \text{ J/g})$$
$$= -5.48 \times 10^4 \text{ J}$$

---

747.   **-2.50×10^4 cal**

Solving this problem is a three-step process. The first step is to reduce the temperature of the steam to the boiling point of water, 100.0°C, using the specific heat of steam: $Q = (mC_p\Delta T)_g$, where Q is the heat energy, m is the mass, C_p is the specific heat of the material, and ΔT is the change in temperature $(T_{final} - T_{initial})$. The second step is to condense the steam to liquid water using the heat of condensation (ΔH_{cond}): $Q = m\Delta H_{cond}$, where m is the mass and $\Delta H_{cond} = -\Delta H_{vap}$, the opposite of the heat of vaporization. The final step is to remove enough heat from the hot water to cool it to the final temperature using the specific heat of water: $Q = (mC_p\Delta T)_l$.

This combination gives you the following equation, with g referring to gas and l referring to liquid:

$$Q = (mC_p\Delta T)_g + m\Delta H_{cond} + (mC_p\Delta T)_l$$
$$= [(40.0 \text{ g})(0.500 \text{ cal/g}^\circ\text{C})(100.0 - 120.0)^\circ\text{C}]_g + (40.0 \text{ g})(-540.0 \text{ cal/g})$$
$$+ [(40.0 \text{ g})(1.00 \text{ cal/g}^\circ\text{C})(25.0 - 100.0)^\circ\text{C}]_l$$
$$= -2.50 \times 10^4 \text{ cal}$$

---

748.   **5.043×10^5 J**

The first thing to do is to raise the temperature of the water to the boiling point, 100.0°C, using the specific heat of water: $Q = (mC_p\Delta T)_l$, where Q is the energy, m is the mass, C_p is the specific heat of the material, and ΔT is the change in temperature $(T_{final} - T_{initial})$. Then boil (vaporize) the water using the heat of vaporization (ΔH_{vap}): $Q = m\Delta H_{vap}$. Last, heat the steam from the boiling point of water to the final temperature using the specific heat of steam: $Q = (mC_p\Delta T)_g$.

This combination gives you the following equation, with g referring to gas and l referring to liquid:

$$Q = (mC_p\Delta T)_l + m\Delta H_{vap} + (mC_p\Delta T)_g$$
$$= [(200.0 \text{ g})(4.18 \text{ J/g}^\circ\text{C})(100.0 - 48.0)^\circ\text{C}]_l + (200.0 \text{ g})(2,257 \text{ J/g})$$
$$+ [(200.0 \text{ g})(2.09 \text{ J/g}^\circ\text{C})(122.0 - 100.0)^\circ\text{C}]_g$$
$$= 5.043 \times 10^5 \text{ J}$$

749. **2.66 × 10⁴ cal**

This problem involves a four-step process. First, raise the temperature of the ice to the melting point, 0.0°C, using the specific heat of ice: $Q = (mC_p\Delta T)_s$, where Q is the energy, m is the mass, C_p is the specific heat of the material, and ΔT is the change in temperature ($T_{final} - T_{initial}$). Then melt the ice using the heat of fusion (ΔH_{fus}): $Q = m\Delta H_{fus}$. Next, raise the temperature of the liquid to the boiling point, 100.0° C, using the specific heat of water: $Q = (mC_p\Delta T)_l$. Finally, use the heat of vaporization, ΔH_{vap}, to convert the liquid water to steam: $Q = m\Delta H_{vap}$.

This combination, with s referring to solid and l referring to liquid, gives you the following equation:

$$Q = \left(mC_p\Delta T\right)_s + m\Delta H_{fus} + \left(mC_p\Delta T\right)_l + m\Delta H_{vap}$$
$$= \left[(36.0 \text{ g})(0.500 \text{ cal/g°C})(0.0-(-40.0))°C\right]_s + (36.0 \text{ g})(80.00 \text{ cal/g})$$
$$+ \left[(36.0 \text{ g})(1.00 \text{ cal/g°C})(100.0-0.0)°C\right]_l + (36.0 \text{ g})(540.0 \text{ cal/g})$$
$$= 2.66 \times 10^4 \text{ cal}$$

750. **–2.20 × 10⁵ J**

Removing energy involves a series of four steps in this case. First, use the specific heat of steam to lower the temperature to the boiling point of water, 100.0°C: $Q = (mC_p\Delta T)_g$ where Q is the energy, m is the mass, C_p is the specific heat of the material, and ΔT is the change in temperature ($T_{final} - T_{initial}$). At the boiling (condensation) point, remove the heat of condensation from the steam: $Q = m\Delta H_{cond}$, where $\Delta H_{cond} = -\Delta H_{vap}$, the opposite of the heat of vaporization. The temperature of the liquid water formed from the condensation must be lowered to the freezing point, 0.0°C, using the specific heat of water: $Q = (mC_p\Delta T)_l$. Finally, freeze the water using the heat of solidification ($\Delta H_{sol} = -\Delta H_{fus}$): $Q = m\Delta H_{sol}$.

This combination, with g referring to gas and l referring to liquid, gives you the following equation:

$$Q = \left(mC_p\Delta T\right)_g + m\Delta H_{cond} + \left(mC_p\Delta T\right)_l + m\Delta H_{sol}$$
$$= \left[(72.0 \text{ g})(2.09 \text{ J/g°C})(100.0-120.0)°C\right]_g + (72.0 \text{ g})(-2,257\text{J/g})$$
$$+ \left[(72.0 \text{ g})(4.18 \text{ J/g°C})(0.0-100.0)°C\right]_l + (72.0 \text{ g})(-334.0 \text{ J/g})$$
$$= -2.20 \times 10^5 \text{J}$$

751. **1.444 × 10⁷ J**

You need five separate steps to solve this problem. In each step, you add a given amount of energy. The amount of energy in the first step, raising the temperature of the ice to the melting point, 0.0°C, depends on the specific heat of ice: $Q = (mC_p\Delta T)_s$, where Q is the energy, m is the mass, C_p is the specific heat of the material, and ΔT is the change in temperature ($T_{final} - T_{initial}$). The second step, melting the ice, uses the heat of fusion of ice: $Q = m\Delta H_{fus}$. Next, to raise the temperature of the water to the boiling point, 100.0°C, use the specific heat of water: $Q = (mC_p\Delta T)_l$. The fourth step, changing the water to steam, involves the heat of vaporization: $Q = m\Delta H_{vap}$. The final step, raising the temperature of the steam, uses the specific heat of steam: $Q = (mC_p\Delta T)_g$.

This combination, with g referring to gas, l referring to liquid, and s referring to solid, gives you the following equation:

$$Q = \left(mC_p\Delta T\right)_s + m\Delta H_{fus} + \left(mC_p\Delta T\right)_l + m\Delta H_{vap} + \left(mC_p\Delta T\right)_g$$

$$= \left[(4{,}536\text{ g})(2.09\text{ J/g°C})(0.0-(-78.0))\text{°C}\right]_s + (4{,}536\text{ g})(334.0\text{ J/g})$$

$$+ \left[(4{,}536\text{ g})(4.18\text{ J/g°C})(100.0-0.0)\text{°C}\right]_l + (4{,}536\text{ g})(2{,}257\text{ J/g})$$

$$+ \left[(4{,}536\text{ g})(2.09\text{ J/g°C})(105.0-100.0)\text{°C}\right]_g$$

$$= 1.444 \times 10^7\text{ J}$$

752.   −385 cal

This problem involves five steps. In each step, you remove a given amount of energy. The amount of energy in the first step, cooling the steam to the condensation point, 100.0°C, depends on the specific heat of steam: $Q = (mC_p\Delta T)_g$ where Q is the energy, m is the mass, C_p is the specific heat of the material, and ΔT is the change in temperature ($T_{final} - T_{initial}$). The second step, condensing the steam to water, utilizes the heat of condensation of steam ($\Delta H_{cond} = -\Delta H_{vap}$): $m\Delta H_{cond}$. Next, use the specific heat of water to lower the water temperature to the freezing point, 0.0°C: $Q = (mC_p\Delta T)_l$. The fourth step, freezing the water, involves the heat of solidification ($\Delta H_{sol} = -\Delta H_{fus}$): $m\Delta H_{sol}$. The final step, lowering the temperature of the ice, uses the specific heat of ice: $Q = (mC_p\Delta T)_s$.

This combination, with g referring to gas, l referring to liquid, and s referring to solid, gives you the following equation:

$$Q = \left(mC_p\Delta T\right)_g + m\Delta H_{cond} + \left(mC_p\Delta T\right)_l + m\Delta H_{sol} + \left(mC_p\Delta T\right)_s$$

$$= \left[(0.500\text{ g})(0.500\text{ cal/g°C})(100.0-150.0)\text{°C}\right]_g + (0.500\text{ g})(-540.0\text{ cal/g})$$

$$+ \left[(0.500\text{ g})(1.00\text{ cal/g°C})(0.0-100.0)\text{°C}\right]_l + (0.500\text{ g})(-80.00\text{ cal/g})$$

$$+ \left[(0.500\text{ g})(0.500\text{ cal/g°C})(100.0-150.0)\text{°C}\right]_s$$

$$= -385\text{ cal}$$

753.   60.0 g

The basic equation relating energy to mass, specific heat, and temperature is

$$Q = mC_p\Delta T$$

where Q is the energy, m is the mass, C_p is the specific heat, and ΔT is the change in temperature. You want to know the mass, so rearrange the equation:

$$m = \frac{Q}{C_P\Delta T}$$

Entering the appropriate values and doing the math gives you

$$m = \frac{172\text{ cal}}{(0.573\text{ cal/g°C})(5.00\text{°C})} = 60.0\text{ g}$$

The sample has a mass of 60.0 g.

754. **4.16 J/g°C**

The basic equation relating energy to mass, specific heat, and temperature is

$$Q = mC_p\Delta T$$

where Q is the energy, m is the mass, C_p is the specific heat, and ΔT is the change in temperature. You want to know the specific heat, so solve the equation for C_p:

$$C_p = \frac{Q}{m\Delta T}$$

Now just enter the numbers and do the math:

$$C_p = \frac{197 \text{ J}}{(22.0 \text{ g})(2.15°\text{C})} = 4.16 \text{ J/g°C}$$

The specific heat of the substance is 4.16 J/g°C.

755. **18.5°C**

The basic equation relating energy to mass, specific heat, and temperature is

$$Q = mC_p\Delta T$$

where Q is the energy, m is the mass, C_p is the specific heat, and ΔT is the change in temperature. You want to find the change in temperature, so solve the equation for ΔT:

$$\Delta T = \frac{Q}{mC_p}$$

Entering the appropriate values gives you

$$\Delta T = \frac{153 \text{ cal}}{(38.1 \text{ g})(0.217 \text{ cal/g°C})} = 18.5°\text{C}$$

The temperature increased by 18.5°C.

756. **46.0°C**

The basic equation relating energy to mass, specific heat, and temperature is

$$Q = mC_p\Delta T$$
$$Q = mC_p(T_f - T_i)$$

where Q is the energy, m is the mass, C_p is the specific heat, and ΔT is the change in temperature (final temperature minus initial temperature). You want to know the final temperature, so solve the equation for T_f:

$$T_f = \frac{Q}{mC_p} + T_i$$

Enter the numbers and complete the calculations:

$$T_f = \frac{1,148 \text{ J}}{(17.35 \text{ g})(2.17 \text{ J/g°C})} + 15.5°\text{C} = 46.0°\text{C}$$

The final temperature is 46.0°C.

757.   **32.0°C**

The basic equation relating energy to mass, specific heat, and temperature is

$$Q = mC_p\Delta T$$
$$Q = mC_p\left(T_f - T_i\right)$$

where Q is the energy, m is the mass, C_p is the specific heat, and ΔT is the change in temperature (final temperature minus initial temperature). You want to know the initial temperature. Solving the equation for T_i gives you

$$T_i = -\frac{Q}{mC_p} + T_f$$

Enter the numbers and do the math:

$$T_i = -\frac{12.35 \text{ cal}}{(19.75 \text{ g})(0.125 \text{ cal/g°C})} + 37.0°C = 32.0°C$$

The sample's initial temperature was 32.0°C.

758.   **9.10 cal**

The basic equation relating energy to mass, specific heat, and temperature is

$$Q = mC_p\Delta T$$

where Q is the energy, m is the mass, C_p is the specific heat, and ΔT is the change in temperature. You want to find the heat (energy), so simply enter the numbers and solve:

$$Q = (3.75 \text{ g})(0.986 \text{ cal/g°C})(2.46°C) = 9.10 \text{ cal}$$

The answer is positive, so the sample absorbed 9.10 calories.

759.   **–0.400 J**

The basic equation relating energy to mass, specific heat, and temperature is

$$Q = mC_p\Delta T$$

where Q is the energy, m is the mass, C_p is the specific heat, and ΔT is the change in temperature. You want to find the heat (energy), so simply enter the numbers and solve:

$$Q = (0.326 \text{ g})(0.896 \text{ J/g°C})(-1.37°C) = -0.400 \text{ J}$$

The sample lost 0.400 J of energy.

Note: Watch the value of ΔT. It's negative because the temperature decreased, which results in a negative heat change.

760.   **1.60 g**

The basic equation relating energy to mass, specific heat, and temperature is

$$\left(mC_p\Delta T\right)_1 = -\left(mC_p\Delta T\right)_2$$

where m is the mass, C_p is the specific heat, and ΔT is the change in temperature for Samples 1 and 2. The negative sign appears on the right side of the equation because Sample 2 lost heat to Sample 1.

You want to find the mass of Sample 2, so solve the equation for m_2:

$$m_2 = \frac{(mC_p\Delta T)_1}{-(C_p\Delta T)_2}$$

Now enter the appropriate values and do the math:

$$m_2 = \frac{\left[(1.35\ \text{g})(2.18\ \text{cal/g°C})(13.9°C)\right]_1}{-\left[(1.36\ \text{cal/g°C})(-18.8°C)\right]_2} = 1.60\ \text{g}$$

Sample 2 has a mass of 1.60 g.

761. 2.60 J/g°C

The basic equation relating energy to mass, specific heat, and temperature is

$$(mC_p\Delta T)_1 = -(mC_p\Delta T)_2$$

where m is the mass, C_p is the specific heat, and ΔT is the change in temperature for Samples 1 and 2. The negative sign appears on the right side of the equation because Sample 2 lost heat to Sample 1.

Rearrange the equation to solve for C_{p1}, the specific heat of Sample 1:

$$C_{p1} = \frac{-(mC_p\Delta T)_2}{(m\Delta T)_1}$$

Entering the appropriate values and doing the math gives you

$$C_{p1} = \frac{-\left[(2.70\ \text{g})(2.15\ \text{J/g°C})(-12.8°C)\right]_2}{\left[(1.42\ \text{g})(20.1°C)\right]_1} = 2.60\ \text{J/g°C}$$

The specific heat of Sample 1 is 2.60 J/g°C.

762. 53.1°C

The basic equation relating energy to mass, specific heat, and temperature is

$$-(mC_p\Delta T)_1 = (mC_p\Delta T)_2$$

where m is the mass, C_p is the specific heat, and ΔT is the change in temperature for Samples 1 and 2. The negative sign on the left side of the equation indicates that Sample 1 lost heat to Sample 2.

You want to find the final temperature of Sample 2, so rewrite ΔT_2 as $T_f - T_i$, the final temperature minus the initial temperature:

$$-(mC_p\Delta T)_1 = \left[mC_p(T_f - T_i)\right]_2$$

Solve the equation for T_{f2}. This equation rearranges to

$$T_{f2} = \frac{-(mC_p\Delta T)_1}{(mC_p)_2} + T_{i2}$$

Entering the appropriate values and doing the calculations gives you

$$T_{f2} = \frac{-\left[(5.13 \text{ g})(0.581 \text{ cal/g°C})(-15.0°C)\right]_1}{\left[(4.19 \text{ g})(0.381 \text{ cal/g°C})\right]_2} + 25.1°C = 53.1°C$$

The final temperature of Sample 2 is 53.1°C.

---

763.   2 kJ

This reaction deals with the conversion of graphite (pencil lead) to diamonds. The basic equation for the heat of reaction is

$$\Delta H°_{rxn} = \sum \Delta H°_f (\text{products}) - \sum \Delta H°_f (\text{reactants})$$

Use the table to look up the standard heat of formation ($\Delta H°_f$) for diamonds and graphite and then enter them in the equation:

$$\Delta H°_{rxn} = \Delta H°_{diamond} - \Delta H°_{graphite}$$
$$= (1 \text{ mol})(2 \text{ kJ/mol}) - (1 \text{ mol})(0 \text{ kJ/mol})$$
$$= 2 \text{ kJ}$$

This reaction requires 2 kJ of heat.

---

764.   −396 kJ

This reaction shows the burning of diamonds. The basic equation for the heat of reaction is

$$\Delta H°_{rxn} = \sum \Delta H°_f (\text{products}) - \sum \Delta H°_f (\text{reactants})$$

Use the table to look up the standard heats of formation ($\Delta H°_f$) for carbon dioxide, diamonds, and oxygen. Enter the numbers in the equation and do the math:

$$\Delta H°_{rxn} = \left(\Delta H°_{carbon\ dioxide}\right) - \left(\Delta H°_{diamond} + \Delta H°_{oxygen}\right)$$
$$= \left[(1 \text{ mol})(-394 \text{ kJ/mol})\right] - \left[(1 \text{ mol})(2 \text{ kJ/mol}) + (1 \text{ mol})(0 \text{ kJ/mol})\right]$$
$$= -396 \text{ kJ}$$

The answer is negative, so this reaction releases 396 kJ of heat.

---

765.   −566 kJ

This reaction shows the combustion of toxic carbon monoxide to form carbon dioxide. The basic equation for the heat of reaction is

$$\Delta H°_{rxn} = \sum \Delta H°_f (\text{products}) - \sum \Delta H°_f (\text{reactants})$$

Use the table to look up the standard heat of formation ($\Delta H°_f$) for each substance involved — carbon dioxide, carbon monoxide, and oxygen — and enter the numbers in the equation:

$$\Delta H°_{rxn} = \left(2\Delta H°_{carbon\ dioxide}\right) - \left(2\Delta H°_{carbon\ monoxide} + \Delta H°_{oxygen}\right)$$
$$= \left[(2 \text{ mol})(-394 \text{ kJ/mol})\right] - \left[(2 \text{ mol})(-111 \text{ kJ/mol}) + (1 \text{ mol})(0 \text{ kJ/mol})\right]$$
$$= -566 \text{ kJ}$$

The answer is negative, so this reaction releases 566 kJ of heat.

766.

–904 kJ

This reaction shows the combustion of ammonia, an important step in the synthesis of nitric acid and many fertilizers. The basic equation for the heat of reaction is

$$\Delta H^{\circ}{}_{rxn} = \sum \Delta H^{\circ}{}_{f}(\text{products}) - \sum \Delta H^{\circ}{}_{f}(\text{reactants})$$

Use the table to determine the standard heats of formation $(\Delta H^{\circ}{}_{f})$ for each of the substances involved — nitrogen oxide, water vapor, ammonia, and oxygen — and enter them in the equation:

$$\Delta H^{\circ}{}_{rxn} = \left[\left(4\Delta H^{\circ}{}_{nitrogen\ oxide} \right) + 6\Delta H^{\circ}{}_{water\ vapor} \right] - \left(4\Delta H^{\circ}{}_{ammonia} + 5\Delta H^{\circ}{}_{oxygen} \right)$$

$$= \left[(4\ \text{mol})(91\ \text{kJ/mol}) + (6\ \text{mol})(-242\ \text{kJ/mol}) \right]$$

$$- \left[(4\ \text{mol})(-46\ \text{kJ/mol}) + (5\ \text{mol})(0\ \text{kJ/mol}) \right]$$

$$= -904\ \text{kJ}$$

The answer is negative, so this reaction releases 904 kJ of heat.

767.

–1,369 kJ

The basic equation for the heat of reaction is

$$\Delta H^{\circ}{}_{rxn} = \sum \Delta H^{\circ}{}_{f}(\text{products}) - \sum \Delta H^{\circ}{}_{f}(\text{reactants})$$

Use the table to determine the standard heats of formation $(\Delta H^{\circ}{}_{f})$ for each substance involved — carbon dioxide, water, ethyl alcohol, and oxygen — and enter them in the equation:

$$\Delta H^{\circ}{}_{rxn} = \left[\left(2\Delta H^{\circ}{}_{carbon\ dioxide} \right) + 3\Delta H^{\circ}{}_{water} \right] - \left(1\Delta H^{\circ}{}_{ethyl\ alcohol} + 3\Delta H^{\circ}{}_{oxygen} \right)$$

$$= \left[(2\ \text{mol})(-394\ \text{kJ/mol}) + (3\ \text{mol})(-286\ \text{kJ/mol}) \right]$$

$$- \left[(1\ \text{mol})(-277\ \text{kJ/mol}) + (3\ \text{mol})(0\ \text{kJ/mol}) \right]$$

$$= -1,369\ \text{kJ}$$

The answer is negative, so this reaction releases 1,369 kJ of heat.

768.

–2,807 kJ

This reaction is the basic reaction known as *respiration*. The basic equation for the heat of reaction is

$$\Delta H^{\circ}{}_{rxn} = \sum \Delta H^{\circ}{}_{f}(\text{products}) - \sum \Delta H^{\circ}{}_{f}(\text{reactants})$$

Use the table to determine the standard heat of formation $(\Delta H^{\circ}{}_{f})$ for each substance involved — carbon dioxide, water, glucose, and oxygen — and enter them in the equation:

$$\Delta H^{\circ}{}_{rxn} = \left[\left(6\Delta H^{\circ}{}_{carbon\ dioxide} \right) + 6\Delta H^{\circ}{}_{water} \right] - \left(1\Delta H^{\circ}{}_{glucose} + 6\Delta H^{\circ}{}_{oxygen} \right)$$

$$= \left[(6\ \text{mol})(-394\ \text{kJ/mol}) + (6\ \text{mol})(-286\ \text{kJ/mol}) \right]$$

$$- \left[(1\ \text{mol})(-1,273\ \text{kJ/mol}) + (6\ \text{mol})(0\ \text{kJ/mol}) \right]$$

$$= -2,807\ \text{kJ}$$

The answer is negative, so this reaction releases 2,807 kJ of heat.

769. **53 kJ/mol**

The basic equation for the heat of reaction is

$$\Delta H^\circ{}_{rxn} = \sum \Delta H^\circ{}_f (\text{products}) - \sum \Delta H^\circ{}_f (\text{reactants})$$

Use the table to determine the standard heat of formation ($\Delta H^\circ{}_f$) for each substance where the value is known — nitrogen oxide and chlorine — and enter these values and the given heat of reaction ($\Delta H^\circ{}_{rxn}$) in the equation. This gives you the following:

$$\Delta H^\circ{}_{rxn} = \left[\left(2\Delta H^\circ{}_{nitrosyl\ chloride} \right) \right] - \left(2\Delta H^\circ{}_{nitrogen\ oxide} + 1\,\Delta H^\circ{}_{chlorine} \right)$$

$$-76 \text{ kJ} = \left[(2 \text{ mol}) \left(\Delta H^\circ{}_{nitrosyl\ chloride} \right) \right] - \left[(2 \text{ mol})(91 \text{ kJ/mol}) + (1 \text{ mol})(0 \text{ kJ/mol}) \right]$$

Now solve for $\Delta H^\circ{}_{nitrosyl\ chloride}$:

$$-76 \text{ kJ} + \left[(2)(91) \right] \text{ kJ} = (2 \text{ mol}) \left(\Delta H^\circ{}_{nitrosyl\ chloride} \right)$$

$$\Delta H^\circ{}_{nitrosyl\ chloride} = 53 \text{ kJ/mol}$$

The standard heat of formation for nitrosyl chloride is 53 kJ/mol.

770. **–105 kJ/mol**

The basic equation for the heat of reaction is

$$\Delta H^\circ{}_{rxn} = \sum \Delta H^\circ{}_f (\text{products}) - \sum \Delta H^\circ{}_f (\text{reactants})$$

Use the table to determine the standard heat of formation ($\Delta H^\circ{}_f$) for each substance where the value is known — carbon dioxide, water vapor, and oxygen — and enter these values and the given heat of reaction ($\Delta H^\circ{}_{rxn}$) in the equation:

$$\Delta H^\circ{}_{rxn} = \left[\left(3\Delta H^\circ{}_{carbon\ dioxide} \right) + 4\Delta H^\circ{}_{water\ vapor} \right] - \left[1\Delta H^\circ{}_{propane} + 5\Delta H^\circ{}_{oxygen} \right]$$

$$-2{,}045 \text{ kJ} = \left[(3 \text{ mol})(-394 \text{ kJ/mol}) + (4 \text{ mol})(-242 \text{ kJ/mol}) \right]$$
$$- \left[(1 \text{ mol}) \left(\Delta H^\circ{}_{propane} \right) + (5 \text{ mol})(0 \text{ kJ/mol}) \right]$$

Then solve for $\Delta H^\circ{}_{propane}$:

$$-2{,}045 \text{ kJ} - \left[3(-394) + 4(-242) \right] \text{ kJ} = -(1 \text{ mol}) \left(\Delta H^\circ{}_{propane} \right)$$

$$\Delta H^\circ{}_{propane} = -105 \text{ kJ/mol}$$

The standard heat of formation for propane is –105 kJ/mol.

771. **73 kJ/mol**

The basic equation for the heat of reaction is

$$\Delta H^\circ{}_{rxn} = \sum \Delta H^\circ{}_f (\text{products}) - \sum \Delta H^\circ{}_f (\text{reactants})$$

Now use the table to determine the standard heat of formation ($\Delta H^\circ{}_f$) for each substance where the value is known — boron oxide, water, and oxygen — and enter these values and the given heat of reaction (ΔH_{rxn}) in the equation:

$$\Delta H^\circ_{rxn} = \left[\left(5\Delta H^\circ_{boron\ oxide} \right) + 9\Delta H^\circ_{water} \right] - \left(2\Delta H^\circ_{pentaborane\text{-}9} + 12\Delta H^\circ_{oxygen} \right)$$

$$-9{,}090.\ kJ = \left[\left(5\ \cancel{mol} \right)\left(-1{,}274\ kJ/\cancel{mol} \right) + \left(9\ \cancel{mol} \right)\left(-286\ kJ/\cancel{mol} \right) \right]$$

$$- \left[\left(2\ mol \right)\left(\Delta H^\circ_{pentaborane\text{-}9} \right) + \left(12\ \cancel{mol} \right)\left(0\ kJ/\cancel{mol} \right) \right]$$

Solve for $\Delta H_{pentaborane\text{-}9}$:

$$-9{,}090.\ kJ - \left[5(-1{,}274) + 9(-286) \right]\ kJ = -\left(2\ mol \right)\left(\Delta H^\circ_{pentaborane\text{-}9} \right)$$

$$\Delta H^\circ_{pentaborane\text{-}9} = 73\ kJ/mol$$

The standard heat of formation for pentaborane-9 is 73 kJ/mol.

772. 170 kJ

Simply add the two thermochemical equations and cancel substances appearing on both sides to get the desired equation. Adding the individual heats of reaction gives you the overall heat of reaction:

$2Cu(s) + \cancel{Cl_2(g)} \rightarrow 2CuCl(s)$	$\Delta H = -36\ kJ$
$CuCl_2(s) \rightarrow \cancel{Cu(s)} + \cancel{Cl_2(g)}$	$\Delta H = 206\ kJ$
$CuCl_2(s) + Cu(s) \rightarrow 2CuCl(s)$	$\Delta H = 170\ kJ$

The heat of reaction is 170 kJ.

773. –512 kJ

Adding the individual heats of reaction gives you the overall heat of reaction.

Double the first thermochemical equation, including its energy, before adding the two equations together, because the fluorine needs to have a coefficient of 2 to match the coefficient in the desired reaction. Cancel any substance that appears in equal amounts on both sides of the reaction arrows.

Here are the calculations, beginning with the doubled first equation:

$2\cancel{H_2(g)} + 2F_2(g) \rightarrow 4HF(g)$	$\Delta H = 2(-542\ kJ)$
$2H_2O(l) \rightarrow 2\cancel{H_2(g)} + O_2(g)$	$\Delta H = 572\ kJ$
$2H_2O(l) + 2F_2(g) \rightarrow O_2(g) + 4HF(g)$	$\Delta H = -512\ kJ$

The heat of reaction is –512 kJ.

774. –566 kJ

Adding the individual heats of reaction gives you the overall heat of reaction.

Double the first thermochemical equation, including its energy, before adding the two equations together, because the CO_2 needs to have a coefficient of 2 to match the coefficient in the desired reaction. Cancel any substance that appears in equal amounts on both sides of the reaction arrows.

Here are the calculations, beginning with the doubled first equation:

$$2C(gr) + 2O_2(g) \rightarrow 2CO_2(g) \qquad\qquad \Delta H = 2(-394 \text{ kJ})$$

$$\underline{2CO(g) \rightarrow 2C(gr) + O_2(g) \qquad\qquad \Delta H = 222 \text{ kJ}}$$

$$2CO(g) + O_2(g) \rightarrow 2CO_2(g) \qquad\qquad \Delta H = -566 \text{ kJ}$$

The heat of reaction is –566 kJ.

---

775. –1,561 kJ

To get the desired equation, simply add the three thermochemical equations and cancel substances that appear in equal amounts on both sides of the reaction arrow. Adding the individual heats of reaction gives you the overall heat of reaction:

$$C_2H_6(g) \rightarrow C_2H_4(g) + H_2(g) \qquad\qquad \Delta H = 136 \text{ kJ}$$

$$H_2(g) + \frac{1}{2}O_2(g) \rightarrow H_2O(l) \qquad\qquad \Delta H = -286 \text{ kJ}$$

$$\underline{C_2H_4(g) + 3O_2(g) \rightarrow 2CO_2(g) + 2H_2O(l) \qquad\qquad \Delta H = -1{,}411 \text{ kJ}}$$

$$C_2H_6(g) + \frac{7}{2}O_2(g) \rightarrow 2CO_2(g) + 3H_2O(l) \qquad\qquad \Delta H = -1{,}561 \text{ kJ}$$

The heat of reaction is –1,561 kJ.

---

776. –925 kJ

Adding the individual heats of reaction gives you the overall heat of reaction.

The desired equation uses hydrogen as a reactant, so reverse the last thermochemical equation to get the hydrogen on the left side, which means reversing the sign on the energy. You also need to multiply the last two equations and their energies by 1/2 to reduce their coefficients to agree with the ones in the final equation.

Here's what you get when you apply these changes and add the equations, canceling any substances that appear in equal amounts on both sides of the reaction arrows:

$$MgO(s) + H_2O(l) \rightarrow Mg(OH)_2(s) \qquad\qquad \Delta H = -37 \text{ kJ}$$

$$Mg(s) + \frac{1}{2}O_2(g) \rightarrow MgO(s) \qquad\qquad \Delta H = \frac{1}{2}(-1{,}204 \text{ kJ})$$

$$\underline{H_2(g) + \frac{1}{2}O_2(g) \rightarrow H_2O(l) \qquad\qquad \Delta H = \frac{1}{2}(-572 \text{ kJ})}$$

$$Mg(s) + H_2(g) + O_2(g) \rightarrow Mg(OH)_2(s) \qquad\qquad \Delta H = -925 \text{ kJ}$$

The heat of reaction is –925 kJ.

---

777. –906 kJ

Adding the individual heats of reaction gives you the overall heat of reaction.

The desired reaction uses ammonia (NH_3) as a reactant, so reverse the first equation to get the ammonia on the left side, which means reversing the sign on the energy; you also need to double the first equation along with its energy, because the desired equation has $4NH_3$, not $2NH_3$. Triple the second equation (to convert $2H_2O$ to $6H_2O$) and double the third equation (to convert N_2 to $2N_2$), along with doubling and tripling their energies.

Here's what you get when you apply these changes and add the equations, canceling any substances that appear in equal amounts on both sides of the reaction arrows:

$$4NH_3(g) \rightarrow 2N_2(g) + 6H_2(g) \qquad \Delta H = 2(92 \text{ kJ})$$
$$6H_2(g) + 3O_2(g) \rightarrow 6H_2O(g) \qquad \Delta H = 3(-484 \text{ kJ})$$
$$2N_2(g) + 2O_2(g) \rightarrow 4NO(g) \qquad \Delta H = 2(181 \text{ kJ})$$
$$\overline{4NH_3(g) + 5O_2(g) \rightarrow 4NO(g) + 6H_2O(g)} \qquad \Delta H = -906 \text{ kJ}$$

The heat of reaction is –906 kJ.

---

778. **256 kJ**

Adding the individual heats of reaction gives you the overall heat of reaction.

To get the sum of the three thermochemical equations to match the desired equation, you need to reverse both the second and third equations (to get the reactants and products on the same side as they're on in the desired equation); this means reversing the signs on their energies. You also need to multiply the first and third equations (including their associated energies) by 1/2 to get the coefficients on HCN and NH_3 to match those in the desired equation.

Here's what you get when you apply these changes and add the equations, canceling any substances that appear in equal amounts on both sides of the reaction arrows:

$$\tfrac{1}{2}H_2(g) + C(gr) + \tfrac{1}{2}N_2(g) \rightarrow HCN(g) \qquad \Delta H = \tfrac{1}{2}(271 \text{ kJ})$$
$$CH_4(g) \rightarrow C(gr) + 2H_2(g) \qquad \Delta H = 75 \text{ kJ}$$
$$NH_3(g) \rightarrow \tfrac{1}{2}N_2(g) + \tfrac{3\,2}{2}H_2(g) \qquad \Delta H = \tfrac{1}{2}(92 \text{ kJ})$$
$$\overline{CH_4(g) + NH_3(g) \rightarrow HCN(g) + 3H_2(g)} \qquad \Delta H = 256 \text{ kJ}$$

The heat of reaction is 256 kJ.

---

779. **–205 kJ**

Adding the individual heats of reaction gives you the overall heat of reaction.

To get the sum of the three thermochemical equations to match the desired equation, you need to double the first equation and its energy. You also need to reverse the third equation, which means reversing the sign on the energy. No changes are needed for the second equation.

Here's what you get when you apply these changes and add the equations, canceling any substances that appear in equal amounts on both sides of the reaction arrows:

$$2H_2(g) + 2Cl_2(g) \rightarrow 4HCl(g) \qquad\qquad \Delta H = 2(-92\ kJ)$$

$$C(gr) + 2Cl_2(g) \rightarrow CCl_4(g) \qquad\qquad \Delta H = -96\ kJ$$

$$CH_4(g) \rightarrow C(gr) + 2H_2(g) \qquad\qquad \Delta H = 75\ kJ$$

$$CH_4(g) + 4Cl_2(g) \rightarrow CCl_4(g) + 4HCl(g) \qquad\qquad \Delta H = -205\ kJ$$

The heat of reaction is –205 kJ.

780.   **–128 kJ**

Adding the individual heats of reaction gives you the overall heat of reaction.

To get the sum of the four thermochemical equations to match the desired equation, you need to reverse the second and fourth equations (to get the CO and CH_3OH on the same side as in the final equation), which also means reversing the sign on their energies. You also need to double the first equation, including its associated energy, because the desired equation has a coefficient of 2 on the H_2.

Here's what you get when you apply these changes and add the equations, canceling any substances that appear in equal amounts on both sides of the reaction arrows:

$$2H_2(g) + O_2(g) \rightarrow 2H_2O(l) \qquad\qquad \Delta H = 2(-286\ kJ)$$

$$CO(g) \rightarrow C(gr) + \tfrac{1}{2}O_2(g) \qquad\qquad \Delta H = 111\ kJ$$

$$C(gr) + O_2(g) \rightarrow CO_2(g) \qquad\qquad \Delta H = -394\ kJ$$

$$CO_2(g) + 2H_2O(l) \rightarrow CH_3OH(l) + \tfrac{3}{2}O_2(g) \qquad\qquad \Delta H = 727\ kJ$$

$$CO(g) + 2H_2(g) \rightarrow CH_3OH(l) \qquad\qquad \Delta H = -128\ kJ$$

The heat of reaction is –128 kJ.

781.   **–49 kJ**

Adding the individual heats of reaction gives you the overall heat of reaction.

To get the sum of the three thermochemical equations to match the desired equation, you need to reverse the first and third equations, which also means reversing the signs on the energy. You also need to multiply the second equation and its associated energy by 1/2 and to multiply the first equation and its associated energy by 3/2.

Here's what you get when you apply these changes and add the equations, canceling any substances that appear in equal amounts on both sides of the reaction arrows:

$$3NO_2(g) \rightarrow 3NO(g) + \tfrac{3}{2}O_2(g) \qquad\qquad \Delta H = \tfrac{3}{2}(173\ kJ)$$

$$N_2(g) + \tfrac{5}{2}O_2(g) + H_2O(l) \rightarrow 2HNO_3(aq) \qquad\qquad \Delta H = \tfrac{1}{2}(-255\ kJ)$$

$$2NO(g) \rightarrow N_2(g) + O_2(g) \qquad\qquad \Delta H = (-181\ kJ)$$

$$3NO_2(g) + H_2O(l) \rightarrow 2HNO_3(aq) + NO(g) \qquad\qquad \Delta H = -49\ kJ$$

The heat of the reaction is –49 kJ.

782. **1,900 torr**

To convert from atmospheres to torr, multiply the given pressure by 760 torr/1 atm:

$$\frac{2.5 \text{ atm}}{1} \times \frac{760 \text{ torr}}{1 \text{ atm}} = 1,900 \text{ torr}$$

783. **76 kPa**

To convert from atmospheres to kilopascals, multiply the given pressure by 101 kPa/1 atm:

$$\frac{0.75 \text{ atm}}{1} \times \frac{101 \text{ kPa}}{1 \text{ atm}} = 76 \text{ kPa}$$

784. **0.695 atm**

To find how many atmospheres are in a certain number of millimeters of mercury, multiply the given pressure by 1 atm/760 mm Hg:

$$\frac{528 \text{ mm Hg}}{1} \times \frac{1 \text{ atm}}{760 \text{ mm Hg}} = 0.695 \text{ atm}$$

785. **2.63 atm**

To find how many atmospheres are in a certain number of pounds per square inch, multiply the given pressure by 1 atm/14.7 psi:

$$\frac{38.7 \text{ psi}}{1} \times \frac{1 \text{ atm}}{14.7 \text{ psi}} = 2.63 \text{ atm}$$

786. **3,800 mm Hg**

To convert from atmospheres to millimeters of mercury, multiply the given pressure by 760 mm Hg/1 atm:

$$\frac{5.00 \text{ atm}}{1} \times \frac{760 \text{ mm Hg}}{1 \text{ atm}} = 3,800 \text{ mm Hg}$$

787. **1,050 mm Hg**

To convert torr to millimeters of mercury, you can convert torr to atmospheres (760 torr = 1 atm) and then convert atmospheres to millimeters of mercury (1 atm = 760 mm Hg):

$$\frac{1,050 \text{ torr}}{1} \times \frac{1 \text{ atm}}{760 \text{ torr}} \times \frac{760 \text{ mm Hg}}{1 \text{ atm}} = 1,050 \text{ mm Hg}$$

Or you can remember that 1 torr = 1 mm Hg:

$$\frac{1,050 \text{ torr}}{1} \times \frac{1 \text{ mm Hg}}{1 \text{ torr}} = 1,050 \text{ mm Hg}$$

788.   **23.4 torr**

To find how many torr are in a certain number of kilopascals, convert kilopascals to atmospheres (by multiplying by 1 atm/101 kPa) and then convert atmospheres to torr (by multiplying by 760 torr/1 atm):

$$\frac{3.11 \, \text{kPa}}{1} \times \frac{1 \, \text{atm}}{101 \, \text{kPa}} \times \frac{760 \, \text{torr}}{1 \, \text{atm}} = 23.4 \, \text{torr}$$

789.   **344 kPa**

To find how many kilopascals are in a certain number of pounds per square inch, convert pounds per square inch to atmospheres (by multiplying by 1 atm/14.7 psi) and then convert atmospheres to kilopascals (by multiplying by 101 kPa/1 atm):

$$\frac{50.0 \, \text{psi}}{1} \times \frac{1 \, \text{atm}}{14.7 \, \text{psi}} \times \frac{101 \, \text{kPa}}{1 \, \text{atm}} = 344 \, \text{kPa}$$

790.   **1.5 L**

When you're given a combination of initial and final pressures and volumes at a constant temperature, the calculations will involve Boyle's law. Draw a little chart to keep track of the quantities and the units:

$P_1 = 3.0$ atm	$P_2 = 10.0$ atm
$V_1 = 5.01$ L	$V_2 = ?$ L

Make sure the units match, substitute the values into the equation from Boyle's law, and solve for V_2, giving your answer two significant figures:

$$P_1 V_1 = P_2 V_2$$
$$(3.0 \, \text{atm})(5.0 \, \text{L}) = (10.0 \, \text{atm}) V_2$$
$$\frac{(3.0 \, \text{atm})(5.0 \, \text{L})}{10.0 \, \text{atm}} = V_2$$
$$1.5 \, \text{L} = V_2$$

Now make sure your answer is reasonable. The pressure is increasing, so the volume should decrease. The new volume is less than the original volume, so this answer makes sense.

791.   **1,070 torr**

When you're given a combination of initial and final pressures and volumes at a constant temperature, the calculations will involve Boyle's law. Draw a little chart to keep track of the quantities and the units:

$P_1 = 765$ torr	$P_2 = ?$ torr
$V_1 = 17.5$ L	$V_2 = 12.5$ L

Make sure the units match, substitute the values into the equation from Boyle's law, and solve for P_2, giving your answer three significant figures:

$$P_1 V_1 = P_2 V_2$$

$$(765 \text{ torr})(17.5 \text{ L}) = P_2 (12.5 \text{ L})$$

$$\frac{(765 \text{ torr})(17.5 \text{ L})}{12.5 \text{ L}} = P_2$$

$$1{,}070 \text{ torr} = P_2$$

Now make sure your answer is reasonable. The volume is decreasing, so the pressure should increase. The new pressure is greater than the original pressure, so this answer makes sense.

792.   **130 L**

When you're given a combination of initial and final pressures and volumes at a constant temperature, the calculations will involve Boyle's law. Draw a little chart to keep track of the quantities and the units:

$P_1 = 0.75$ atm	$P_2 = 0.25$ atm
$V_1 = 44.8$ L	$V_2 = ?$ L

Make sure the units match, substitute the values into the equation from Boyle's law, and solve for V_2, giving your answer two significant figures:

$$P_1 V_1 = P_2 V_2$$

$$(0.75 \text{ atm})(44.8 \text{ L}) = (0.25 \text{ atm})V_2$$

$$\frac{(0.75 \text{ atm})(44.8 \text{ L})}{0.25 \text{ atm}} = V_2$$

$$130 \text{ L} = V_2$$

Last, make sure your answer is reasonable. The pressure is decreasing, so the volume should increase. The new volume is larger than the original volume, so this answer makes sense.

793.   **1.11 atm**

When you're given a combination of initial and final pressures and volumes at a constant temperature, the calculations will involve Boyle's law. Draw a little chart to keep track of the quantities and the units:

$P_1 = 1.75$ atm	$P_2 = ?$ atm
$V_1 = 547$ mL	$V_2 = 861$ mL

Make sure the units match, substitute the values into the equation from Boyle's law, and solve for P_2, giving your answer three significant figures:

$$P_1V_1 = P_2V_2$$

$$(1.75 \text{ atm})(547 \text{ mL}) = P_2(861 \text{ mL})$$

$$\frac{(1.75 \text{ atm})(547 \text{ mL})}{861 \text{ mL}} = P_2$$

$$1.11 \text{ atm} = P_2$$

Now make sure your answer is reasonable. The volume is increasing, so the pressure should decrease. The new pressure is lower than the original pressure, so this answer makes sense.

794. **99.9 mL**

When you're given a combination of initial and final pressures and volumes at a constant temperature, the calculations will involve Boyle's law. Draw a little chart to keep track of the quantities and the units:

$P_1 = 95.0$ kPa	$P_2 = 211$ kPa
$V_1 = ?$ mL	$V_2 = 45.0$ mL

Make sure the units match, substitute the values into the equation from Boyle's law, and solve for V_1, giving your answer three significant figures:

$$P_1V_1 = P_2V_2$$

$$(95.0 \text{ kPa})V_1 = (211 \text{ kPa})(45.0 \text{ mL})$$

$$V_1 = \frac{(211 \text{ kPa})(45.0 \text{ mL})}{95.0 \text{ kPa}}$$

$$V_1 = 99.9 \text{ mL}$$

Now make sure your answer is reasonable. The pressure increased, so the volume should decrease. The original volume is larger than the final volume, so this answer makes sense.

795. **555 torr**

When you're given a combination of initial and final pressures and volumes at a constant temperature, the calculations will involve Boyle's law. Draw a little chart to keep track of the quantities and the units:

$P_1 = ?$ torr	$P_2 = 5.10$ atm
$V_1 = 2{,}645$ mL	$V_2 = 379$ mL

Next, make sure the units match. The answer needs to be in torr, so you can change the atmospheres to torr now:

$$P_2 = \frac{5.10 \text{ atm}}{1} \times \frac{760 \text{ torr}}{1 \text{ atm}} = 3{,}876 \text{ torr}$$

Then substitute the values into the equation from Boyle's law and solve for P_1, giving your answer three significant figures:

$$P_1V_1 = P_2V_2$$
$$P_1(2{,}645 \text{ mL}) = (3{,}876 \text{ torr})(379 \text{ mL})$$
$$P_1 = \frac{(3{,}876 \text{ torr})(379 \text{ mL})}{2{,}645 \text{ mL}}$$
$$P_1 = 555 \text{ torr}$$

Your other option is to use this equation to find P_1 in atmospheres and convert to torr at the end.

Last, make sure your answer is reasonable. The volume decreased, so the pressure should increase. The original pressure, 555 torr, is less than the final pressure, 3,876 torr, so this answer makes sense.

---

796. **1,580 mL**

When you're given a combination of initial and final pressures and volumes at a constant temperature, the calculations will involve Boyle's law. Draw a little chart to keep track of the quantities and the units:

$P_1 = 1{,}020$ torr	$P_2 = 7{,}660$ torr
$V_1 = ?$ mL	$V_2 = 0.210$ L

Next, make sure the units match. The answer needs to be in milliliters, so you can change the liters to milliliters now:

$$V_2 = \frac{0.21 \text{ L}}{1} \times \frac{1{,}000 \text{ mL}}{1 \text{ L}} = 210. \text{ mL}$$

Then substitute the values into the equation from Boyle's law and solve for V_1, giving your answer three significant figures:

$$P_1V_1 = P_2V_2$$
$$(1{,}020 \text{ torr})V_1 = (7{,}660 \text{ torr})(210. \text{ mL})$$
$$V_1 = \frac{(7{,}660 \text{ torr})(210. \text{ mL})}{1{,}020 \text{ torr}}$$
$$V_1 = 1{,}580 \text{ mL}$$

Your other option is to use this equation to find V_1 in liters and convert to milliliters at the end.

Last, make sure your answer makes sense. The pressure is increasing, so the volume should decrease. The original volume is 1,580 mL, or 1.58 L, which is larger than the final volume, 0.210 L, so this answer makes sense.

---

797. **0.568 atm**

When you're given a combination of initial and final pressures and volumes at a constant temperature, the calculations will involve Boyle's law. Draw a little chart to keep track of the quantities and the units:

$P_1 = ?$ atm	$P_2 = 720.$ torr
$V_1 = 1.00$ L	$V_2 = 600.$ mL

Next, make sure the units match. In this problem, the volumes are in milliliters and liters, and the pressures are in torr and atmospheres. The volume conversion is easy when you remember that 1.00 L = 1,000 mL; the other conversion requires changing torr to atmospheres using 1 atm = 760 torr:

$$V_1 = 1.00 \text{ L} = 1,000 \text{ mL}$$

$$P_2 = \frac{720.\text{ torr}}{1} \times \frac{1 \text{ atm}}{760 \text{ torr}} = 0.947 \text{ atm}$$

Now substitute the values into the equation from Boyle's law and solve for P_1, giving your answer three significant figures:

$$P_1 V_1 = P_2 V_2$$

$$P_1(1,000 \text{ mL}) = (0.947 \text{ atm})(600. \text{ mL})$$

$$P_1 = \frac{(0.947 \text{ atm})(600. \text{ mL})}{1,000 \text{ mL}}$$

$$P_1 = 0.568 \text{ atm}$$

Now make sure your answer is reasonable. The volume decreased, so the pressure should increase. The original pressure is smaller than the final pressure, so this answer makes sense.

798.   0.35 L

When you're given a combination of initial and final pressures and volumes at a constant temperature, the calculations will involve Boyle's law. Draw a little chart to keep track of the quantities and the units:

$P_1 = 1.50$ atm	$P_2 = 540$ torr
$V_1 = ?$ L	$V_2 = 730$ mL

Next, make sure the units match. In this problem, the pressures are in atmospheres and torr, and the volumes are in milliliters and liters. Completing the pressure conversion first will allow the units to cancel when you place the pressures into the equation from Boyle's law:

$$P_1 = \frac{1.50 \text{ atm}}{1} \times \frac{760 \text{ torr}}{1 \text{ atm}} = 1,140 \text{ torr}$$

Substitute the values into the equation and solve for V_1:

$$P_1 V_1 = P_2 V_2$$

$$(1,140 \text{ torr})V_1 = (540 \text{ torr})(730 \text{ mL})$$

$$V_1 = \frac{(540 \text{ torr})(730 \text{ mL})}{1,140 \text{ torr}}$$

$$V_1 = 350 \text{ mL}$$

Give the final answer in liters, with two significant figures:

$$V_1 = \frac{350 \, \text{mL}}{1} \times \frac{1 \, \text{L}}{1,000 \, \text{mL}} = 0.35 \, \text{L}$$

Make sure your answer is reasonable. The pressure decreased, so the volume should increase. The original volume is smaller than the final volume, so this answer makes sense.

799. 7.0×10^2 mL

When you're given a combination of initial and final pressures and volumes at a constant temperature, the calculations will involve Boyle's law. Draw a little chart to keep track of the quantities and the units:

$P_1 = 920$ mm Hg	$P_2 = 2.5$ atm
$V_1 = ?$ mL	$V_2 = 0.34$ L

Next, make sure the units match. In this problem, the pressures are in millimeters of mercury and atmospheres, and the volumes are in liters and milliliters. Do the pressure conversion first so the units will cancel when you enter them in the equation from Boyle's law:

$$P_2 = \frac{2.5 \, \text{atm}}{1} \times \frac{760 \, \text{mm Hg}}{1 \, \text{atm}} = 1,900 \, \text{mm Hg}$$

Then substitute the values into the equation and solve for V_1:

$$P_1 V_1 = P_2 V_2$$
$$(920 \, \text{mm Hg}) V_1 = (1,900 \, \text{mm Hg})(0.34 \, \text{L})$$
$$V_1 = \frac{(1,900 \, \text{mm Hg})(0.34 \, \text{L})}{920 \, \text{mm Hg}}$$
$$V_1 = 0.70 \, \text{L}$$

Change the answer from liters to milliliters, giving the answer two significant figures:

$$V_1 = \frac{0.70 \, \text{L}}{1} \times \frac{1,000 \, \text{mL}}{1 \, \text{L}} = 700 \, \text{mL} = 7.0 \times 10^2 \, \text{mL}$$

Last, make sure your answer is reasonable. The pressure increased, so the volume should decrease. The original volume is larger than the final volume, so this answer makes sense.

800. 8.4 L

When you're given a combination of initial and final volumes and temperatures at a constant pressure, the calculations will involve Charles's law. Draw a little chart to keep track of the quantities and the units:

$V_1 = 4.2$ L	$V_2 = ?$ L
$T_1 = 200.$ K	$T_2 = 400.$ K

Next, make sure the temperatures are in kelvins. They are, so substitute the values into the equation from Charles's law and solve for V_2, giving your answer two significant figures:

$$\frac{V_1}{T_1} = \frac{V_2}{T_2}$$

$$\frac{4.2 \text{ L}}{200. \text{ K}} = \frac{V_2}{400. \text{ K}}$$

$$(4.2 \text{ L})(400. \text{ K}) = (200. \text{ K})V_2$$

$$\frac{(4.2 \text{ L})(400. \cancel{\text{K}})}{200. \cancel{\text{K}}} = V_2$$

$$V_2 = 8.4 \text{ L}$$

This answer makes sense. Volume and temperature are directly proportional, so if the temperature increases, then the volume must increase, too.

---

801.   **240 mL**

When you're given a combination of initial and final volumes and temperatures at a constant pressure, the calculations will involve Charles's law. Draw a little chart to keep track of the quantities and the units:

$V_1 = ?$ mL	$V_2 = 473$ mL
$T_1 = 20.$ K	$T_2 = 40.$ K

Next, make sure the temperatures are in kelvins. They are, so substitute the values into the equation from Charles's law and solve for V_1, giving your answer two significant figures:

$$\frac{V_1}{T_1} = \frac{V_2}{T_2}$$

$$\frac{V_1}{20. \text{ K}} = \frac{473 \text{ mL}}{40. \text{ K}}$$

$$V_1(40. \text{ K}) = (20. \text{ K})(473 \text{ mL})$$

$$V_1 = \frac{(20. \cancel{\text{K}})(473 \text{ mL})}{40. \cancel{\text{K}}}$$

$$V_1 = 240 \text{ mL}$$

This answer makes sense. Volume and temperature are directly proportional, so if the original temperature was lower than the final temperature, then the original volume was smaller than the final volume, too.

---

802.   **3.62 L**

When you're given a combination of initial and final volumes and temperatures at a constant pressure, the calculations will involve Charles's law. Draw a little chart to keep track of the quantities and the units:

$V_1 = ?$ L	$V_2 = 2.50$ L
$T_1 = 323$ K	$T_2 = 223$ K

Next, make sure the temperatures are in kelvins. They are, so substitute the values into the equation from Charles's law and solve for the V_1, giving your answer three significant figures:

$$\frac{V_1}{T_1} = \frac{V_2}{T_2}$$

$$\frac{V_1}{323\ K} = \frac{2.50\ L}{223\ K}$$

$$V_1(223\ K) = (323\ K)(2.50\ L)$$

$$V_1 = \frac{(323\ K)(2.50\ L)}{223\ K}$$

$$V_1 = 3.62\ L$$

This answer makes sense. Because volume and temperature are directly proportional, an initial temperature that's higher than the final temperature indicates that the initial volume was larger than the final volume.

803.   **5.00 L**

When you're given a combination of initial and final volumes and temperatures at a constant pressure, the calculations will involve Charles's law. Draw a little chart to keep track of the quantities and the units:

$V_1 = 20.0\ L$	$V_2 = ?\ L$
$T_1 = 300.\ K$	$T_2 = 75.0\ K$

Next, make sure the temperatures are in kelvins. They are, so substitute the values into the equation from Charles's law and solve for V_2, giving your answer three significant figures:

$$\frac{V_1}{T_1} = \frac{V_2}{T_2}$$

$$\frac{20.0\ L}{300.\ K} = \frac{V_2}{75.0\ K}$$

$$(20.0\ L)(75.0\ K) = (300.\ K)V_2$$

$$\frac{(20.0\ L)(75.0\ K)}{300.\ K} = V_2$$

$$V_2 = 5.00\ L$$

This answer makes sense. The temperature is decreasing, so the volume should be decreasing, because at a constant pressure, volume and temperature are directly proportional.

804.   **43.3 mL**

When you're given a combination of initial and final volumes and temperatures at a constant pressure, the calculations will involve Charles's law. Draw a little chart to keep track of the quantities and the units:

$V_1 = 50.0$ mL	$V_2 = ?$ mL
$T_1 = 100.0°C$	$T_2 = 50.0°C$

Next, make sure the temperatures are in kelvins. To convert to kelvins, add 273.2 to the Celsius temperatures:

$$T_1 = 100.0°C + 273.2 = 373.2 \text{ K}$$

$$T_2 = 50.0°C + 273.2 = 323.2 \text{ K}$$

Substitute the values into the equation from Charles's law and solve for V_2, giving your answer three significant figures:

$$\frac{V_1}{T_1} = \frac{V_2}{T_2}$$

$$\frac{50.0 \text{ mL}}{373.2 \text{ K}} = \frac{V_2}{323.2 \text{ K}}$$

$$(50.0 \text{ mL})(323.2 \text{ K}) = (373.2 \text{ K})V_2$$

$$\frac{(50.0 \text{ mL})(323.2 \text{ K})}{373.2 \text{ K}} = V_2$$

$$V_2 = 43.3 \text{ mL}$$

Volume and pressure are directly proportional, so when the temperature decreases, the volume also decreases. The final volume of 43.3 mL is less than 50.0 mL, so the answer is reasonable.

---

805.   **114 K**

When you're given a combination of initial and final volumes and temperatures at a constant pressure, the calculations will involve Charles's law. Draw a little chart to keep track of the quantities and the units:

$V_1 = 350.$ mL	$V_2 = 0.100$ L
$T_1 = 127°C$	$T_2 = ?$ K

Next, make sure the temperature is in kelvins. To convert the Celsius temperature into kelvins, add 273:

$$T_1 = 127°C + 273 = 400. \text{ K}$$

The volumes also need to be in the same unit, so convert the liters into milliliters (or alternatively, the milliliters into liters):

$$V_2 = \frac{0.100 \text{ L}}{1} \times \frac{1,000 \text{ mL}}{1 \text{ L}} = 100. \text{ mL}$$

Next, substitute the values into the equation from Charles's law and solve for T_2:

$$\frac{V_1}{T_1} = \frac{V_2}{T_2}$$

$$\frac{350.\ \text{mL}}{400.\ \text{K}} = \frac{100.\ \text{mL}}{T_2}$$

$$(350.\ \text{mL})T_2 = (400.\ \text{K})(100.\ \text{mL})$$

$$T_2 = \frac{(400.\ \text{K})(100.\ \cancel{\text{mL}})}{350.\ \cancel{\text{mL}}}$$

$$T_2 = 114\ \text{K}$$

In this problem, the volume is decreasing, so the temperature must also decrease. The final temperature of 114 K is definitely lower than 400. K, so this answer makes sense.

806.   519 K

When you're given a combination of initial and final volumes and temperatures at a constant pressure, the calculations will involve Charles's law. Draw a little chart to keep track of the quantities and the units:

$V_1 = 981$ mL	$V_2 = 1{,}520$ mL
$T_1 = 335$ K	$T_2 = ?$ K

Next, make sure the temperature is in kelvins. It is, so substitute the values into the equation from Charles's law and solve for T_2, giving your answer three significant figures:

$$\frac{V_1}{T_1} = \frac{V_2}{T_2}$$

$$\frac{981\ \text{mL}}{335\ \text{K}} = \frac{1{,}520\ \text{mL}}{T_2}$$

$$(981\ \text{mL})T_2 = (335\ \text{K})(1{,}520\ \text{mL})$$

$$T_2 = \frac{(335\ \text{K})(1{,}520\ \cancel{\text{mL}})}{981\ \cancel{\text{mL}}}$$

$$T_2 = 519\ \text{K}$$

The volume increased in this problem, so the temperature should also increase. The final temperature of 519 K is greater than 335 K, so the answer is reasonable.

807.   1,150 K

When you're given a combination of initial and final volumes and temperatures at a constant pressure, the calculations will involve Charles's law. Draw a little chart to keep track of the quantities and the units:

$V_1 = 2.63$ L	$V_2 = 627$ mL
$T_1 = ?$ K	$T_2 = 275$ K

Next, make sure the temperature is in kelvins, as it is here. Also, both volumes need to have the same units so they'll cancel during the calculations. Converting liters to milliliters requires multiplying by 1,000 mL/1 L:

$$V_1 = \frac{2.63 \; \cancel{L}}{1} \times \frac{1{,}000 \text{ mL}}{1 \; \cancel{L}} = 2{,}630 \text{ mL}$$

Substitute the values into the equation from Charles's law and solve for T_1, giving your answer three significant figures:

$$\frac{V_1}{T_1} = \frac{V_2}{T_2}$$

$$\frac{2{,}630 \text{ mL}}{T_1} = \frac{627 \text{ mL}}{275 \text{ K}}$$

$$(2{,}630 \text{ mL})(275 \text{ K}) = T_1(627 \text{ mL})$$

$$\frac{(2{,}630 \; \cancel{mL})(275 \text{ K})}{627 \; \cancel{mL}} = T_1$$

$$T_1 = 1{,}150 \text{ K}$$

In this problem, the volume is decreasing, so the temperature must also decrease. The final temperature of 275 K is definitely smaller than 1,150 K, so this answer makes sense.

808.   **2,140 K**

When you're given a combination of initial and final volumes and temperatures at a constant pressure, the calculations will involve Charles's law. Draw a little chart to keep track of the quantities and the units:

$V_1 = 90.0 \text{ mL}$	$V_2 = 10.0 \text{ mL}$
$T_1 = ? \text{ K}$	$T_2 = -35°C$

Next, make sure the temperature is in kelvins. It isn't, so add 273 to the Celsius temperature:

$$T_2 = -35°C + 273 = 238 \text{ K}$$

Substitute the values into the equation from Charles's law and solve for T_1, giving the answer three significant figures:

$$\frac{V_1}{T_1} = \frac{V_2}{T_2}$$

$$\frac{90.0 \text{ mL}}{T_1} = \frac{10.0 \text{ mL}}{238 \text{ K}}$$

$$(90.0 \text{ mL})(238 \text{ K}) = T_1(10.0 \text{ mL})$$

$$\frac{(90.0 \; \cancel{mL})(238 \text{ K})}{10.0 \; \cancel{mL}} = T_1$$

$$T_1 = 2{,}140 \text{ K}$$

In this problem, the volume is decreasing, so the temperature must also decrease. The final temperature of 238 K is definitely smaller than 2,140 K, so this answer makes sense.

809.   **1,500 K**

When you're given a combination of initial and final volumes and temperatures at a constant pressure, the calculations will involve Charles's law. Draw a little chart to keep track of the quantities and the units:

$V_1 = 750$ mL	$V_2 = 3.0$ L
$T_1 = 95°C$	$T_2 = ?$ K

Next, make sure the temperature is in kelvins. It isn't, so add 273 to the Celsius temperature:

$$T_1 = 95°C + 273 = 368 \text{ K}$$

You also need to make sure that the volume units are the same, so change the liters to milliliters by multiplying by 1,000 mL/1 L (or alternatively, change the milliliters to liters):

$$V_2 = \frac{3.0 \, \cancel{L}}{1} \times \frac{1,000 \text{ mL}}{1 \, \cancel{L}} = 3,000 \text{ mL}$$

Substitute the values into the equation from Charles's law and solve for T_2, giving your answer two significant figures:

$$\frac{V_1}{T_1} = \frac{V_2}{T_2}$$

$$\frac{750 \text{ mL}}{368 \text{ K}} = \frac{3,000 \text{ mL}}{T_2}$$

$$(750 \text{ mL})T_2 = (368 \text{ K})(3,000 \text{ mL})$$

$$T_2 = \frac{(368 \text{ K})(3,000 \, \cancel{mL})}{750 \, \cancel{mL}}$$

$$T_2 = 1,500 \text{ K}$$

In this problem, the volume is increasing, so the temperature should also increase. The final temperature of 1,500 K is definitely larger than 368 K, so this answer makes sense.

810.   1,350 torr

When you're given a combination of initial and final pressures and temperatures at a constant volume, the calculations will involve Gay-Lussac's law. Draw a little chart to keep track of the quantities and the units:

$P_1 = 900.$ torr	$P_2 = ?$ torr
$T_1 = 300.$ K	$T_2 = 450.$ K

Next, make sure the temperatures are in kelvins. They are, so substitute the values into the equation from Gay-Lussac's law and solve for P_2, giving your answer three significant figures:

$$\frac{P_1}{T_1} = \frac{P_2}{T_2}$$

$$\frac{900. \text{ torr}}{300. \text{ K}} = \frac{P_2}{450. \text{ K}}$$

$$(900. \text{ torr})(450. \text{ K}) = (300. \text{ K})P_2$$

$$\frac{(900. \text{ torr})(450. \, \cancel{K})}{300. \, \cancel{K}} = P_2$$

$$P_2 = 1,350 \text{ torr}$$

In this problem, the temperature is increasing, so the pressure should increase. The final pressure of 1,350 torr is greater than 900. torr, so this answer makes sense.

811. 1.5 atm

When you're given a combination of initial and final pressures and temperatures at a constant volume, the calculations will involve Gay-Lussac's law. Draw a little chart to keep track of the quantities and the units:

$P_1 = 2.5$ atm	$P_2 = ?$ atm
$T_1 = 500.$ K	$T_2 = 300.$ K

Next, make sure the temperatures are in kelvins. They are, so substitute the values into the equation from Gay-Lussac's law and solve for P_2, giving your answer two significant figures:

$$\frac{P_1}{T_1} = \frac{P_2}{T_2}$$

$$\frac{2.5 \text{ atm}}{500. \text{ K}} = \frac{P_2}{300. \text{ K}}$$

$$(2.5 \text{ atm})(300. \text{ K}) = (500. \text{ K})P_2$$

$$\frac{(2.5 \text{ atm})(300. \text{K})}{500. \text{K}} = P_2$$

$$P_2 = 1.5 \text{ atm}$$

In this problem, the temperature is decreasing, so the pressure should decrease. The final pressure of 1.5 atm is less than 2.5 atm, so this answer makes sense.

812. 1,280 torr

When you're given a combination of initial and final pressures and temperatures at a constant volume, the calculations will involve Gay-Lussac's law. Draw a little chart to keep track of the quantities and the units:

$P_1 = ?$ torr	$P_2 = 675$ torr
$T_1 = 425$ K	$T_2 = 225$ K

Next, make sure the temperatures are in kelvins. They are, so substitute the values into the equation from Gay-Lussac's law and solve for P_1, giving your answer three significant figures:

$$\frac{P_1}{T_1} = \frac{P_2}{T_2}$$

$$\frac{P_1}{425 \text{ K}} = \frac{675 \text{ torr}}{225 \text{ K}}$$

$$P_1(225 \text{ K}) = (425 \text{ K})(675 \text{ torr})$$

$$P_1 = \frac{(425 \text{ K})(675 \text{ torr})}{225 \text{ K}}$$

$$P_1 = 1,280 \text{ torr}$$

In this question, the temperature is decreasing, so the pressure should decrease. The initial pressure of 1,280 torr is greater than 675 torr, so the answer makes sense.

813. **5.05 atm**

When you're given a combination of initial and final pressures and temperatures at a constant volume, the calculations will involve Gay-Lussac's law. Draw a little chart to keep track of the quantities and the units:

P_1 = ? atm	P_2 = 8.10 atm
T_1 = 315 K	T_2 = 505 K

Next, make sure the temperatures are in kelvins. They are, so substitute the values into the equation from Gay-Lussac's law and solve for P_1, giving your answer three significant figures:

$$\frac{P_1}{T_1} = \frac{P_2}{T_2}$$

$$\frac{P_1}{315 \text{ K}} = \frac{8.10 \text{ atm}}{505 \text{ K}}$$

$$P_1(505 \text{ K}) = (315 \text{ K})(8.10 \text{ atm})$$

$$P_1 = \frac{(315 \text{ K})(8.10 \text{ atm})}{505 \text{ K}}$$

$$P_1 = 5.05 \text{ atm}$$

The temperature in this question is increasing, so the pressure should increase. The answer checks, because the initial pressure of 5.05 atm is less than the final pressure of 8.10 atm.

814. **420 kPa**

When you're given a combination of initial and final pressures and temperatures at a constant volume, the calculations will involve Gay-Lussac's law. Draw a little chart to keep track of the quantities and the units:

P_1 = 470 kPa	P_2 = ? kPa
T_1 = 310 K	T_2 = 280 K

Next, make sure the temperatures are in kelvins. They are, so substitute the values into the equation from Gay-Lussac's law and solve for P_2, giving your answer two significant figures:

$$\frac{P_1}{T_1} = \frac{P_2}{T_2}$$

$$\frac{470 \text{ kPa}}{310 \text{ K}} = \frac{P_2}{280 \text{ K}}$$

$$(470 \text{ kPa})(280 \text{ K}) = (310 \text{ K})P_2$$

$$\frac{(470 \text{ kPa})(280 \text{ K})}{310 \text{ K}} = P_2$$

$$P_2 = 420 \text{ kPa}$$

To check this answer, compare the initial temperatures. The temperature is decreasing, so the pressure should decrease. The final pressure of 420 kPa is less than 470 kPa, so this answer makes sense.

815. 1,760 K

When you're given a combination of initial and final pressures and temperatures at a constant volume, the calculations will involve Gay-Lussac's law. Draw a little chart to keep track of the quantities and the units:

P_1 = 780. mm Hg	P_2 = 1,280 mm Hg
T_1 = 801°C	T_2 = ? K

Next, make sure the temperature is in kelvins. It isn't, so add 273 to the Celsius temperature:

$$T_1 = 801°C + 273 = 1,074 \text{ K}$$

Substitute the values into the equation from Gay-Lussac's law and solve for T_2, giving your answer three significant figures:

$$\frac{P_1}{T_1} = \frac{P_2}{T_2}$$

$$\frac{780. \text{ mm Hg}}{1,074 \text{K}} = \frac{1,280 \text{ mm Hg}}{T_2}$$

$$(780. \text{ mm Hg})T_2 = (1,074 \text{ K})(1,280 \text{ mm Hg})$$

$$T_2 = \frac{(1,074 \text{ K})(1,280 \text{ mm Hg})}{780. \text{ mm Hg}}$$

$$T_2 = 1,760 \text{ K}$$

The pressure is increasing, so the temperature should increase. The final temperature of 1,760 K is approximately 1,490°C, which is greater than 801°C, so this answer makes sense.

816. 660.°C

When you're given a combination of initial and final pressures and temperatures at a constant volume, the calculations will involve Gay-Lussac's law. Draw a little chart to keep track of the quantities and the units:

P_1 = 1.00 atm	P_2 = 2.50 atm
T_1 = 100.°C	T_2 = ? °C

Next, make sure the temperature is in kelvins. It isn't, so add 273 to the Celsius temperature:

$$T_1 = 100.°C + 273 = 373 \text{ K}$$

Substitute the values into the equation from Gay-Lussac's law and solve for T_2, giving your answer three significant figures:

$$\frac{P_1}{T_1} = \frac{P_2}{T_2}$$

$$\frac{1.00 \text{ atm}}{373 \text{ K}} = \frac{2.50 \text{ atm}}{T_2}$$

$$(1.00 \text{ atm})T_2 = (373 \text{ K})(2.50 \text{ atm})$$

$$T_2 = \frac{(373 \text{ K})(2.50 \text{ atm})}{1.00 \text{ atm}}$$

$$T_2 = 933 \text{ K}$$

The final temperature needs to be in degrees Celsius, so subtract 273 from the Kelvin temperature:

$$T_2 = 933 \text{ K} - 273 = 660.^\circ\text{C}$$

To check the answer, note that the pressure increased, so the temperature should increase as well. The final temperature of 660.°C is greater than 100.°C, so this answer makes sense.

817. **235 K**

When you're given a combination of initial and final pressures and temperatures at a constant volume, the calculations will involve Gay-Lussac's law. Draw a little chart to keep track of the quantities and the units:

P_1 = 428 torr	P_2 = 1.25 atm
T_1 = ? K	T_2 = 521 K

The pressures aren't in the same unit, so convert one unit to the other. Here's the conversion from torr to atmospheres:

$$P_1 = \frac{428 \text{ torr}}{1} \times \frac{1 \text{ atm}}{760 \text{ torr}} = 0.563 \text{ atm}$$

The temperature is already in kelvins, so substitute the values into the equation from Gay-Lussac's law and solve for T_1, giving your answer three significant figures:

$$\frac{P_1}{T_1} = \frac{P_2}{T_2}$$

$$\frac{0.563 \text{ atm}}{T_1} = \frac{1.25 \text{ atm}}{521 \text{ K}}$$

$$(0.563 \text{ atm})(521 \text{ K}) = T_1(1.25 \text{ atm})$$

$$\frac{(0.563 \text{ atm})(521 \text{ K})}{1.25 \text{ atm}} = T_1$$

$$235 \text{ K} = T_1$$

The pressure is increasing (going from 0.563 atm to 1.25 atm), so the temperature should increase as well. The temperature goes from 235 K to 521 K, so this answer is reasonable.

818. −76°C

When you're given a combination of initial and final pressures and temperatures at a constant volume, the calculations will involve Gay-Lussac's law. Draw a little chart to keep track of the quantities and the units:

$P_1 = 44.1$ psi	$P_2 = 2{,}028$ torr
$T_1 = ?$ °C	$T_2 = 175$ K

The pressures aren't in the same unit, so convert one unit to the other. Here's the conversion from pounds per square inch to torr:

$$P_1 = \frac{44.1 \text{ psi}}{1} \times \frac{1 \text{ atm}}{14.7 \text{ psi}} \times \frac{760 \text{ torr}}{1 \text{ atm}} = 2{,}280 \text{ torr}$$

The given temperature is already in kelvins, so substitute the values into the equation from Gay-Lussac's law and solve for T_1, giving your answer three significant figures:

$$\frac{P_1}{T_1} = \frac{P_2}{T_2}$$

$$\frac{2{,}280 \text{ torr}}{T_1} = \frac{2{,}028 \text{ torr}}{175 \text{ K}}$$

$$(2{,}280 \text{ torr})(175 \text{ K}) = T_1(2{,}028 \text{ torr})$$

$$\frac{(2{,}280 \text{ torr})(175 \text{ K})}{2{,}028 \text{ torr}} = T_1$$

$$197 \text{ K} = T_1$$

The answer needs to be in degrees Celsius, so take the temperature in kelvins and subtract 273:

$$T_1 = 197 \text{ K} - 273 = -76°\text{C}$$

In this problem, the pressure decreases from 2,280 torr to 2,028 torr, so the temperature should also decrease. You see the temperature decreasing from 197 K to 175 K, so this answer is reasonable.

819. 391°C

When you're given a combination of initial and final pressures and temperatures at a constant volume, the calculations will involve Gay-Lussac's law. Draw a little chart to keep track of the quantities and the units:

$P_1 = 0.612$ atm	$P_2 = 34.9$ in. Hg
$T_1 = 75$°C	$T_2 = ?$ °C

First, convert the initial temperature from degrees Celsius to kelvins by adding 273:

$$T_1 = 75°\text{C} + 273 = 348 \text{ K}$$

The pressures aren't in the same unit, so convert one unit to the other. Here's the conversion from inches of mercury to atmospheres (1 atm = 29.9 in. Hg):

$$P_1 = \frac{0.612 \text{ atm}}{1} \times \frac{29.9 \text{ in. Hg}}{1 \text{ atm}} = 18.3 \text{ in. Hg}$$

Substitute the values into the equation from Gay-Lussac's law and solve for T_2, giving your answer three significant figures:

$$\frac{P_1}{T_1} = \frac{P_2}{T_2}$$

$$\frac{18.3 \text{ in. Hg}}{348 \text{ K}} = \frac{34.9 \text{ in. Hg}}{T_2}$$

$$(18.3 \text{ in. Hg})T_2 = (348 \text{ K})(34.9 \text{ in. Hg})$$

$$T_2 = \frac{(348 \text{ K})(34.9 \text{ in. Hg})}{18.3 \text{ in. Hg}}$$

$$T_2 = 664 \text{ K}$$

The final temperature needs to be in degrees Celsius, so subtract 273 from the Kelvin temperature:

$$T_2 = 664 \text{ K} - 273 = 391°C$$

To check the answer, note that the pressure increases, so the temperature should increase as well. The final temperature of 391°C is greater than the initial temperature of 75°C, so the final answer makes sense.

820. 1.47 atm

When you're given a combination of initial and final pressures, volumes, and temperatures for a gas sample, the calculations will involve the combined gas law. The units match and the temperatures are in kelvins, so substitute the given values into the equation and solve for P_2, giving your answer three significant figures:

$$\frac{P_1 V_1}{T_1} = \frac{P_2 V_2}{T_2}$$

$$\frac{(2.00 \text{ atm})(75.0 \text{ mL})}{223 \text{ K}} = \frac{P_2(125 \text{ mL})}{273 \text{ K}}$$

$$(2.00 \text{ atm})(75.0 \text{ mL})(273 \text{ K}) = (223 \text{ K})(P_2)(125 \text{ mL})$$

$$\frac{(2.00 \text{ atm})(75.0 \text{ mL})(273 \text{ K})}{(223 \text{ K})(125 \text{ mL})} = P_2$$

$$1.47 \text{ atm} = P_2$$

The final pressure is 1.47 atm.

821. 7.7 L

When you're given a combination of initial and final pressures, volumes, and temperatures for a gas sample, the calculations will involve the combined gas law. The units match and the temperatures are in kelvins, so substitute the given values into the equation and solve for V_2, giving your answer two significant figures:

$$\frac{P_1V_1}{T_1} = \frac{P_2V_2}{T_2}$$

$$\frac{(970 \text{ torr})(4.1 \text{ L})}{273 \text{ K}} = \frac{(760 \text{ torr})V_2}{400. \text{ K}}$$

$$(970 \text{ torr})(4.1 \text{ L})(400. \text{ K}) = (273 \text{ K})(760 \text{ torr})V_2$$

$$\frac{(970 \cancel{\text{torr}})(4.1 \text{ L})(400. \cancel{\text{K}})}{(273 \cancel{\text{K}})(760 \cancel{\text{torr}})} = V_2$$

$$7.7 \text{ L} = V_2$$

The final volume is 7.7 L.

---

822. **1.7 atm**

When you're given a combination of initial and final pressures, volumes, and temperatures for a gas sample, the calculations will involve the combined gas law. The units match and the temperatures are in kelvins, so substitute the given values into the equation and solve for P_1, giving your answer two significant figures:

$$\frac{P_1V_1}{T_1} = \frac{P_2V_2}{T_2}$$

$$\frac{P_1(327 \text{ mL})}{323 \text{ K}} = \frac{(5.1 \text{ atm})(188 \text{ mL})}{544 \text{ K}}$$

$$P_1(327 \text{ mL})(544 \text{ K}) = (323 \text{ K})(5.1 \text{ atm})(188 \text{ mL})$$

$$P_1 = \frac{(323 \text{ K})(5.1 \text{ atm})(188 \cancel{\text{mL}})}{(327 \cancel{\text{mL}})(544 \cancel{\text{K}})}$$

$$P_1 = 1.7 \text{ atm}$$

The initial pressure is 1.7 atm.

---

823. **320 mL**

When you're given a combination of initial and final pressures, volumes, and temperatures for a gas sample, the calculations will involve the combined gas law. The units match and the temperatures are in kelvins, so substitute the given values into the equation and solve for V_1, giving your answer two significant figures:

$$\frac{P_1V_1}{T_1} = \frac{P_2V_2}{T_2}$$

$$\frac{(280 \text{ mm Hg})V_1}{520 \text{ K}} = \frac{(760 \text{ mm Hg})(50. \text{ mL})}{220 \text{ K}}$$

$$(280 \text{ mm Hg})(V_1)(220 \text{ K}) = (520 \text{ K})(760 \text{ mm Hg})(50. \text{ mL})$$

$$V_1 = \frac{(520 \cancel{\text{K}})(760 \cancel{\text{mm Hg}})(50. \text{ mL})}{(280 \cancel{\text{mm Hg}})(220 \cancel{\text{K}})}$$

$$V_1 = 320 \text{ mL}$$

The initial volume is 320 mL.

824. **1.68 atm**

When you're given a combination of initial and final pressures, volumes, and temperatures for a gas sample, the calculations will involve the combined gas law. Draw a little chart to keep track of the quantities and the units:

P_1 = 4.23 atm	P_2 = ? atm
V_1 = 1,870 mL	V_2 = 6.01 L
T_1 = 293 K	T_2 = 373 K

The volumes aren't in the same units, so convert the milliliters to liters or the liters to milliliters. Here's the conversion for milliliters to liters:

$$V_1 = \frac{1{,}870 \text{ mL}}{1} \times \frac{1 \text{ L}}{1{,}000 \text{ mL}} = 1.87 \text{ L}$$

The temperatures are already in kelvins, so substitute the given values into the equation for the combined gas law and solve for P_2, giving your answer three significant figures:

$$\frac{P_1 V_1}{T_1} = \frac{P_2 V_2}{T_2}$$

$$\frac{(4.23 \text{ atm}) \times (1.87 \text{ L})}{293 \text{ K}} = \frac{P_2 (6.01 \text{ L})}{373 \text{ K}}$$

$$(4.23 \text{ atm})(1.87 \text{ L})(373 \text{ K}) = (293 \text{ K})(P_2)(6.01 \text{ L})$$

$$\frac{(4.23 \text{ atm})(1.87 \text{ L})(373 \text{ K})}{(293 \text{ K})(6.01 \text{ L})} = P_2$$

$$1.68 \text{ atm} = P_2$$

The final pressure is 1.68 atm.

825. **2.1 L**

Remember that standard temperature is 273 K and standard pressure is 1 atm (***Note:*** The pressure is an exact number, so it doesn't affect the number of significant figures in the answer). With this info, you have a combination of initial and final pressures, volumes, and temperatures for a gas sample, so the calculations will involve the combined gas law. Draw a little chart to keep track of the quantities and the units:

P_1 = 450 torr	P_2 = 1 atm
V_1 = 10.0 L	V_2 = ? L
T_1 = 773 K	T_2 = 273 K

The pressures aren't in the same units, so convert atmospheres to torr: P_2 = 1 atm = 760 torr. The temperatures are already in kelvins, so substitute the values into the equation for the combined gas law and solve for V_2, giving your answer two significant figures:

$$\frac{P_1V_1}{T_1} = \frac{P_2V_2}{T_2}$$

$$\frac{(450 \text{ torr})(10.0 \text{ L})}{773 \text{ K}} = \frac{(760 \text{ torr})V_2}{273 \text{ K}}$$

$$(450 \text{ torr})(10.0 \text{ L})(273 \text{ K}) = (773 \text{ K})(760 \text{ torr})V_2$$

$$\frac{(450 \text{ torr})(10.0 \text{ L})(273 \text{ K})}{(773 \text{ K})(760 \text{ torr})} = V_2$$

$$2.1 \text{ L} = V_2$$

The final volume will be 2.1 L.

---

826. **16 atm**

When you're given a combination of initial and final pressures, volumes, and temperatures for a gas sample, the calculations will involve the combined gas law. Draw a little chart to keep track of the quantities and the units:

$P_1 = ?$ atm	$P_2 = 5.5$ atm
$V_1 = 0.75$ L	$V_2 = 2,647$ mL
$T_1 = 25°C$	$T_2 = 100.°C$

Next, make sure the temperatures are in kelvins. They aren't, so add 273 to the Celsius temperatures:

$T_1 = 25°C + 273 = 298$ K

$T_2 = 100.°C + 273 = 373$ K

Also, you need to convert one of the volume units so that the units match. Here's the conversion from liters to milliliters:

$$V_1 = \frac{0.75 \text{ L}}{1} \times \frac{1{,}000 \text{ mL}}{1 \text{ L}} = 750 \text{ mL}$$

Substitute the values into the equation for the combined gas law and solve for P_1, giving your answer two significant figures:

$$\frac{P_1V_1}{T_1} = \frac{P_2V_2}{T_2}$$

$$\frac{P_1(750 \text{ mL})}{298 \text{ K}} = \frac{(5.5 \text{ atm})(2{,}647 \text{ mL})}{373 \text{ K}}$$

$$P_1(750 \text{ mL})(373 \text{ K}) = (298 \text{ K})(5.5 \text{ atm})(2{,}647 \text{ mL})$$

$$P_1 = \frac{(298 \text{ K})(5.5 \text{ atm})(2{,}647 \text{ mL})}{(750 \text{ mL})(373 \text{ K})}$$

$$P_1 = 16 \text{ atm}$$

The original pressure was 16 atm.

827. 1,220 mL

When you're given a combination of initial and final pressures, volumes, and temperatures for a gas sample, the calculations will involve the combined gas law. Draw a little chart to keep track of the quantities and the units:

$P_1 = 3.00$ atm	$P_2 = 3,862$ mm Hg
$V_1 = ?$ mL	$V_2 = 594$ mL
$T_1 = 300.°C$	$T_2 = 200.°C$

Next, make sure the temperatures are in kelvins. They aren't, so add 273 to the Celsius temperatures given:

$$T_1 = 300.°C + 273 = 573 \text{ K}$$
$$T_2 = 200.°C + 273 = 473 \text{ K}$$

Also, the pressures need to be in the same unit. Here's the conversion from atmospheres to millimeters of mercury:

$$P_1 = \frac{3.00 \text{ atm}}{1} \times \frac{760 \text{ mm Hg}}{1 \text{ atm}} = 2,280 \text{ mm Hg}$$

Substitute the values into the equation for the combined gas law and solve for V_1, giving your answer three significant figures:

$$\frac{P_1 V_1}{T_1} = \frac{P_2 V_2}{T_2}$$

$$\frac{(2,280 \text{ mm Hg})V_1}{573 \text{ K}} = \frac{(3,862 \text{ mm Hg})(594 \text{ mL})}{473 \text{ K}}$$

$$(2,280 \text{ mm Hg})(V_1)(473 \text{ K}) = (573 \text{ K})(3,862 \text{ mm Hg})(594 \text{ mL})$$

$$V_1 = \frac{(573 \text{ K})(3,862 \text{ mm Hg})(594 \text{ mL})}{(2,280 \text{ mm Hg})(473 \text{ K})}$$

$$V_1 = 1,220 \text{ mL}$$

The original volume was 1,220 mL.

828. 330 K

When you're given a combination of initial and final pressures, volumes, and temperatures for a gas sample, the calculations will involve the combined gas law. Draw a little chart to keep track of the quantities and the units:

$P_1 = 970$ torr	$P_2 = 0.78$ atm
$V_1 = 220.$ mL	$V_2 = 2.4$ L
$T_1 = 50.$ K	$T_2 = ?$ K

You need to convert the pressure and volume units so they'll cancel later. Here are the conversions from atmospheres to torr and from liters to milliliters:

$$P_2 = \frac{0.78 \text{ atm}}{1} \times \frac{760 \text{ torr}}{1 \text{ atm}} = 592.8 \text{ torr}$$

$$V_2 = \frac{2.4 \text{ L}}{1} \times \frac{1,000 \text{ mL}}{1 \text{ L}} = 2,400 \text{ mL}$$

Next, substitute the given and converted values into the equation for the combined gas law and solve for T_2, giving your answer two significant figures:

$$\frac{P_1 V_1}{T_1} = \frac{P_2 V_2}{T_2}$$

$$\frac{(970 \text{ torr})(220. \text{ mL})}{50. \text{ K}} = \frac{(592.8 \text{ torr})(2,400 \text{ mL})}{T_2}$$

$$(970 \text{ torr})(220. \text{ mL}) T_2 = (50. \text{ K})(592.8 \text{ torr})(2,400 \text{ mL})$$

$$T_2 = \frac{(50. \text{ K})(592.8 \text{ torr})(2,400 \text{ mL})}{(970 \text{ torr})(220. \text{ mL})}$$

$$T_2 = 330 \text{ K}$$

The gas's final temperature is 330 K.

---

829. **518 K**

When you're given a combination of initial and final pressures, volumes, and temperatures for a gas sample, the calculations will involve the combined gas law. Draw a little chart to keep track of the quantities and the units:

$P_1 = 35.3 \text{ psi}$	$P_2 = 6.18 \text{ atm}$
$V_1 = 10.0 \text{ L}$	$V_2 = 4,290 \text{ mL}$
$T_1 = ? \text{ K}$	$T_2 = 572 \text{ K}$

You need to convert the pressure and the volume units so they'll cancel later. Here are the conversions from pounds per square inch to atmospheres and from milliliters to liters:

$$P_1 = \frac{35.3 \text{ psi}}{1} \times \frac{1 \text{ atm}}{14.7 \text{ psi}} = 2.40 \text{ atm}$$

$$V_2 = \frac{4,290 \text{ mL}}{1} \times \frac{1 \text{ L}}{1,000 \text{ mL}} = 4.29 \text{ L}$$

The temperature is already in kelvins, so substitute the values into the equation for the combined gas law and solve for T_1, giving your answer three significant figures:

$$\frac{P_1 V_1}{T_1} = \frac{P_2 V_2}{T_2}$$

$$\frac{(2.40 \text{ atm})(10.0 \text{ L})}{T_1} = \frac{(6.18 \text{ atm})(4.29 \text{ L})}{572 \text{ K}}$$

$$(2.40 \text{ atm})(10.0 \text{ L})(572 \text{ K}) = T_1 (6.18 \text{ atm})(0.429 \text{ L})$$

$$\frac{(2.40 \text{ atm})(10.0 \text{ L})(572 \text{ K})}{(6.18 \text{ atm})(4.29 \text{ L})} = T_1$$

$$518 \text{ K} = T_1$$

The initial temperature was 518 K.

830. 560. L

To use Avogadro's law, you need to know the number of moles of the gas, so use the molar mass of helium to convert from grams to moles. Then multiply by 22.4 L/1 mol to find the volume of the gas:

$$\frac{100. \text{ g He}}{1} \times \frac{1 \text{ mol He}}{4.0026 \text{ g He}} \times \frac{22.4 \text{ L He}}{1 \text{ mol He}} = 560. \text{ L He}$$

831. 12.7 L

To use Avogadro's law, you need to know the number of moles of the gas, so use the molar mass of carbon dioxide to convert from grams to moles. Then multiply by 22.4 L/1 mol to find the volume of the gas:

$$\frac{25.0 \text{ g CO}_2}{1} \times \frac{1 \text{ mol CO}_2}{(12.0107 + 2 \times 15.9994) \text{ g CO}_2} \times \frac{22.4 \text{ L CO}_2}{1 \text{ mol CO}_2} = 12.7 \text{ L CO}_2$$

832. 2.46×10^{23} molecules

Convert the volume of the nitrogen gas to moles using 22.4 L = 1 mol (at standard temperature and pressure). Then multiply by Avogadro's number to find the number of molecules:

$$\frac{9.14 \text{ L N}_2}{1} \times \frac{1 \text{ mol N}_2}{22.4 \text{ L N}_2} \times \frac{6.022 \times 10^{23} \text{ molecules N}_2}{1 \text{ mol N}_2} = 2.46 \times 10^{23} \text{ molecules N}_2$$

833. 27.14 g

First, determine the number of moles of methane gas using 22.4 L = 1 mol (at standard temperature and pressure). The multiply by the molar mass of CH_4 to find the mass of the sample:

$$\frac{37.89 \text{ L CH}_4}{1} \times \frac{1 \text{ mol CH}_4}{22.4 \text{ L CH}_4} \times \frac{(12.0107 + 4 \times 1.00794) \text{ g CH}_4}{1 \text{ mol CH}_4} = 27.14 \text{ g CH}_4$$

834. 110. L

Avogadro's law states that the volume of a gas is proportional to the number of moles present (assuming the pressure and temperature are constant). One way of using Avogadro's law is through the following relationship:

$$\frac{V_1}{n_1} = \frac{V_2}{n_2}$$

Here are the given values from the question:

$V_1 = 55.0$ L	$V_2 = ?$ L
$n_1 = 2.10$ mol	$n_2 = ?$ mol

Calculate the number of moles of NO_2 produced using the mole ratio from the balanced equation. You have 2 mol of NO_2 for every mole of N_2O_4:

$$n_2 = \frac{2.10 \ \cancel{mol \ N_2O_4}}{1} \times \frac{2 \ mol \ NO_2}{1 \ \cancel{mol \ N_2O_4}} = 4.20 \ mol \ NO_2$$

Enter your numbers in Avogadro's relationship and solve for V_2, the volume of NO_2 gas produced. Give your answer three significant figures.

$$\frac{V_1}{n_1} = \frac{V_2}{n_2}$$

$$\frac{55.0 \ L}{2.10 \ mol} = \frac{V_2}{4.20 \ mol}$$

$$(55.0 \ L)(4.20 \ mol) = (2.10 \ mol)V_2$$

$$\frac{(55.0 \ L)(4.20 \ \cancel{mol})}{2.10 \ \cancel{mol}} = V_2$$

$$110. \ L = V_2$$

The reaction produces 110. L of NO_2.

835.   **18.6 L**

Avogadro's law states that the volume of a gas is proportional to the number of moles present (assuming the pressure and temperature are constant). One way of using Avogadro's law is through the following relationship:

$$\frac{V_1}{n_1} = \frac{V_2}{n_2}$$

Here are the given values from the question:

$V_1 = 12.4 \ L$	$V_2 = ? \ L$
$n_1 = 0.296 \ mol$	$n_2 = ? \ mol$

Calculate the number of moles produced using the mole ratio from the balanced equation. You have 3 mol of O_2 for every 2 mol of O_3:

$$n_2 = \frac{0.296 \ \cancel{mol \ O_3}}{1} \times \frac{3 \ mol \ O_2}{2 \ \cancel{mol \ O_3}} = 0.444 \ mol \ O_2$$

Next, substitute the values into Avogadro's relationship and solve for V_2, the volume of O_2 gas produced. Give your answer three significant figures.

$$\frac{V_1}{n_1} = \frac{V_2}{n_2}$$

$$\frac{12.4 \ L}{0.296 \ mol} = \frac{V_2}{0.444 \ mol}$$

$$(12.4 \ L)(0.444 \ mol) = (0.296 \ mol)V_2$$

$$\frac{(12.4 \ L)(0.444 \ \cancel{mol})}{0.296 \ \cancel{mol}} = V_2$$

$$18.6 \ L = V_2$$

The reaction produces 18.6 L of O_2.

836. **13.1 atm**

After reading the question, list all the given information. You know the volume, number of moles of the gas, and temperature, and you want to find the pressure, so using the ideal gas law makes sense.

$P = ?$ atm
$V = 12.5$ L
$n = 5.00$ mol
$T = 400.$ K

Rearrange the ideal gas law to solve for P, and then enter the given values and $R = 0.0821$ L·atm/K·mol in the equation. Give your answer three significant figures:

$$PV = nRT$$

$$P = \frac{nRT}{V}$$

$$P = \frac{(5.00 \ \text{mol})\left(0.0821 \frac{L \cdot atm}{K \cdot mol}\right)(400. \ K)}{12.5 \ L}$$

$$P = 13.1 \ \text{atm}$$

The pressure of the gas sample is 13.1 atm.

837. **196 L**

After reading the question, list all the given information. You know the pressure, number of moles of the gas, and temperature, and you want to find the volume, so using the ideal gas law makes sense.

$P = 0.747$ atm
$V = ?$ L
$n = 8.91$ mol
$T = 200.$ K

Rearrange the ideal gas law to solve for V, and then enter the given values and $R = 0.0821$ L·atm/K·mol in the equation. Give your answer three significant figures:

$$PV = nRT$$

$$V = \frac{nRT}{P}$$

$$V = \frac{(8.91 \ \text{mol})\left(0.0821 \frac{L \cdot atm}{K \cdot mol}\right)(200. \ K)}{0.747 \ \text{atm}}$$

$$V = 196 \ \text{L}$$

The volume of the gas is 196 L.

---

838. **0.0613 mol**

After reading the question, list all the given information. You know the pressure, volume, and temperature, and you want to find the number of moles, so using the ideal gas law makes sense.

$P = 3.00$ atm
$V = 500.0$ mL
$n = ?$ mol
$T = 298$ K

Rearrange the ideal gas law to solve for n, the number of moles, and then enter the given values (converting milliliters to liters) and $R = 0.0821$ L·atm/K·mol in the equation. Give your answer three significant figures:

$$PV = nRT$$

$$\frac{PV}{RT} = n$$

$$\frac{(3.00 \text{ atm})(0.5000 \text{ L})}{\left(0.0821 \frac{\text{L} \cdot \text{atm}}{\text{K} \cdot \text{mol}}\right)(298 \text{ K})} = n$$

$$0.0613 \text{ mol} = n$$

The gas sample contains 0.0613 mol of helium.

---

839. **230 K**

After reading the question, list all the given information. You know the pressure, volume, and number of moles of the gas, and you want to find the temperature, so using the ideal gas law makes sense.

$P = 4.1$ atm
$V = 2.8$ L
$n = 0.60$ mol
$T = ?$ K

Rearrange the ideal gas law to solve for the temperature T, and then enter the given values and $R = 0.0821$ L·atm/K·mol in the equation. Give your answer two significant figures:

$$PV = nRT$$

$$\frac{PV}{nR} = T$$

$$\frac{(4.1 \text{ atm})(2.8 \text{ L})}{(0.60 \text{ mol})\left(0.0821 \frac{\text{L} \cdot \text{atm}}{\text{K} \cdot \text{mol}}\right)} = T$$

$$230 \text{ K} = T$$

The gas's temperature is 230 K.

840. 1,060 L

After reading the question, list all the given information. You know the pressure, number of moles of the gas, and temperature, and you want to find the volume, so using the ideal gas law makes sense.

$P = 1.50$ atm
$V = ?$ L
$n = 40.6$ mol
$T = 205°C$

Convert the temperature from degrees Celsius to kelvins:

$T = 205°C + 273 = 478$ K

Rearrange the ideal gas law to solve for V, and then enter the given and calculated values and $R = 0.0821$ L·atm/K·mol in the equation. Give your answer three significant figures:

$$PV = nRT$$

$$V = \frac{nRT}{P}$$

$$V = \frac{(40.6 \, \text{mol})\left(0.0821 \, \frac{\text{L} \cdot \text{atm}}{\text{K} \cdot \text{mol}}\right)(478 \, \text{K})}{1.50 \, \text{atm}}$$

$$V = 1,060 \text{ L}$$

The volume of the carbon dioxide gas is 1,060 L.

841. 195 K

After reading the question, list all the given information. You know the pressure, volume, and number of moles of the gas, and you want to find the temperature, so using the ideal gas law makes sense.

$P = 6.98$ psi
$V = 11.2$ L
$n = 0.333$ mol
$T = ?$ K

Convert the pressure to atmospheres by multiplying the pounds per square inch by 1 atm/14.7 psi:

$$P = \frac{6.98 \text{ psi}}{1} \times \frac{1 \text{ atm}}{14.7 \text{ psi}} = 0.475 \text{ atm}$$

Rearrange the ideal gas law to solve for T, and then enter the given and calculated values and $R = 0.0821$ L·atm/K·mol in the equation. Give your answer three significant figures:

$$PV = nRT$$

$$\frac{PV}{nR} = T$$

$$\frac{(0.475 \text{ atm})(11.2 \text{ L})}{(0.333 \text{ mol})\left(0.0821 \dfrac{\text{L} \cdot \text{atm}}{\text{K} \cdot \text{mol}}\right)} = T$$

$$195 \text{ K} = T$$

The temperature of the gas sample is 195 K.

842. **0.0944 mol**

After reading the question, list all the given information. You know the pressure, volume, and temperature, and you want to find the number of moles of the gas, so using the ideal gas law makes sense.

$P = 672$ torr
$V = 3{,}050$ mL
$n = ?$ mol
$T = 75°C$

Convert the pressure, volume, and temperature into units that will cancel with the units in the gas constant R: atmospheres, liters, and kelvins:

$$P = \frac{672 \text{ torr}}{1} \times \frac{1 \text{ atm}}{760 \text{ torr}} = 0.884 \text{ atm}$$

$$V = \frac{3{,}050 \text{ mL}}{1} \times \frac{1 \text{ L}}{1{,}000 \text{ mL}} = 3.05 \text{ L}$$

$$T = 75°C + 273 = 348 \text{ K}$$

Rearrange the ideal gas law to solve for n, the number of moles, and then enter the given and calculated values and $R = 0.0821$ L·atm/K·mol in the equation. Give your answer three significant figures:

$$PV = nRT$$

$$\frac{PV}{RT} = n$$

$$\frac{(0.884 \text{ atm})(3.05 \text{ L})}{\left(0.0821 \dfrac{\text{L} \cdot \text{atm}}{\text{K} \cdot \text{mol}}\right)(348 \text{ K})} = n$$

$$0.0944 \text{ mol} = n$$

The sample contains 0.0944 mol of oxygen gas.

843. **2.43 atm**

After reading the question, list all the given information. You know the volume, number of grams of the gas (which you can convert to moles), and temperature, and you want to find the pressure, so using the ideal gas law makes sense.

$$P = ? \text{ atm}$$

$$V = 75.0 \text{ mL}$$

$$n = ? \text{ mol}$$

$$m = 0.618 \text{ g}$$

$$T = 255 \text{ K}$$

Convert the volume from milliliters to liters (so that the unit will cancel with the volume unit in the gas constant R). Also convert the grams to moles, using twice the molar mass of chlorine because there are two chlorine atoms per molecule of chlorine gas:

$$V = \frac{75.0 \text{ mL}}{1} \times \frac{1 \text{ L}}{1,000 \text{ mL}} = 0.0750 \text{ L}$$

$$n = \frac{0.618 \text{ g Cl}_2}{1} \times \frac{1 \text{ mol Cl}_2}{\left(2 \times 35.453 \text{ g Cl}_2\right)} = 0.00872 \text{ mol Cl}_2$$

After all that, rearrange the equation for the ideal gas law to solve for P, and then enter the given and calculated values and $R = 0.0821$ L·atm/K·mol in the equation. Give your answer three significant figures:

$$PV = nRT$$

$$P = \frac{nRT}{V}$$

$$P = \frac{\left(0.00872 \text{ mol}\right)\left(0.0821 \frac{\text{L·atm}}{\text{K·mol}}\right)\left(255 \text{ K}\right)}{0.0750 \text{ L}}$$

$$P = 2.43 \text{ atm}$$

The pressure of the chlorine gas is 2.43 atm.

844.   16 g/mol

To find the molar mass given the number of grams, the pressure, the volume, and the temperature, you need to find the number of moles of the gas. Identify the given information first:

$$P = 0.90 \text{ atm}$$

$$V = 555 \text{ L}$$

$$n = ? \text{ mol}$$

$$m = 261 \text{ g}$$

$$T = 373 \text{ K}$$

Solve the ideal gas law for n, and then enter the given values and $R = 0.0821$ L·atm/ K·mol in the equation:

$$PV = nRT$$

$$\frac{PV}{RT} = n$$

$$\frac{(0.90 \text{ atm})(555 \text{ L})}{\left(0.0821 \dfrac{\text{L} \cdot \text{atm}}{\text{K} \cdot \text{mol}}\right)(373 \text{ K})} = n$$

$$16.3 \text{ mol} = n$$

To find the gas's molar mass, divide the number of grams by the number of moles. Give your answer two significant figures (because the pressure has only two significant figures):

$$MM = \frac{m}{n} = \frac{261 \text{ g}}{16.3 \text{ mol}} = 16 \text{ g/mol}$$

The molar mass is 16 g/mol.

---

845. **44.0 g/mol**

To find the molar mass given the number of grams, the pressure, the volume, and the temperature, you need to find the number of moles of the gas. Identify the given information first:

$P = 1.05$ atm
$V = 11.3$ L
$n = ?$ mol
$m = 12.15$ g
$T = 523$ K

Solve the ideal gas law for n, the number of moles, and then enter the given values and $R = 0.0821$ L·atm/K·mol in the equation:

$$PV = nRT$$

$$\frac{PV}{RT} = n$$

$$\frac{(1.05 \text{ atm})(11.3 \text{ L})}{\left(0.0821 \dfrac{\text{L} \cdot \text{atm}}{\text{K} \cdot \text{mol}}\right)(523 \text{ K})} = n$$

$$0.276 \text{ mol} = n$$

To find the gas's molar mass, divide the number of grams by the number of moles. Give your answer three significant figures:

$$MM = \frac{m}{n} = \frac{12.15 \text{ g}}{0.276 \text{ mol}} = 44.0 \text{ g/mol}$$

The molar mass of the gas is 44.0 g/mol.

---

846. **847 torr**

Given the individual partial pressures, using Dalton's law is the best way to solve this problem. Enter the partial pressures in the equation and do the math:

$$P_{total} = P_{N_2} + P_{O_2} + P_{He}$$
$$= 255 \text{ torr} + 491 \text{ torr} + 101 \text{ torr}$$
$$= 847 \text{ torr}$$

847. **3.38 atm**

Given the individual partial pressures, using Dalton's law is the best way to solve this problem. First, convert the torr to atmospheres:

$$P_{Cl_2} = \frac{567 \text{ torr}}{1} \times \frac{1 \text{ atm}}{760 \text{ torr}} = 0.746 \text{ atm}$$

$$P_{F_2} = \frac{843 \text{ torr}}{1} \times \frac{1 \text{ atm}}{760 \text{ torr}} = 1.11 \text{ atm}$$

Then enter your numbers into the formula for Dalton's law, giving your answer two decimal places:

$$P_{total} = P_{Ar} + P_{Cl_2} + P_{F_2}$$
$$= 1.52 \text{ atm} + 0.746 \text{ atm} + 1.11 \text{ atm}$$
$$= 3.38 \text{ atm}$$

848. **2.99 atm**

Given the ratio of the partial pressures (7:3:2) and the total pressure, set up an algebraic equation to determine the partial pressure of any of the gases:

$$P_{total} = P_x + P_y + P_z$$
$$P_{total} = 7x + 3x + 2x$$
$$5.12 \text{ atm} = 12x$$
$$0.427 \text{ atm} = x$$

Now substitute in the value of x and calculate the partial pressure of P_x, giving your answer three significant figures:

$$P_x = 7x$$
$$= 7(0.427 \text{ atm})$$
$$= 2.99 \text{ atm}$$

849. **1.63 atm**

Given the total pressure and some partial pressures, using Dalton's law is the best way to solve this problem. Substitute in the given values and solve for the partial pressure of O_2, giving your answer two decimal places:

$$P_{total} = P_{SO_3} + P_{SO_2} + P_{O_2}$$
$$4.42 \text{ atm} = 1.77 \text{ atm} + 1.02 \text{ atm} + P_{O_2}$$
$$1.63 \text{ atm} = P_{O_2}$$

850.   493 torr

The number of moles is proportional to the pressure of the gas (from the ideal gas law), so you can use the mole fraction (χ_n) to solve for the partial pressure of the propane. First, find the total number of moles in the sample by adding the given values:

$$n_{total} = 0.822 \text{ mol} + 0.282 \text{ mol} + 0.550 \text{ mol} = 1.654 \text{ mol}$$

Next, multiply the mole fraction of propane by the total pressure to get the partial pressure due to propane. Give your answer three significant figures:

$$P_{propane} = \chi_{propane} \times P_{total}$$

$$= \frac{n_{propane}}{n_{total}} \times P_{total}$$

$$= \frac{0.550 \text{ mol}}{1.654 \text{ mol}} \times 1{,}482 \text{ torr}$$

$$= 493 \text{ torr}$$

851.   1.31 atm

Each gas exerts a partial pressure proportional to the amount of gas present, so first determine the number of moles of N_2:

$$n_{N_2} = \frac{15.0 \text{ g } N_2}{1} \times \frac{1 \text{ mol } N_2}{(14.0067 \times 2) \text{ g } N_2}$$

$$= 0.535 \text{ mol } N_2$$

Then you can use the ideal gas law to determine the partial pressure due to the nitrogen gas. Don't forget to convert the temperature to kelvins by adding 273 to the Celsius temperature:

$$T = 25°C + 273 = 298 \text{ K}$$

Then enter the numbers in the ideal gas law and solve for the partial pressure of N_2. Give your answer three significant figures:

$$PV = n_{N_2}RT$$

$$P_{N_2} = \frac{n_{N_2}RT}{V}$$

$$P_{N_2} = \frac{(0.535 \text{ mol})\left(0.0821 \frac{L \cdot atm}{K \cdot mol}\right)(298 \text{ K})}{10.0 \text{ } L}$$

$$P_{N_2} = 1.31 \text{ atm}$$

852.   2.84 atm

Each gas exerts a partial pressure proportional to the amount of gas present, so first determine the number of moles of each gas. Divide the given masses by the molar masses of N_2 and O_2:

$$n_{N_2} = \frac{15.0 \text{ g N}_2}{1} \times \frac{1 \text{ mol N}_2}{(14.0067 \times 2) \text{ g N}_2} = 0.535 \text{ mol N}_2$$

$$n_{O_2} = \frac{20.0 \text{ g O}_2}{1} \times \frac{1 \text{ mol O}_2}{(15.9994 \times 2) \text{ g O}_2} = 0.625 \text{ mol O}_2$$

Then use the ideal gas law, $PV = nRT$, to determine the partial pressure of each gas. Don't forget to convert the temperature to kelvins by adding 273 to the Celsius temperature:

$$T = 25°C + 273 = 298 \text{ K}$$

Now substitute the values for N_2 into the ideal gas law and solve for the pressure (remember that the gas constant $R = 0.0821$ L·atm/K·mol):

$$P_{N_2}V = n_{N_2}RT$$

$$P_{N_2} = \frac{n_{N_2}RT}{V}$$

$$P_{N_2} = \frac{(0.535 \text{ mol})\left(0.0821 \frac{\text{L} \cdot \text{atm}}{\text{K} \cdot \text{mol}}\right)(298 \text{ K})}{10.0 \text{ L}}$$

$$P_{N_2} = 1.31 \text{ atm}$$

Use the ideal gas law again to find the partial pressure of O_2:

$$P_{O_2}V = n_{O_2}RT$$

$$P_{O_2} = \frac{n_{O_2}RT}{V}$$

$$P_{O_2} = \frac{(0.625 \text{ mol})\left(0.0821 \frac{\text{L} \cdot \text{atm}}{\text{K} \cdot \text{mol}}\right)(298 \text{ K})}{10.0 \text{ L}}$$

$$P_{O_2} = 1.53 \text{ atm}$$

Then use Dalton's law. Determine the total pressure by adding together the two partial pressures, giving your final answer two decimal places:

$$P_{\text{total}} = P_{N_2} + P_{O_2}$$
$$= 1.31 \text{ atm} + 1.53 \text{ atm}$$
$$= 2.84 \text{ atm}$$

---

853.   **731.8 torr and 10.7 g NH$_4$NO$_2$**

Find the partial pressure due to the nitrogen gas using Dalton's law:

$$P_{\text{total}} = P_{N_2} + P_{\text{water vapor}}$$

$$P_{\text{total}} - P_{\text{water vapor}} = P_{N_2}$$

$$757.0 \text{ torr} - 25.21 \text{ torr} = P_{N_2}$$

$$731.8 \text{ torr} = P_{N_2}$$

Next, calculate the number of moles of ammonium nitrate using the ideal gas law, $PV = nRT$ (remember that the gas constant $R = 0.0821$ L·atm/K·mol):

$$PV = nRT$$

$$\frac{PV}{RT} = n$$

$$\frac{(731.8 \text{ torr})(4.25 \, \cancel{L})}{\left(0.0821 \, \frac{\cancel{L} \cdot \text{atm}}{K \cdot \text{mol}}\right)(26°C)} = n$$

The units need to cancel, so convert the torr to atmospheres by multiplying by 1 atm/760 torr and convert the 26°C to kelvins by adding 273:

$$\frac{\left(\frac{731.8 \, \cancel{\text{torr}}}{1} \times \frac{1 \, \text{atm}}{760 \, \cancel{\text{torr}}}\right)(4.25 \, \cancel{L})}{\left(0.0821 \, \frac{\cancel{L} \cdot \cancel{\text{atm}}}{K \cdot \text{mol}}\right)(26 + 273 \, K)} = n$$

$$0.167 \text{ mol} = n$$

Last, use stoichiometry to find the mass of ammonium nitrite using the mole ratio and the molar mass of ammonium nitrite. Give your answer three significant figures:

$$m = \frac{0.167 \, \cancel{\text{mol } N_2}}{1} \times \frac{1 \, \text{mol } NH_4NO_2}{1 \, \cancel{\text{mol } N_2}}$$

$$\times \frac{(14.0067 + (4 \times 1.00794) + 14.0067 + (2 \times 15.9994)) \text{g } NH_4NO_2}{1 \, \text{mol } NH_4NO_2}$$

$$m = 10.7 \text{ g } NH_4NO_2$$

854.　**742.2 torr and 0.485 g KCl**

First, write and balance the chemical equation that represents the reaction. Here's the unbalanced equation:

$$KClO_3(g) \rightarrow KCl(g) + O_2(g)$$

Balance the oxygen by multiplying the $KClO_3$ by 2 and the O_2 by 3:

$$2KClO_3(g) \rightarrow KCl(g) + 3O_2(g)$$

And then balance the potassium and the chlorine by multiplying the KCl by 2:

$$2KClO_3(g) \rightarrow 2KCl(g) + 3O_2(g)$$

Now find the partial pressure due to the oxygen gas using Dalton's law:

$$P_{total} = P_{O_2} + P_{water \, vapor}$$

$$P_{total} - P_{water \, vapor} = P_{O_2}$$

$$762.0 \text{ torr} - 19.83 \text{ torr} = P_{O_2}$$

$$742.2 \text{ torr} = P_{O_2}$$

Next, calculate the number of moles of oxygen gas using the ideal gas law, $PV = nRT$, where R is a constant equal to 0.0821 L·atm/K·mol:

$$PV = nRT$$

$$\frac{PV}{RT} = n$$

$$\frac{(742.6 \text{ torr})(242 \text{ mL})}{\left(0.0821 \frac{\text{L} \cdot \text{atm}}{\text{K} \cdot \text{mol}}\right)(22°\text{C})} = n$$

The units need to cancel, so convert the torr to atmospheres by multiplying by 1 atm/760 torr, convert the milliliters to liters by multiplying by 1 L/1,000 mL, and convert the 22°C to kelvins by adding 273:

$$\frac{\left(\frac{742.6 \text{ torr}}{1} \times \frac{1 \text{ atm}}{760 \text{ torr}}\right)\left(\frac{242 \text{ mL}}{1} \times \frac{1 \text{ L}}{1,000 \text{ mL}}\right)}{\left(0.0821 \frac{\text{L} \cdot \text{atm}}{\text{K} \cdot \text{mol}}\right)(22 + 273 \text{ K})} = n$$

$$9.76 \times 10^{-3} \text{ mol} = n$$

Last, use stoichiometry to find the mass of potassium chloride using the mole ratio and the molar mass of potassium chloride. Give your answer three significant figures:

$$m = \frac{9.76 \times 10^{-3} \text{ mol } O_2}{1} \times \frac{2 \text{ mol KCl}}{3 \text{ mol } O_2} \times \frac{(39.0983 + 35.453)\text{g KCl}}{1 \text{ mol KCl}}$$

$$m = 0.485 \text{ g KCl}$$

855. **O_2, because it has the smaller molar mass**

According to Graham's law, the gas with the smaller molar mass effuses/diffuses faster, so the gas with the larger molar mass must diffuse at a slower rate. O_2 has a molar mass of about 2(16 g/mol) = 32 g/mol, and argon has a molar mass of about 39 g/mol, so the oxygen gas must effuse faster.

856. **CO_2, because it has the larger molar mass**

According to Graham's law, the gas with the smaller molar mass effuses/diffuses faster, so the gas with the larger molar mass must effuse at a slower rate. The molar mass of CO_2 is about 12 g/mol + 2(16 g/mol) = 44 g/mol, and the molar mass of NH_3 is about 14 g/mol + 3(1 g/mol) = 17 g/mol, so the carbon dioxide must effuse more slowly.

857. **SO_2 effuses 1.0261 times as fast as ClO_2.**

According to Graham's law, the gas with the smaller molar mass, SO_2, will effuse faster. To find the ratio of the rates of the two gases, substitute one molar mass *(MM)* of each gas into the formula:

$$\frac{\text{rate}_{SO_2}}{\text{rate}_{ClO_2}} = \sqrt{\frac{MM_{ClO_2}}{MM_{SO_2}}}$$

$$= \sqrt{\frac{35.453 + 2(15.9994)}{32.065 + 2(15.9994)}}$$

$$= \sqrt{\frac{67.452}{64.063}}$$

$$= 1.0261$$

The answer must be greater than 1, because you're comparing the rate of the faster gas to the rate of the slower one. If your answer is less than 1, you likely placed the wrong gas in the denominator, and the answer is the reciprocal to the one you obtained.

858. **122 L HCl**

At standard temperature and pressure, you can use the molar volume of a gas for stoichiometry, because Avogadro's law proportionally relates the volume to the number of moles of a gas. Using 22.4 L = 1 mol of a gas and the mole ratio from the balanced equation, you can convert from liters H_2 to liters HCl using the following map:

liters H_2 → moles H_2 → moles HCl → liters HCl

Here are the calculations:

$$V = \frac{60.9 \text{ L } H_2}{1} \times \frac{1 \text{ mol } H_2}{22.4 \text{ L } H_2} \times \frac{2 \text{ mol HCl}}{1 \text{ mol } H_2} \times \frac{22.4 \text{ L HCl}}{1 \text{ mol HCl}}$$

$$= 122 \text{ L HCl}$$

Or you can use the relationship that for every liter of H_2 consumed, 2 L of HCl is formed:

$$V = \frac{60.9 \text{ L } H_2}{1} \times \frac{2 \text{ L HCl}}{1 \text{ L } H_2} = 122 \text{ L HCl}$$

859. **1,790 g Hg**

First, write the reaction:

$$HgO \rightarrow Hg + O_2$$

Now balance it. To balance the oxygen atoms, multiply the HgO by 2:

$$2HgO \rightarrow Hg + O_2$$

Next, balance the mercury atoms by multiplying the Hg on the products side by 2:

$$2HgO \rightarrow 2Hg + O_2$$

Verify that the atoms on each side of the arrow are equal to each other.

Next, calculate the mass of mercury using molar volume (22.4 L/mol), the mole ratio (from the balanced chemical equation), and molar mass (from the atomic masses on the periodic table):

$$m = \frac{100. \text{ L } O_2}{1} \times \frac{1 \text{ mol } O_2}{22.4 \text{ L } O_2} \times \frac{2 \text{ mol Hg}}{1 \text{ mol } O_2} \times \frac{200.59 \text{ g Hg}}{1 \text{ mol Hg}}$$

$$= 1,790 \text{ g Hg}$$

860. **284 L O_2**

First, write the reaction:

$$H_2O_2 \rightarrow H_2O + O_2$$

Then balance it. To balance the oxygen atoms, multiply the water by 2; this gives you an even number of oxygen atoms on the products side:

$$H_2O_2 \rightarrow 2H_2O + O_2$$

Next, balance the hydrogen atoms by multiplying the hydrogen peroxide on the reactant side by 2:

$$2H_2O_2 \rightarrow 2H_2O + O_2$$

Verify that the atoms on each side of the arrow are equal to each other.

Next, calculate the volume of oxygen gas produced using molar volume (22.4 L/mol), the mole ratio (from the balanced chemical equation), and molar mass (from the atomic masses on the periodic table):

$$V = \frac{862 \text{ g H}_2O_2}{1} \times \frac{1 \text{ mol H}_2O_2}{(1.00794 \times 2 + 15.9994 \times 2) \text{ g H}_2O_2} \times \frac{1 \text{ mol O}_2}{2 \text{ mol H}_2O_2} \times \frac{22.4 \text{ L O}_2}{1 \text{ mol O}_2}$$

$$= 284 \text{ L O}_2$$

861. 33.3 L NH₃

First, write the reaction:

$$N_2 + H_2 \rightarrow NH_3$$

The balance it. Multiply the ammonia by 2 to balance the two atoms of nitrogen on the reactants side:

$$N_2 + H_2 \rightarrow 2NH_3$$

Multiply the hydrogen on the reactants side by 3 to balance the six atoms of hydrogen on the product side:

$$N_2 + 3H_2 \rightarrow 2NH_3$$

Verify that the atoms on each side of the arrow are equal to each other.

Next, find the limiting reactant by comparing the number of moles of each reactant:

$$n_{N_2} = \frac{25.0 \text{ L N}_2}{1} \times \frac{1 \text{ mol N}_2}{22.4 \text{ L N}_2} = 1.12 \text{ mol N}_2$$

$$n_{H_2} = \frac{50.0 \text{ L H}_2}{1} \times \frac{1 \text{ mol H}_2}{22.4 \text{ L H}_2} = 2.23 \text{ mol H}_2$$

You need three times as much H_2 as N_2 according to the mole ratio from the balanced reaction, so the H_2 is the limiting reactant. Next, solve for the volume of NH_3:

liters $H_2 \rightarrow$ moles $H_2 \rightarrow$ moles $NH_3 \rightarrow$ liters NH_3

$$V_{NH_3} = \frac{50.0 \text{ L H}_2}{1} \times \frac{1 \text{ mol H}_2}{22.4 \text{ L H}_2} \times \frac{2 \text{ mol NH}_3}{3 \text{ mol H}_2} \times \frac{22.4 \text{ L NH}_3}{1 \text{ mol NH}_3}$$

$$= 33.3 \text{ L NH}_3$$

862. the substance dissolved

In a solution, the *solute* is the substance dissolved. It's normally the component of a solution present in a lesser quantity.

863. the substance doing the dissolving

In a solution, the *solvent* is the substance doing the dissolving. It's normally the component of a solution present in a greater quantity.

864. nitrogen

Air is a solution of gases. The major component (solvent) is nitrogen, which comprises about 80% of air.

865. iron

High-carbon steel — actually, all steel — is a solution containing carbon and other elements dissolved in iron. You can consider all alloys to be solutions. For example, sterling silver is copper dissolved in silver.

866. water

Regular, unflavored vodka is mostly alcohol and water. Because less than half of the vodka (40%) is alcohol, water (the major component) must be the solvent.

867. a saturated solution

The solution above the undissolved solid is a *saturated* solution. No more solid can dissolve in it.

868. an unsaturated solution

The solution above the undissolved solid was a saturated solution. Separating this solution and then adding more solvent lowers the concentration to something less than saturated: *unsaturated*.

869. a supersaturated solution

The solution must have been supersaturated. A supersaturated solution is only partially stable; such solutions spontaneously break down, rejecting any solute in excess of what's present in the saturated solution. If the excess solute is a solid, it appears as a precipitate, a solid separating from the solution.

870. volume of the solution

Molarity (M) is the moles of solute per liter (volume) of solution. Multiplying the molarity of the solution by its volume in liters yields the moles of solute.

871. **mass of the solvent**

Molality (m) is the moles of solute per kilogram (mass) of solvent. Multiplying the molality of the solution by the solvent's mass in kilograms yields the moles of solute.

872. **1.2×10^{-2} ppt Pb^{2+}**

To give the concentration of a solution in terms of parts per thousand, express the quantity of solute over the quantity of solvent in the same units and multiply the result by 1,000. The amount of lead is in grams, so you need to determine the grams of solution; the grams of solution is the product of the volume of the solution (mL) and its density (g/mL).

$$\frac{\left(1.5\times10^{-2}\text{ g Pb}^{2+}\right)}{\left(1{,}275\text{ mL}\right)\left(\dfrac{1.00\text{ g}}{\text{mL}}\right)}(1{,}000)=1.2\times10^{-2}\text{ ppt Pb}^{2+}$$

The given concentration of the lead is expressed to two significant figures, which limits the final answer to two significant figures.

873. **2.6 ppm DDT**

To give the concentration of a solution in terms of parts per million, express the quantity of solute over the quantity of solvent in the same units and multiply the result by 1,000,000, or 10^6. The amount of DDT is in grams, so you need to determine the grams of solution; the grams of solution is the product of the volume of the solution (mL) and its density (g/mL).

$$\frac{\left(5.5\times10^{-3}\text{ g DDT}\right)}{\left(2{,}125\text{ mL}\right)\left(\dfrac{1.00\text{ g}}{\text{mL}}\right)}(10^{6})=2.6\text{ ppm DDT}$$

The given mass of the DDT has two significant figures, which limits the final answer to two significant figures.

874. **11 ppb Hg**

To express the concentration of a solution in terms of parts per billion, express the quantity of solute over the quantity of solvent in the same units and multiply the result by 1,000,000,000, or 10^9.

In this case, a convenient unit for the quantity is grams. To convert the moles of mercury to grams, multiply by the molar mass of mercury from the periodic table (200.6 g/mol). To find the quantity of solvent, multiply the volume of the solution by its density. Note that to use the density, you first need to convert the volume (liters) to milliliters. (Another option is to convert the density to g/L.)

$$\frac{\left(8.5\times10^{-8}\text{ mol Hg}\right)\left(\dfrac{200.6\text{ g Hg}}{\text{mol Hg}}\right)}{\left(1.5\text{ L}\right)\left(\dfrac{1{,}000\text{ mL}}{\text{L}}\right)\left(\dfrac{1.00\text{ g}}{\text{mL}}\right)}(10^{9})=11\text{ ppb Hg}$$

The given concentration of the mercury has two significant figures, which limits the final answer to two significant figures.

875. **0.080 M NaNO$_3$**

To determine the molarity of a solution, divide the moles of the solute by the liters of solution:

$$\frac{0.200 \text{ mol NaNO}_3}{2.5 \text{ L}} = 0.080 \text{ M NaNO}_3$$

The given volume (2.5 L) has two significant figures, which limits the answer to two significant figures.

876. **0.07246 M NaOH**

To determine the molarity of a solution, divide the moles of the solute by the liters of solution. For this problem, you need to convert the milliliters to liters.

$$\frac{0.1250 \text{ mol NaOH}}{\left(1{,}725 \text{ mL}\right)\left(\dfrac{1 \text{ L}}{1{,}000 \text{ mL}}\right)} = 0.07246 \text{ M NaOH}$$

The given quantities have four significant figures, which limits the answer to four significant figures.

877. **0.30 M NaCl**

The molarity is the moles of solute divided by the liters of solution. In this case, you need to use the molar mass (58 g/mol) to convert the grams of NaCl to moles. Begin the problem by expressing the given information as the grams of NaCl over the volume of the solution (L). From this setup, you can see that only the molar mass is needed to convert to the appropriate units, because mol/L = M.

$$\frac{\left(26 \text{ g NaCl}\right)}{1.5 \text{ L}}\left(\frac{1 \text{ mol NaCl}}{58 \text{ g NaCl}}\right) = 0.30 \text{ M NaCl}$$

The given values all have two significant figures, so the answer has two significant figures.

878. **0.58 mol LiCl**

Multiplying the molarity of a solution by its volume (in liters) gives you the moles of solute. In this type of problem, expressing the molarity of the solution in terms of its definition (mol/L) is useful:

$$\left(\frac{0.16 \text{ mol LiCl}}{L}\right)\left(3.5 \text{ L}\right) = 0.58 \text{ mol LiCl}$$

The given values all have two significant figures, so the answer has two significant figures.

879. **0.61 L**

Convert the mass (15 g) to moles using the molar mass of sulfuric acid (98 g/mol) and then multiply the result by the reciprocal of the molarity (1/M = L/mol):

$$\left(15\,g\,H_2SO_4\right)\left(\frac{1\,mol\,H_2SO_4}{98\,g\,H_2SO_4}\right)\left(\frac{1\,L}{0.25\,mol\,H_2SO_4}\right) = 0.61\,L$$

The given values all have two significant figures, so the answer has two significant figures.

880. **94.0 g MgCl$_2$**

First, determine the moles of MgCl$_2$ by multiplying the molarity of the solution by its volume (in liters). In this type of problem, expressing the molarity of the solution in terms of its definition (mol/L) is useful. Multiplying the moles of MgCl$_2$ by its molar mass yields the grams of solute.

$$\left(\frac{0.564\,mol\,MgCl_2}{L}\right)\left(1.75\,L\right)\left(\frac{95.2\,g\,MgCl_2}{mol\,MgCl_2}\right) = 94.0\,g\,MgCl_2$$

The given values all have three significant figures, so the answer has three significant figures.

881. **8.0 L**

You want the volume of the diluted solution, so rearrange the equation $M_1V_1 = M_2V_2$ to solve for V_2. Then enter the numbers and do the math:

$$V_2 = \frac{M_1V_1}{M_2} = \frac{\left(12\,M\,HCl\right)\left(1.0\,L\right)}{\left(1.5\,M\,HCl\right)} = 8.0\,L$$

The given values all have two significant figures, so the answer has two significant figures.

882. **3.8 L**

The question asks for the volume of the diluted solution, so rearrange the equation $M_1V_1 = M_2V_2$ to solve for V_2. Then enter the numbers:

$$V_2 = \frac{M_1V_1}{M_2} = \frac{\left(15\,M\,NaOH\right)\left(0.50\,L\right)}{\left(2.0\,M\,NaOH\right)} = 3.8\,L$$

The given values all have two significant figures, so the answer has two significant figures.

883. **1.4 M HNO$_3$**

You want the molarity of the diluted solution, so rearrange the equation $M_1V_1 = M_2V_2$ to solve for M_2. Then enter the numbers and do the math:

$$M_2 = \frac{M_1 V_1}{V_2} = \frac{(6.0 \text{ M HNO}_3)(1.5 \text{ L})}{(6.5 \text{ L})} = 1.4 \text{ M HNO}_3$$

The given values all have two significant figures, so the answer has two significant figures.

884.   **0.60 M H₃PO₄**

The question asks for the molarity of the diluted solution, so rearrange the equation $M_1 V_1 = M_2 V_2$ to solve for M_2. Both volumes are in milliliters, so you don't need to convert the volume units. However, you do need to remember that the final volume, V_2, is the sum of the two volumes given in the problem.

$$M_2 = \frac{M_1 V_1}{V_2} = \frac{(6.0 \text{ M H}_3\text{PO}_4)(15 \text{ mL})}{(135 + 15) \text{ mL}} = 0.60 \text{ M H}_3\text{PO}_4$$

The given values have a minimum of two significant figures, so the answer has two significant figures.

885.   **0.42 L HCl**

You want the volume of the starting solution, so rearrange the equation $M_1 V_1 = M_2 V_2$ to solve for V_1. Then enter the numbers and do the math:

$$V_1 = \frac{M_2 V_2}{M_1} = \frac{(1.0 \text{ M HCl})(5.0 \text{ L})}{(12 \text{ M HCl})} = 0.42 \text{ L}$$

The given values all have two significant figures, so the answer has two significant figures.

886.   **1.9×10^2 mL**

The question asks for the volume of the starting solution, so rearrange the equation $M_1 V_1 = M_2 V_2$ to solve for V_1. Both the given volume and the volume you want to find are in milliliters, so you don't need to convert the volume units.

$$V_1 = \frac{M_2 V_2}{M_1} = \frac{(3.0 \text{ M HNO}_3)(875 \text{ mL})}{(14 \text{ M HNO}_3)} = 1.9 \times 10^2 \text{ mL}$$

The given values have a minimum of two significant figures, so the answer has two significant figures.

887.   **6.49 L**

You want the volume of the diluted solution, so rearrange the equation $M_1 V_1 = M_2 V_2$ to find V_2:

$$V_2 = \frac{M_1 V_1}{M_2} = \frac{(1.00 \text{ M NaCl})(1.00 \text{ L})}{(0.154 \text{ M NaCl})} = 6.49 \text{ L}$$

The given values all have three significant figures, so the answer has three significant figures.

888. **20.0 L**

The question asks for the volume of the diluted solution, so rearrange the equation $M_1V_1 = M_2V_2$ to find V_2:

$$V_2 = \frac{M_1V_1}{M_2} = \frac{(1.00 \text{ M CaCl}_2)(5.00 \text{ L})}{(0.250 \text{ M CaCl}_2)} = 20.0 \text{ L}$$

The given values all have three significant figures, so the answer has three significant figures.

889. **0.20 m KBr**

Molality (m) is the moles of solute divided by the kilograms of solvent:

$$\frac{0.25 \text{ mol KBr}}{1.25 \text{ kg}} = 0.20 \text{ m KBr}$$

The given values have a minimum of two significant figures, so the answer has two significant figures.

890. **7.9×10^{-3} m Ca(NO$_3$)$_2$**

Molality (m) is the moles of solute divided by the kilograms of solvent. The mass of the solvent is in grams, so convert the grams of water to kilograms.

$$\left(\frac{0.014 \text{ mol Ca(NO}_3)_2}{1,775 \text{ g}}\right)\left(\frac{1,000 \text{ g}}{1 \text{ kg}}\right) = 7.9 \times 10^{-3} \text{ m Ca(NO}_3)_2$$

The given values have a minimum of two significant figures, so the answer has two significant figures.

891. **0.384 m HCl**

Molality (m) is the moles of solute divided by the kilograms of solvent. Use the molar mass of HCl (36.5 g/mol) to convert the grams of the solute, HCl, to moles. The grams cancel out, so you don't have to convert grams to kilograms.

$$\left(\frac{10.5 \text{ g HCl}}{0.750 \text{ kg}}\right)\left(\frac{1 \text{ mol HCl}}{36.5 \text{ g HCl}}\right) = 0.384 \text{ m HCl}$$

The given values all have three significant figures, so the answer has three significant figures.

892. **0.24 mol ZnCl$_2$**

The question asks for the moles of the solute. Molality (m) is the moles of solute divided by the kilograms of solvent; therefore, the number of moles present is the molality of the solution (0.16 m) multiplied by the kilograms of solvent (1.5 kg). Enter the numbers and do the math:

$$\left(\frac{0.16 \text{ mol } ZnCl_2}{kg}\right)\left(\frac{1.5 \text{ kg}}{1}\right) = 0.24 \text{ mol } ZnCl_2$$

The given values have two significant figures, so the answer has two significant figures.

893.   **6.9 g $CdCl_2$**

You want to find the mass of the solute. Molality is the moles of solute divided by the kilograms of solvent; therefore, the number of moles present is the molality of the solution (0.015 m) multiplied by the kilograms of the solvent (2.5 kg). Multiplying the number of moles by the molar mass gives you the mass of the solute in grams.

$$\left(\frac{0.015 \text{ mol } CdCl_2}{kg}\right)\left(\frac{2.5 \text{ kg}}{1}\right)\left(\frac{183.317 \text{ g } CdCl_2}{\text{mol } CdCl_2}\right) = 6.9 \text{ g } CdCl_2$$

The given values have a minimum of two significant figures; therefore, the answer has two significant figures.

894.   **3.94 g $C_6H_{12}O_6$**

The question asks for the mass of the solute, glucose, in grams. Molality is the moles of solute divided by the kilograms of solvent; therefore, the number of moles present is the molality of the solution (0.125 m) multiplied by the kilograms of the solvent, water. Multiplying the number of moles by the molar mass gives you the amount of the solute in grams. Note that for the units to cancel correctly, you need to convert the grams of solvent (175 g) to kilograms; this conversion appears in the second and third terms in the following equation:

$$\left(\frac{0.125 \text{ mol } C_6H_{12}O_6}{1 \text{ kg}}\right)\left(\frac{175 \text{ g}}{1}\right)\left(\frac{1 \text{ kg}}{1,000 \text{ g}}\right)\left(\frac{180.2 \text{ g } C_6H_{12}O_6}{1 \text{ mol } C_6H_{12}O_6}\right) = 3.94 \text{ g } C_6H_{12}O_6$$

The given values have a minimum of three significant figures, so the answer has three significant figures.

895.   **3**

Calcium nitrate is a strong electrolyte, so in solution, it separates into calcium ions and nitrate ions. Based on the formula $Ca(NO_3)_2$, calcium nitrate gives you one calcium ion and two nitrate ions, for a total of three ions. The formation of three ions means that the van't Hoff factor, i, is 3.

896.   **It's lower.**

Weak electrolytes produce fewer ions than similar strong electrolytes; therefore, the van't Hoff factor for weak electrolytes is always less than the van't Hoff factor for a similar strong electrolyte. Nonelectrolytes can't produce any ions in solution, so they always have a van't Hoff factor of 1.

897. **2.79°C**

The general equation for freezing point depression is $\Delta T = iK_f m$. For nonelectrolytes, the van't Hoff factor, i, is 1 because a nonelectrolyte can produce only one molecule in solution. Enter the numbers and do the math:

$$\Delta T = iK_f m$$

$$= (1)\left(\frac{1.86°C}{m}\right)(1.50\, m)$$

$$= 2.79°C$$

The given values all have three significant figures, so the answer has three significant figures. (The van't Hoff number has no effect on significant figures because it's an exact number in this case.)

A freezing point depression of 2.79°C means that the solution freezes 2.79°C below the normal freezing point of the solvent.

898. **23.4°C**

The general equation for freezing point depression is $\Delta T = iK_f m$. For nonelectrolytes, the van't Hoff factor, i, is 1.

$$\Delta T = iK_f m$$

$$= (1)\left(\frac{1.86°C}{m}\right)(12.6\, m)$$

$$= 23.4°C$$

The given values all have three significant figures, so the answer has three significant figures. (The van't Hoff number has no effect on significant figures because it's an exact number in this case.)

A freezing point depression of 23.4°C means that the solution freezes 23.4°C below the normal freezing point of the solvent.

899. **0.102°C**

The general equation for boiling point elevation is $\Delta T = iK_b m$. For nonelectrolytes, the van't Hoff factor, i, is 1. To determine the molality of the solution, m, divide the moles of solute (isopropyl alcohol) by the kilograms of solvent (water). Here are the calculations:

$$\Delta T = iK_b m$$

$$= (1)\left(\frac{0.512°C}{1\, m}\right)\left(\frac{0.200\, mol}{1.00\, kg}\right)\left(\frac{1\, m}{\left(\frac{1\, mol}{1\, kg}\right)}\right)$$

$$= 0.102°C$$

The given values all have three significant figures, so the answer has three significant figures. (The van't Hoff number has no effect on significant figures because it's an exact number in this case.)

A boiling point elevation of 0.102°C means that the solution boils 0.102°C above the normal boiling point of the solvent.

900. **0.108°C**

The general equation for boiling point elevation is $\Delta T = iK_b m$. For nonelectrolytes, the van't Hoff factor, i, is 1. To determine the molality of the solution, m, divide the moles of solute (propyl alcohol) by the kilograms of solvent (water). Here are the calculations:

$$\Delta T = iK_b m$$

$$= (1)\left(\frac{0.512°C}{1\,m}\right)\left(\frac{0.370\,\text{mol}}{1.75\,\text{kg}}\right)\left(\frac{1\,m}{\left(\frac{1\,\text{mol}}{1\,\text{kg}}\right)}\right)$$

$$= 0.108°C$$

The given values all have three significant figures, so the answer has three significant figures. (The van't Hoff number has no effect on significant figures because it's an exact number in this case.)

A boiling point elevation of $0.108°C$ means that the solution boils $0.108°C$ above the normal boiling point of the solvent.

901. **0.915°C**

The general equation for freezing point depression is $\Delta T = iK_f m$. For nonelectrolytes, the van't Hoff factor, i, is 1. To determine the molality of the solution, m, convert the grams of ethyl alcohol to moles and divide the result by the kilograms of the solvent. Here are the calculations:

$$\Delta T = iK_f m$$

$$= (1)\left(\frac{1.86°C}{m}\right)\left(\frac{17.0\,\text{g}\,C_2H_5OH}{0.750\,\text{kg}}\right)\left(\frac{1\,\text{mol}\,C_2H_5OH}{46.1\,\text{g}\,C_2H_5OH}\right)\left(\frac{1\,m}{\left(\frac{\text{mol}\,C_2H_5OH}{\text{kg}}\right)}\right)$$

$$= 0.915°C$$

The given values all have three significant figures, so the answer has three significant figures. (The van't Hoff number has no effect on significant figures because it's an exact number in this case.)

A freezing point depression of $0.915°C$ means that the solution freezes $0.915°C$ below the normal freezing point of the solvent.

902. **100.673°C**

The general equation for boiling point elevation is $\Delta T = iK_b m$. For nonelectrolytes, the van't Hoff factor, i, is 1. To determine the molality of the solution, m, convert the grams of propylene glycol to moles and divide the result by the kilograms of the solvent. Here are the calculations:

$$\Delta T = iK_b m$$

$$= (1) \left(\frac{0.512°C}{1\,m} \right) \left(\frac{125\,g}{1.25\,kg} \right) \left(\frac{1\,mol}{76.1\,g} \right) \left(\frac{1\,m}{\left(\frac{1\,mol}{1\,kg} \right)} \right)$$

$$= 0.673°C$$

The given values all have three significant figures, so the answer has three significant figures. (The van't Hoff number has no effect on significant figures because it's an exact number in this case.)

A boiling point elevation of 0.673°C means that the solution boils 0.673°C above the normal boiling point of the solvent, which is water. This solution will boil at 100.000°C + 0.673°C = 100.673°C.

---

903. **0.904°C**

The general equation for boiling point elevation is $\Delta T = iK_b m$. Hydrogen bromide is a strong electrolyte producing two ions when put into solution; therefore, the van't Hoff factor, i, is 2. To determine the molality of the solution, m, convert the grams of hydrogen bromide to moles and divide the result by the kilograms of the solvent.

$$\Delta T = iK_b m$$

$$= (2) \left(\frac{0.512°C}{1\,m} \right) \left(\frac{125\,g}{1.75\,kg} \right) \left(\frac{1\,mol}{80.9\,g} \right) \left(\frac{1\,m}{\left(\frac{1\,mol}{1\,kg} \right)} \right)$$

$$= 0.904°C$$

The given values all have three significant figures, so the answer has three significant figures. (The van't Hoff number has no effect on significant figures because it's an exact number in this case.)

A boiling point elevation of 0.904°C means that the solution boils 904°C above the normal boiling point of the solvent.

---

904. **−0.969°C**

The general equation is for freezing point depression is $\Delta T = iK_f m$. For nonelectrolytes, the van't Hoff factor, i, is 1. To determine the molality of the solution, m, convert the grams of methyl alcohol to moles and divide the result by the kilograms of the solvent. Here are the calculations:

$$\Delta T = iK_f m$$

$$= (1) \left(\frac{1.86°C}{1\,m} \right) \left(\frac{25.0\,g\,CH_3OH}{1.50\,kg} \right) \left(\frac{1\,mol\,CH_3OH}{32.0\,g\,CH_3OH} \right) \left(\frac{1\,m}{\left(\frac{1\,mol\,CH_3OH}{1\,kg} \right)} \right)$$

$$= 0.969°C$$

The given values all have three significant figures, so the answer has three significant figures. (The van't Hoff number has no effect on significant figures because it's an exact number in this case.)

A freezing point depression of 0.969°C means that the solution freezes 0.969°C below the normal freezing point of the solvent, which is water. Therefore, the freezing point is

$$0°C - 0.969°C = -0.969°C$$

905.   **–3.69°C**

The general equation for freezing point depression is $\Delta T = iK_f m$. Rubidium chloride is a strong electrolyte producing two ions when put into solution, so the van't Hoff factor is 2. To determine the molality of the solution, m, convert the grams of rubidium chloride to moles and divide the result by the kilograms of the solvent. Here are the calculations:

$$\Delta T = iK_f m$$

$$= (2)\left(\frac{1.86°C}{1\,m}\right)\left(\frac{15.0\,g\,RbCl}{0.125\,kg}\right)\left(\frac{1\,mol\,RbCl}{121\,g\,RbCl}\right)\left(\frac{1\,m}{\left(\frac{1\,mol\,RbCl}{1\,kg}\right)}\right)$$

$$= 3.69°C$$

The given values all have three significant figures, so the answer has three significant figures. (The van't Hoff number has no effect on significant figures because it's an exact number in this case.)

A freezing point depression of 3.69°C means that the solution freezes 3.69°C below the normal freezing point of the solvent, which is water. This solution will freeze at

$$0°C - 3.69°C = -3.69°C$$

906.   **83.40°C**

The general equation for boiling point elevation is $\Delta T = iK_b m$. For nonelectrolytes, the van't Hoff factor, i, is 1. To determine the molality of the solution, m, convert the grams of hexane to moles and divide the result by the kilograms of the solvent. Here are the calculations:

$$\Delta T = iK_b m$$

$$= (1)\left(\frac{2.79°C}{m}\right)\left(\frac{125\,g}{1.50\,kg}\right)\left(\frac{1\,mol}{86.2\,g}\right)\left(\frac{1\,m}{\left(\frac{1\,mol}{1\,kg}\right)}\right)$$

$$= 2.70°C$$

The given values all have three significant figures, so the answer has three significant figures. (The van't Hoff number has no effect on significant figures because it's an exact number in this case.)

A boiling point elevation of 2.70°C means that the solution boils 2.70°C above the normal boiling point of the solvent, which is cyclohexane. Thus, the boiling point is

$$80.70°C + 2.70°C = 83.40°C$$

907.

6.12 atm

The general equation for osmotic pressure is $\Pi V = inRT$ or $\Pi = iMRT$, where Π is the osmotic pressure, V is the volume, i is the van't Hoff factor, n is the number of moles, R is the gas constant, T is the temperature in kelvins, and M is the molarity.

The problem gives you the number of moles, so use the $\Pi V = inRT$ form. For nonelectrolytes, the van't Hoff factor is 1. The temperature must be in kelvins, so convert from degrees Celsius: $T = °C + 273 = 25 + 273 = 298$ K.

Solve the equation $\Pi V = inRT$ for Π, enter the appropriate values, and do the math:

$$\Pi = \frac{inRT}{V}$$

$$= \frac{(1)(0.125 \text{ mol})\left(0.0821 \frac{L \cdot atm}{mol \cdot K}\right)(298 \text{ K})}{(0.500 \text{ L})}$$

$$= 6.12 \text{ atm}$$

The values in the calculation all have three significant figures, so the answer has three significant figures. (The van't Hoff number has no effect on significant figures because it's an exact number in this case.)

908.

30.6 atm

The general equation for osmotic pressure is $\Pi V = inRT$ or $\Pi = iMRT$, where Π is the osmotic pressure, V is the volume, i is the van't Hoff factor, n is the number of moles, R is the gas constant, T is the temperature, and M is the molarity.

You're given the molarity, so use $\Pi = iMRT$. For nonelectrolytes, the van't Hoff factor is 1. The temperature must be in kelvins, so convert from degrees Celsius: $T = °C + 273 = 25 + 273 = 298$ K.

Enter the appropriate values in the equation and do the math. Expressing the molarity in terms of its definition (mol/L) can help you set up the problem correctly.

$$\Pi = iMRT$$

$$= (1)\left(\frac{1.25 \text{ mol}}{L}\right)\left(0.0821 \frac{L \cdot atm}{mol \cdot K}\right)(298 \text{ K})$$

$$= 30.6 \text{ atm}$$

The values in the calculation all have three significant figures, so the answer has three significant figures. (The van't Hoff number has no effect on significant figures because it's an exact number in this case.)

909.

1.75 atm

The general equation for osmotic pressure is $\Pi V = inRT$ or $\Pi = iMRT$, where Π is the osmotic pressure, V is the volume, i is the van't Hoff factor, n is the number of moles, R is the gas constant, T is the temperature, and M is the molarity.

You can find the number of moles from the mass and molar mass of the solute, so use the $\Pi V = inRT$ form. For nonelectrolytes, the van't Hoff factor is 1. The temperature must be in kelvins, so convert from degrees Celsius: $T = °C + 273 = 45 + 273 = 318$ K. To convert the mass of the solute to moles, multiply by the molar mass.

Solve the equation $\Pi V = inRT$ for Π, enter the appropriate values, and do the math:

$$\Pi = \frac{inRT}{V}$$

$$= \frac{(1)\left[(5.00\,\cancel{g})\left(\frac{\cancel{mol}}{272\,\cancel{g}}\right)\right]\left(0.0821\,\frac{\cancel{L}\cdot atm}{\cancel{mol}\cdot\cancel{K}}\right)(318\,\cancel{K})}{(0.275\,\cancel{L})}$$

$$= 1.75\text{ atm}$$

The values all have three significant figures, so the answer has three significant figures. (The van't Hoff number has no effect on significant figures because it's an exact number in this case.)

910. 3.48×10^3 g/mol

The general equation for osmotic pressure is $\Pi V = inRT$ or $\Pi = iMRT$, where Π is the osmotic pressure, V is the volume, i is the van't Hoff factor, n is the number of moles, R is the gas constant, T is the temperature, and M is the molarity.

To find the molar mass, you first need to calculate the number of moles, n, so use the $\Pi V = inRT$ form of the equation. For nonelectrolytes, the van't Hoff factor is 1. The temperature must be in kelvins, so convert from degrees Celsius: $T = {}^\circ C + 273 = 37 + 273 = 310.$ K.

Rearrange $\Pi V = inRT$ and solve for n to find the number of moles:

$$n = \frac{\Pi V}{iRT}$$

$$= \frac{(3.00\times10^{-3}\,\cancel{atm})(1.00\,\cancel{L})}{(1)\left(0.0821\frac{\cancel{L}\cdot\cancel{atm}}{mol\cdot\cancel{K}}\right)(310.\,\cancel{K})}$$

$$= 1.1787\times10^{-4}\text{ mol}$$

To get the molar mass of the polymer, divide the mass of the sample by the number of moles:

$$\frac{0.410\text{ g}}{1.1787\times10^{-4}\text{ mol}} = 3.48\times10^3\text{ g/mol}$$

The values in the calculation all have three significant figures, so the answer has three significant figures. (The van't Hoff number has no effect on significant figures because it's an exact number in this case.) Extra numbers appear in the calculations for the intermediate value (moles) to minimize errors due to intermediate rounding.

911.   17.7 atm

The general equation for osmotic pressure is $\Pi V = inRT$ or $\Pi = iMRT$, where Π is the osmotic pressure, V is the volume, i is the van't Hoff factor, n is the number of moles, R is the gas constant, T is the temperature, and M is the molarity.

You can find the number of moles, n, by multiplying the grams of calcium nitrate by its molar mass, so use the $\Pi V = inRT$ form. This is a strong electrolyte that produces three ions per molecule in solution (one calcium ion and two nitrate ions), so the van't Hoff factor is 3. The temperature, 37°C, must be in kelvins, so do the temperature conversion: $T = {}^\circ C + 273 = 37 + 273 = 310.$ K.

To finish the problem, rearrange the equation and enter the appropriate values:

$$\Pi = \frac{inRT}{V}$$

$$= \frac{(3)\left[(4.75\,\cancel{g})\left(\frac{mol}{164\,\cancel{g}}\right)\right]\left(0.0821\frac{\cancel{L}\cdot atm}{mol\cdot \cancel{K}}\right)(310.\,\cancel{K})}{(125\,\cancel{mL})}\left(\frac{1{,}000\,\cancel{mL}}{\cancel{L}}\right)$$

$$= 17.7\ atm$$

The values in the calculation all have three significant figures, so the answer has three significant figures. (The van't Hoff number has no effect on significant figures because it's an exact number in this case.)

912. HNO_3

Nitric acid (HNO_3) is a strong acid. It's one of the most commonly encountered strong acids, along with hydrochloric acid (HCl), hydrobromic acid (HBr), hydroiodic acid (HI), chloric acid ($HClO_3$), perchloric acid ($HClO_4$), and sulfuric acid (H_2SO_4).

913. KOH

Potassium hydroxide (KOH) is a commonly encountered strong base, as are sodium hydroxide ($NaOH$) and calcium hydroxide ($Ca(OH)_2$). Other strong bases are lithium hydroxide ($LiOH$), rubidium hydroxide ($RbOH$), cesium hydroxide ($CsOH$), strontium hydroxide ($Sr(OH)_2$), and barium hydroxide, ($Ba(OH)_2$).

914. HNO_2

HNO_2 is nitrous acid. In most cases, if the formula begins with H, the compound is an acid. Water (H_2O) is a common exception. Seven acids (HNO_3, HCl, HBr, HI, $HClO_3$, $HClO_4$, and H_2SO_4) are strong acids — assume all other acids are weak unless told otherwise.

915. NH_3

Ammonia (NH_3) and similar compounds are common weak bases. Similar substances, including amines such as methylamine (CH_3NH_2), are organic bases that are also weak.

916. CH_3OH

Methanol, CH_3OH, is not an acid or a base. Metal hydroxides, such as $NaOH$, are well-known bases. However, when the –OH group is attached to a nonmetal, the hydroxides are often neutral compounds. The presence of very electronegative elements can cause these otherwise neutral compounds to be acidic.

917. CH_3NH_2

The compound methylamine (CH_3NH_2) is similar to ammonia (NH_3), and like ammonia, it's a weak base. Many compounds containing nitrogen and carbon are weak bases.

918. Cl^-

To determine the conjugate base of any acid, simply remove one H^+. In this case, $HCl - H^+$ leaves Cl^-. Never remove more than one H^+ to determine the conjugate base.

919. $HC_2H_3O_2$

To determine the conjugate acid of any base, simply add one H^+. In this case, $C_2H_3O_2^- + H^+$ forms $HC_2H_3O_2$. Never add more than one H^+ to determine the conjugate acid.

920. NH_4^+

To determine the conjugate acid of any base, simply add one H^+. In this case, $NH_3 + H^+$ forms NH_4^+. Never add more than one H^+ to determine the conjugate acid.

921. NH_2^-

To determine the conjugate base of any acid, simply remove one H^+. In this case, $NH_3 - H^+$ leaves NH_2^-. Never remove more than one H^+ to determine the conjugate base.

922. $H_2PO_4^-$

To determine the conjugate base of any acid, simply remove one H^+. In this case, $H_3PO_4 - H^+$ leaves $H_2PO_4^-$. Never remove more than one H^+ to determine the conjugate base.

923. $H_3SO_4^+$

To determine the conjugate acid of any base, simply add one H^+. In this case, $H_2SO_4 + H^+$ forms $H_3SO_4^+$.

Note: In aqueous solutions, H_2SO_4 is a strong acid; however, stronger acids, such as $HClO_4$, exist. The stronger acid is capable of transferring an H^+ not only to a base but also to a weaker acid. This occurs in many aqueous solutions where an acid transfers an H^+ to the base H_2O to produce the conjugate acid H_3O^+.

924. HPO_4^{2-}

The ion HPO_4^{2-} may either lose a hydrogen ion to form PO_4^{3-} or gain a hydrogen ion to form $H_2PO_4^-$. The ability to lose a hydrogen ion makes it a conjugate acid, and the ability to gain a hydrogen ion makes it a conjugate base.

925. **2.0**

To determine the pH of a solution, you need to know the hydrogen ion concentration. For a strong acid such as HNO_3 (nitric acid), the hydrogen ion concentration is equal to the concentration of the acid. Here, $[H^+] = 0.01$ M. Inserting this concentration in the pH relationship gives you the following:

$$pH = -\log[H^+] = -\log[0.01] = 2.0$$

In this case, converting the concentration to scientific notation, 10^{-2}, and taking the negative of the exponent also gives the pH. Because this is an acid, the pH should be below 7, which it is.

If this compound were a weak acid, you'd have to use the K_a of the acid to determine the hydrogen ion concentration. K_a is the equilibrium constant for the ionization of a weak acid.

926.   **3.0**

To determine the pOH of a solution, you need to know the hydroxide ion concentration. For a strong base such as KOH (potassium hydroxide), the hydroxide ion concentration is equal to the concentration of the base. In this case, [OH⁻] = 0.001 M. Inserting this concentration in the pOH relationship gives you the following:

$$pOH = -\log[OH^-] = -\log[0.001] = 3.0$$

In this case, converting the concentration to scientific notation, 10^{-3}, and taking the negative of the exponent also gives you the pOH. Because this is an acid, the pOH should be below 7, which it is.

If this compound were a weak base, you'd have to use the K_b of the base to determine the hydroxide ion concentration. K_b is the equilibrium constant for the ionization of a weak base.

927.   **12.18**

To determine the pH of a base solution, it's helpful to find the pOH of the solution first. To determine the pOH, find the hydroxide ion concentration. For a strong base such as NaOH (sodium hydroxide), the hydroxide ion concentration is equal to the concentration of the base. Here, [OH⁻] = 0.015 M. Inserting this concentration in the pOH relationship gives you the following:

$$pOH = -\log[OH^-] = -\log[0.015] = 1.82$$

The last step is to find the pH using the relationship $pK_w = pH + pOH$, where $pK_w = 14.000$:

$$pH = pK_w - pOH = 14.000 - 1.82 = 12.18$$

Another way to solve the problem is to use the relationship $K_w = [H^+][OH^-] = 1.00 \times 10^{-14}$:

$$[H^+] = \frac{K_w}{[OH^-]} = \frac{1.00 \times 10^{-14}}{1.5 \times 10^{-2}} = 6.6667 \times 10^{-13} \text{ M}$$

$$pH = -\log[H^+] = -\log[6.6667 \times 10^{-13}] = 12.18$$

Because this solution is a base, the pH should be above 7, which it is.

If this compound were a weak base, you'd have to use the K_b of the base to determine the hydroxide ion concentration. K_b is the equilibrium constant for the ionization of a weak base.

928.   **11.40**

To determine the pOH of an acid solution, it's helpful to find the pH of the solution first. To determine the pH, find the hydrogen ion concentration. For a strong acid such as HCl (hydrochloric acid), the hydrogen ion concentration is equal to the concentration of the acid. In this case, $[H+] = 0.0025$ M. Inserting this concentration in the pH relationship gives you the following:

$$pH = -\log\left[H^+\right] = -\log\left[0.0025\right] = 2.60$$

The last step is to find the pOH using the relationship pK_w = pH + pOH, where pK_w = 14.000:

$$pOH = 14.000 - pH = 14.000 - 2.60 = 11.40$$

Another way to solve the problem is to use the relationship $K_w = [H^+][OH^-] = 1.00 \times 10^{-14}$, where K_w is the autoionization equilibrium constant for water:

$$\left[OH^-\right] = \frac{K_w}{\left[H^+\right]} = \frac{1.00 \times 10^{-14}}{2.5 \times 10^{-3}} = 4.0 \times 10^{-12} \text{ M}$$

$$pOH = -\log\left[OH^-\right] = -\log\left[4.0 \times 10^{-12}\right] = 11.40$$

Because this is an acid, the pH should be below 7 and the pOH should be above 7, which they are.

If this compound were a weak acid, you'd have to use the acid's ionization constant, K_a, to determine the hydrogen ion concentration.

929.   **11.8**

To determine the pH of a base solution, it's helpful to find the pOH of the solution first. To determine the pOH, find the hydroxide ion concentration. For a strong base such as $Ba(OH)_2$ (barium hydroxide), the hydroxide ion concentration is double the concentration of the base (double because of the subscript 2). In this case, $[OH^-] = 2(0.003) = 0.006$ M. Inserting this concentration in the pOH relationship gives you the following:

$$pOH = -\log\left[OH^-\right] = -\log\left[0.006\right] = 2.2$$

The last step is to find the pH using the relationship pK_w = pH + pOH, where pK_w = 14.000:

$$pH = pK_w - pOH = 14.000 - 2.2 = 11.8$$

Because this compound is a base, the pH should be above 7, which it is.

You can also solve the problem by using the relationship $K_w = [H^+][OH^-] = 1.00 \times 10^{-14}$:

$$\left[H^+\right] = \frac{K_w}{\left[OH^-\right]} = \frac{1.00 \times 10^{-14}}{0.006} = 1.6667 \times 10^{-12} \text{ M}$$

$$pH = -\log\left[H^+\right] = -\log\left[1.6667 \times 10^{-12}\right] = 11.8$$

If this compound were a weak base, you'd have to use the base's ionization constant, K_b, to determine the hydroxide ion concentration.

930. 11.30

To determine the pH of a base solution, it's helpful to find the pOH of the solution first. To determine the pOH, find the hydroxide ion concentration. For a strong base such as $Sr(OH)_2$ (strontium hydroxide), the hydroxide ion concentration is double the concentration of the base (double because of the subscript 2). In this case, $[OH^-] = 2(1.0 \times 10^{-3}) = 2.0 \times 10^{-3}$ M. Inserting this concentration in the pOH relationship gives you the following:

$$pOH = -\log\left[OH^-\right] = -\log\left[2.0 \times 10^{-3}\right] = 2.70$$

The last step is to find the pH using the relationship $pK_w = pH + pOH$, where $pK_w = 14.000$:

$$pH = 14.000 - pOH = 14.000 - 2.70 = 11.30$$

Because this is a base, the pH should be above 7 and the pOH should be below 7, which they are.

You can also solve the problem by using the relationship $K_w = [H^+][OH^-] = 1.00 \times 10^{-14}$. In this case,

$$\left[H^+\right] = \frac{K_w}{\left[OH^-\right]} = \frac{1.00 \times 10^{-14}}{2.0 \times 10^{-3}} = 5.0 \times 10^{-12} \text{ M}$$

$$pH = -\log\left[H^+\right] = -\log\left[5.0 \times 10^{-12}\right] = 11.30$$

If this compound were a weak base, you'd have to use the base's ionization constant, K_b, to determine the hydroxide ion concentration.

931. 0

To determine the pH of a solution, you need to know the hydrogen ion concentration. For a strong acid such as HBr (hydrobromic acid), the hydrogen ion concentration is equal to the concentration of the acid. Here, $[H^+] = 1.0$ M. Inserting this concentration in the pH relationship gives you the following:

$$pH = -\log\left[H^+\right] = -\log[1.0] = 0$$

In this case, converting the concentration to scientific notation, 10^0, and taking the negative of the exponent also gives you the pH.

Because the compound is an acid, the pH should be below 7, which it is.

If this compound were a weak acid, you'd need to use the acid's ionization constant, K_a, to determine the hydrogen ion concentration.

932. 15.18

To determine the pH of a base solution, it's helpful to find the pOH of the solution first. To determine the pOH, you need the hydroxide ion concentration. For a strong base such as NaOH (sodium hydroxide), the hydroxide ion concentration is equal to the concentration of the base. In this case, $[OH^-] = 15$ M. Inserting this concentration in the pOH relationship gives you the following:

$$pOH = -\log\left[OH^-\right] = -\log[15] = -1.18$$

The last step is to find the pH using the relationship $pK_w = pH + pOH$, where $pK_w = 14.000$:

$$pH = 14.000 - pOH = 14.000 - (-1.18) = 15.18$$

You can also solve the problem by using the relationship $K_w = [H^+][OH^-] = 1.00 \times 10^{-14}$:

$$\left[H^+ \right] = \frac{K_w}{\left[OH^- \right]} = \frac{1.00 \times 10^{-14}}{15} = 6.6667 \times 10^{-16} \text{ M}$$

$$pH = -\log \left[H^+ \right] = -\log \left[6.6667 \times 10^{-16} \right] = 15.18$$

Because this compound is a base, the pH should be above 7, which it is. You have a very high concentration of a strong base, so it's possible for the pH to be above 14.

If this compound were a weak base, you'd have to use the base's ionization constant, K_b, to determine the hydroxide ion concentration.

933.   **2.39**

Acetic acid isn't one of the strong acids (HNO_3, HCl, HBr, HI, $HClO_3$, $HClO_4$, and H_2SO_4), so it must be a weak acid. Weak acid calculations need an acid equilibrium constant, K_a. The generic form of every K_a problem is

$$CA \rightleftharpoons H^+ + CB$$

$$K_a = \frac{\left[H^+ \right]\left[CB \right]}{\left[CA \right]}$$

where CA refers to the conjugate acid and CB refers to the conjugate base. In this case, the equilibrium equation and the change in the concentrations are

$$HC_2H_3O_2 \rightleftharpoons H^+ + C_2H_3O_2^-$$
$$1.0 - x \quad\quad +x \quad +x$$

Entering this information in the K_a expression gives you the following:

$$K_a = \frac{\left[H^+ \right]\left[C_2H_3O_2^- \right]}{\left[HC_2H_3O_2 \right]}$$

$$1.7 \times 10^{-5} = \frac{\left[x \right]\left[x \right]}{\left[1.0 - x \right]}$$

This is a quadratic equation, and you can solve it as such. But before doing the math, think about the problem logically. If the change in the denominator is insignificant, you can drop the $-x$ and make the problem easier to solve.

A simple check is to compare the exponent on the K to the exponent on the concentration (in scientific notation). In this case, the exponent on the K is –5, and the exponent on the concentration is 0. If the exponent on the K is at least 3 less than the exponent on the concentration (as it is in this case), you can assume that $-x$ in the denominator is insignificant. That means you can rewrite the K equation as

$$1.7 \times 10^{-5} = \frac{\left[x \right]\left[x \right]}{\left[1.0 \right]}$$

$$x = 4.1 \times 10^{-3} = \left[H^+ \right]$$

The x value is sufficiently small that $1.0 - x \approx 1.0$, which validates the assumption.

Finally, use the pH definition:

$$pH = -\log\left[H^+\right] = -\log\left[4.1 \times 10^{-3}\right] = 2.39$$

934. 1.66

Nitrous acid isn't one of the strong acids (HNO_3, HCl, HBr, HI, $HClO_3$, $HClO_4$, and H_2SO_4), so it must be a weak acid. Weak acid calculations need an acid equilibrium constant, K_a. The generic form of every K_a problem is

$$CA \rightleftharpoons H^+ + CB$$

$$K_a = \frac{\left[H^+\right]\left[CB\right]}{\left[CA\right]}$$

where CA refers to the conjugate acid and CB refers to the conjugate base. In this case, the equilibrium equation and the change in the concentrations are

$$HNO_2 \rightleftharpoons H^+ + NO_2^-$$
$$1.0 - x \quad +x \quad +x$$

Entering this information in the K_a expression gives you the following:

$$K_a = \frac{\left[H^+\right]\left[NO_2^-\right]}{\left[HNO_2\right]}$$

$$5.0 \times 10^{-4} = \frac{[x][x]}{[1.0 - x]}$$

This is a quadratic equation, and you can solve it as such. But before doing the math, think about the problem logically. If the change in the denominator is insignificant, you can drop the $-x$ and make the problem easier to solve.

A simple check is to compare the exponent on the K to the exponent on the concentration (in scientific notation). In this case, the exponent on the K is -4, and the exponent on the concentration is 0. If the exponent on the K is at least 3 less than the exponent on the concentration (as it is in this case), you can assume that $-x$ in the denominator is insignificant. That means you can rewrite the K equation as

$$5.0 \times 10^{-4} = \frac{[x][x]}{[1.0]}$$

$$x = 2.2 \times 10^{-2} = \left[H^+\right]$$

The x value is sufficiently small that $1.0 - x \approx 1.0$, which validates the assumption.

Finally, use the pH definition:

$$pH = -\log\left[H^+\right] = -\log\left[2.2 \times 10^{-2}\right] = 1.66$$

935. 2.38

Ammonia isn't one of the strong bases (NaOH, KOH, LiOH, RbOH, CsOH, $Ca(OH)_2$, $Sr(OH)_2$, and $Ba(OH)_2$), so it must be a weak base. Weak base calculations need a base equilibrium constant, K_b. The generic form of every K_b problem is

$$CB \rightleftharpoons OH^- + CA$$

$$K_b = \frac{\left[OH^-\right]\left[CA\right]}{\left[CB\right]}$$

where CA refers to the conjugate acid and CB refers to the conjugate base. For balancing purposes, water may be present in the equilibrium chemical equation; however, water is also the solvent, so it shouldn't be in the equilibrium expression.

In this case, the equilibrium equation and the change in the concentrations are

$$NH_3 + H_2O \rightleftharpoons OH^- + NH_4^+$$
$$1.0 - x \qquad\quad +x \quad +x$$

Entering this information in the K_b expression gives you the following:

$$K_b = \frac{\left[OH^-\right]\left[NH_4^+\right]}{\left[NH_3\right]}$$

$$1.8 \times 10^{-5} = \frac{[x][x]}{[1.0-x]}$$

This is a quadratic equation, and you can solve it as such. But before doing the math, think about the problem logically. If the change in the denominator is insignificant, you can drop the $-x$ and make the problem easier to solve.

A simple check is to compare the exponent on the K to the exponent on the concentration (in scientific notation). In this case, the exponent on the K is -5, and the exponent on the concentration is 0. If the exponent on the K is at least 3 less than the exponent on the concentration (as it is in this case), you can assume that $-x$ in the denominator is insignificant. That means you can rewrite the K equation as

$$1.8 \times 10^{-5} = \frac{[x][x]}{[1.0]}$$

$$x = 4.2 \times 10^{-3} = \left[OH^-\right]$$

The x value is sufficiently small that $1.0 - x \approx 1.0$, which validates the assumption.

Finally, use the pOH definition:

$$pOH = -\log\left[OH^-\right] = -\log\left[4.2 \times 10^{-3}\right] = 2.38$$

---

936.   **5.10**

Hydrocyanic acid isn't one of the strong acids (HNO_3, HCl, HBr, HI, $HClO_3$, $HClO_4$, and H_2SO_4), so it must be a weak acid. Weak acid calculations need an acid equilibrium constant, K_a. The generic form of every K_a problem is

$$CA \rightleftharpoons H^+ + CB$$

$$K_a = \frac{\left[H^+\right]\left[CB\right]}{\left[CA\right]}$$

where CA refers to the conjugate acid and CB refers to the conjugate base. In this case, the equilibrium equation and the change in the concentrations are

$$HCN \rightleftharpoons H^+ + CN^-$$
$$0.100 - x \quad +x \quad +x$$

Entering this information in the K_a expression gives you the following:

$$K_a = \frac{\left[H^+\right]\left[CN^-\right]}{\left[HCN\right]}$$

$$6.2\times10^{-10} = \frac{[x][x]}{[0.100-x]}$$

This is a quadratic equation, and you can solve it as such. But before doing the math, think about the problem logically. If the change in the denominator is insignificant, you can drop the $-x$ and make the problem easier to solve.

A simple check is to compare the exponent on the K to the exponent on the concentration (in scientific notation). In this case, the exponent on the K is -10, and the exponent on the concentration is -1. If the exponent on the K is at least 3 less than the exponent on the concentration (as it is in this case), you can assume that $-x$ in the denominator is insignificant. That means you can rewrite the K equation as

$$6.2\times10^{-10} = \frac{[x][x]}{[0.100]}$$

$$x = 7.9\times10^{-6} = \left[H^+\right]$$

The x value is sufficiently small that $1.0 - x \approx 1.0$, which validates the assumption.

Finally, use the pH definition:

$$pH = -\log\left[H^+\right] = -\log\left[7.9\times10^{-6}\right] = 5.10$$

---

937.   2.74

Acetic acid isn't one of the strong acids (HNO_3, HCl, HBr, HI, $HClO_3$, $HClO_4$, and H_2SO_4), so it must be a weak acid. Weak acid calculations need an acid equilibrium constant, K_a. The generic form of every K_a problem is

$$CA \rightleftharpoons H^+ + CB$$

$$K_a = \frac{\left[H^+\right]\left[CB\right]}{\left[CA\right]}$$

where CA refers to the conjugate acid and CB refers to the conjugate base. In this case, the equilibrium equation and the change in the concentrations are

$$HC_2H_3O_2 \rightleftharpoons H^+ + C_2H_3O_2^-$$
$$0.20-x \qquad +x \qquad +x$$

Entering this information in the K_a expression gives you the following:

$$K_a = \frac{\left[H^+\right]\left[C_2H_3O_2^-\right]}{\left[HC_2H_3O_2\right]}$$

$$1.7\times10^{-5} = \frac{[x][x]}{[0.20-x]}$$

This is a quadratic equation, and you can solve it as such. But before doing the math, think about the problem logically. If the change in the denominator is insignificant, you can drop the $-x$ and make the problem easier to solve.

A simple check is to compare the exponent on the K to the exponent on the concentration (in scientific notation). In this case, the exponent on the K is –5, and the exponent on the concentration is –1. If the exponent on the K is at least 3 less than the exponent on the concentration (as it is in this case), you can assume that $-x$ in the denominator is insignificant. That means you can rewrite the K equation as

$$1.7 \times 10^{-5} = \frac{[x][x]}{[0.20]}$$

$$x = 1.8 \times 10^{-3} = [H^+]$$

The x value is sufficiently small that $1.0 - x \approx 1.0$, which validates the assumption.

Finally, use the pH definition:

$$pH = -\log[H^+] = -\log[1.8 \times 10^{-3}] = 2.74$$

---

938.   **2.85**

Periodic acid isn't one of the strong acids (HNO_3, HCl, HBr, HI, $HClO_3$, $HClO_4$, and H_2SO_4), so it must be a weak acid. Weak acid calculations need an acid equilibrium constant, K_a. The generic form of every K_a problem is

$$CA \rightleftharpoons H^+ + CB$$

$$K_a = \frac{[H^+][CB]}{[CA]}$$

where CA refers to the conjugate acid and CB refers to the conjugate base. In this case, the equilibrium equation and the change in the concentrations are

$$HIO_4 \rightleftharpoons H^+ + IO_4^-$$
$$0.0015 - x \quad +x \quad +x$$

Entering this information in the K_a expression gives you the following:

$$K_a = \frac{[H^+][IO_4^-]}{[HIO_4]}$$

$$2.8 \times 10^{-2} = \frac{[x][x]}{[0.0015 - x]}$$

This is a quadratic equation, and you can solve it as such. But before doing the math, think about the problem logically. If the change in the denominator is insignificant, you can drop the $-x$ and make the problem easier to solve.

A simple check is to compare the exponent on the K to the exponent on the concentration (in scientific notation). In this case, the exponent on the K is –2, and the exponent on the concentration is –3. If the exponent on the K is at least 3 less than the exponent on the concentration, you can assume that $-x$ in the denominator is insignificant. That doesn't occur in this case, so you have to solve this equation as a quadratic.

Write the quadratic equation in standard form:

$$(2.8 \times 10^{-2})(0.0015 - x) = x^2$$

$$(4.2 \times 10^{-5}) - (2.8 \times 10^{-2})x = x^2$$

$$x^2 + (2.8 \times 10^{-2})x - (4.2 \times 10^{-5}) = 0$$

Then solve for x using the quadratic formula. Here, $a = 1$, $b = 2.8 \times 10^{-2}$, and $c = -4.2 \times 10^{-5}$:

$$x = \frac{-b \pm \sqrt{b^2 - 4ac}}{2a}$$

$$= \frac{-\left(2.8 \times 10^{-2}\right) \pm \sqrt{\left(2.8 \times 10^{-2}\right)^2 - 4(1)\left(-4.2 \times 10^{-5}\right)}}{2(1)}$$

$$= 1.4 \times 10^{-3} = \left[H^+\right]$$

You can neglect the negative root because negative concentrations aren't possible.

Finally, use the pH definition:

$$pH = -\log\left[H^+\right] = -\log\left[1.4 \times 10^{-3}\right] = 2.85$$

939. 2.68

Chlorous acid isn't one of the strong acids (HNO_3, HCl, HBr, HI, $HClO_3$, $HClO_4$, and H_2SO_4), so it must be a weak acid. Weak acid calculations need an acid equilibrium constant, K_a. The generic form of every K_a problem is

$$CA \rightleftharpoons H^+ + CB$$

$$K_a = \frac{\left[H^+\right]\left[CB\right]}{\left[CA\right]}$$

where CA refers to the conjugate acid and CB refers to the conjugate base. In this case, the equilibrium equation and the change in the concentrations are

$$HClO_2 \rightleftharpoons H^+ + ClO_2^-$$
$$0.0025 - x \quad +x \quad +x$$

Entering this information in the K_a expression gives you the following:

$$K_a = \frac{\left[H^+\right]\left[ClO_2^-\right]}{\left[HClO_2\right]}$$

$$1.1 \times 10^{-2} = \frac{[x][x]}{[0.0025 - x]}$$

This is a quadratic equation, and you can solve it as such. But before doing the math, think about the problem logically. If the change in the denominator is insignificant, you can drop the $-x$ and make the problem easier to solve.

A simple check is to compare the exponent on the K to the exponent on the concentration (in scientific notation). In this case, the exponent on the K is -2, and the exponent on the concentration is -3. If the exponent on the K is at least 3 less than the exponent on the concentration, you can assume that $-x$ in the denominator is insignificant. That doesn't occur in this case, so you have to solve this equation as a quadratic.

Write the quadratic equation in standard form:

$$\left(1.1 \times 10^{-2}\right)(0.0025 - x) = x^2$$

$$2.75 \times 10^{-5} - \left(1.1 \times 10^{-2}\right)x = x^2$$

$$x^2 + \left(1.1 \times 10^{-2}\right)x - \left(2.75 \times 10^{-5}\right) = 0$$

Then solve for x using the quadratic formula. Here, $a = 1$, $b = 1.1 \times 10^{-2}$, and $c = -2.75 \times 10^{-5}$:

$$x = \frac{-b \pm \sqrt{b^2 - 4ac}}{2a}$$

$$= \frac{-\left(1.1 \times 10^{-2}\right) \pm \sqrt{\left(1.1 \times 10^{-2}\right)^2 - 4(1)\left(-2.75 \times 10^{-5}\right)}}{2(1)}$$

$$= 2.1 \times 10^{-3} = \left[H^+\right]$$

You can neglect the negative root because negative concentrations aren't possible.

Finally, use the pH definition:

$$pH = -\log\left[H^+\right] = -\log\left[2.1 \times 10^{-3}\right] = 2.68$$

---

940.   4.39

Pyridine isn't one of the strong bases (NaOH, KOH, LiOH, RbOH, CsOH, $Ca(OH)_2$, $Sr(OH)_2$, and $Ba(OH)_2$), so it must be a weak base. Weak base calculations need a base equilibrium constant, K_b. The generic form of every K_b problem is

$$CB \rightleftharpoons OH^- + CA$$

$$K_b = \frac{\left[OH^-\right]\left[CA\right]}{\left[CB\right]}$$

where CA refers to the conjugate acid and CB refers to the conjugate base. For balancing purposes, water may be present in the equilibrium chemical equation; however, water is also the solvent, so it shouldn't be in the equilibrium expression.

In this case, the equilibrium equation and the change in the concentrations are

$$C_5H_5N + H_2O \rightleftharpoons OH^- + C_5H_5NH^+$$
$$1.0 - x +x +x$$

Entering this information in the K_b expression gives you the following:

$$K_b = \frac{\left[OH^-\right]\left[C_5H_5NH^+\right]}{\left[C_5H_5N\right]}$$

$$1.7 \times 10^{-9} = \frac{\left[x\right]\left[x\right]}{\left[1.0 - x\right]}$$

This is a quadratic equation, and you can solve it as such. But before doing the math, think about the problem logically. If the change in the denominator is insignificant, you can drop the $-x$ and make the problem easier to solve.

A simple check is to compare the exponent on the K to the exponent on the concentration (in scientific notation). In this case, the exponent on the K is -9, and the exponent on the concentration is 0. If the exponent on the K is at least 3 less than the exponent on the concentration (as it is in this case), you can assume that $-x$ in the denominator is insignificant. That means you can rewrite the K equation as

$$1.7 \times 10^{-9} = \frac{\left[x\right]\left[x\right]}{\left[1.0\right]}$$

$$x = 4.1 \times 10^{-5} = \left[OH^-\right]$$

The x value is sufficiently small that $1.0 - x \approx 1.0$, which validates the assumption.

Finally, use the pOH definition:

$$\text{pOH} = -\log\left[\text{OH}^-\right] = -\log\left[4.1 \times 10^{-5}\right] = 4.39$$

---

941.   **11.20**

Ammonia isn't one of the strong bases (NaOH, KOH, LiOH, RbOH, CsOH, Ca(OH)$_2$, Sr(OH)$_2$, and Ba(OH)$_2$), so it must be a weak base. Weak base calculations need a base equilibrium constant, K_b. The generic form of every K_b problem is

$$CB \rightleftharpoons OH^- + CA$$

$$K_b = \frac{\left[OH^-\right]\left[CA\right]}{\left[CB\right]}$$

where CA refers to the conjugate acid and CB refers to the conjugate base. For balancing purposes, water may be present in the equilibrium chemical equation; however, water is also the solvent, so it shouldn't be in the equilibrium expression.

In this case, the equilibrium equation and the change in the concentrations are

$$NH_3 + H_2O \rightleftharpoons OH^- + NH_4^+$$

$$0.15 - x \qquad +x \quad +x$$

Entering this information in the K_b expression gives you the following:

$$K_b = \frac{\left[OH^-\right]\left[NH_4^+\right]}{\left[NH_3\right]}$$

$$1.8 \times 10^{-5} = \frac{[x][x]}{[0.15 - x]}$$

This is a quadratic equation, and you can solve it as such. But before doing the math, think about the problem logically. If the change in the denominator is insignificant, you can drop the $-x$ and make the problem easier to solve.

A simple check is to compare the exponent on the K to that of the concentration (in scientific notation). In this case, the exponent on the K is −5, and the exponent on the concentration is −1. If the exponent on the K is at least 3 less than the exponent on the concentration (as it is in this case), you can assume that $-x$ in the denominator is insignificant. That means you can rewrite the K equation as

$$1.8 \times 10^{-5} = \frac{[x][x]}{[0.15]}$$

$$x = 1.6 \times 10^{-3} = \left[OH^-\right]$$

The x value is sufficiently small that $1.0 - x \approx 1.0$, which validates the assumption.

Find the pOH using the pOH definition:

$$\text{pOH} = -\log\left[OH^-\right] = -\log\left[1.6 \times 10^{-3}\right] = 2.80$$

Finally, find the pH using the relationship $pK_w = \text{pH} + \text{pOH}$, where $pK_w = 14.000$:

$$\text{pH} = pK_w - \text{pOH} = 14.000 - 2.80 = 11.20$$

942. 12.56

Methylamine isn't one of the strong bases (NaOH, KOH, LiOH, RbOH, CsOH, Ca(OH)$_2$, Sr(OH)$_2$, and Ba(OH)$_2$), so it must be a weak base. Weak base calculations need a base equilibrium constant, K_b. The generic form of every K_b problem is

$$CB \rightleftharpoons OH^- + CA$$

$$K_b = \frac{\left[OH^-\right]\left[CA\right]}{\left[CB\right]}$$

where CA refers to the conjugate acid and CB refers to the conjugate base. For balancing purposes, water may be present in the equilibrium chemical equation; however, water is also the solvent, so it shouldn't be in the equilibrium expression.

In this problem, the equilibrium equation and the change in the concentrations are

$$CH_3NH_2 + H_2O \rightleftharpoons OH^- + CH_3NH_3^+$$
$$2.5 - x \qquad\qquad +x \quad +x$$

Entering this information in the K_b expression gives you the following:

$$K_b = \frac{\left[OH^-\right]\left[CH_3NH_3^+\right]}{\left[CH_3NH_2\right]}$$

$$5.2 \times 10^{-4} = \frac{[x][x]}{[2.5 - x]}$$

This is a quadratic equation, and you can solve it as such. But before doing the math, think about the problem logically. If the change in the denominator is insignificant, you can drop the $-x$ and make the problem easier to solve.

A simple check is to compare the exponent on the K to the exponent on the concentration (in scientific notation). In this case, the exponent on the K is –4, and the exponent on the concentration is 0. If the exponent on the K is at least 3 less than the exponent on the concentration (as it is in this case), you can assume that $-x$ in the denominator is insignificant. That means you can rewrite the K equation as

$$5.2 \times 10^{-4} = \frac{[x][x]}{[2.5]}$$

$$x = 3.6 \times 10^{-2} = \left[OH^-\right]$$

The x value is sufficiently small that $1.0 - x \approx 1.0$, which validates the assumption.

Use the pOH definition to find the pOH:

$$pOH = -\log\left[OH^-\right] = -\log\left[3.6 \times 10^{-2}\right] = 1.44$$

Finally, find the pH using the relationship $pK_w = pH + pOH$, where $pK_w = 14.000$:

$$pH = pK_w - pOH = 14.000 - 1.44 = 12.56$$

943. 1.38

Chlorous acid isn't one of the strong acids (HNO$_3$, HCl, HBr, HI, HClO$_3$, HClO$_4$, and H$_2$SO$_4$), so it must be a weak acid. Weak acid calculations need an acid equilibrium constant, K_a. The generic form of every K_a problem is

$$CA \rightleftharpoons H^+ + CB$$

$$K_a = \frac{[H^+][CB]}{[CA]}$$

where CA refers to the conjugate acid and CB refers to the conjugate base. In this case, the equilibrium equation and the change in the concentrations are

$$HClO_2 \rightleftharpoons H^+ + ClO_2^-$$
$$0.0025 - x \quad +x \quad +x$$

Entering this information in the K_a expression gives you the following:

$$K_a = \frac{[H^+][ClO_2^-]}{[HClO_2]}$$

$$1.1 \times 10^{-2} = \frac{[x][x]}{[0.20 - x]}$$

This is a quadratic equation, and you can solve it as such. But before doing the math, think about the problem logically. If the change in the denominator is insignificant, you can drop the $-x$ and make the problem easier to solve.

A simple check is to compare the exponent on the K to the exponent on the concentration (in scientific notation). In this case, the exponent on the K is –2, and the exponent on the concentration is –1. If the exponent on the K is at least 3 less than the exponent on the concentration, you can assume that $-x$ in the denominator is insignificant. That doesn't occur in this case, so you have to solve this equation as a quadratic.

Write the quadratic equation in standard form:

$$(1.1 \times 10^{-2})(0.20 - x) = x^2$$

$$(2.2 \times 10^{-3}) - (1.1 \times 10^{-2})x = x^2$$

$$x^2 + (1.1 \times 10^{-2})x - (2.2 \times 10^{-3}) = 0$$

Then solve for x using the quadratic formula. Here, $a = 1$, $b = 1.1 \times 10^{-2}$, and $c = -2.2 \times 10^{-3}$:

$$x = \frac{-b \pm \sqrt{b^2 - 4ac}}{2a}$$

$$= \frac{-(1.1 \times 10^{-2}) \pm \sqrt{(1.1 \times 10^{-2})^2 - 4(1)(-2.2 \times 10^{-3})}}{2(1)}$$

$$= 4.2 \times 10^{-2} = [H^+]$$

You can neglect the negative root because negative concentrations aren't possible.

Finally, find the pH using the pH definition:

$$pH = -\log[H^+] = -\log[4.2 \times 10^{-2}] = 1.38$$

944.   2.68

Cyanic acid isn't one of the strong acids (HNO_3, HCl, HBr, HI, $HClO_3$, $HClO_4$, and H_2SO_4), so it must be a weak acid. Weak acid calculations need an acid equilibrium constant, K_a. The generic form of every K_a problem is

$$CA \rightleftharpoons H^+ + CB$$

$$K_a = \frac{[H^+][CB]}{[CA]}$$

where CA refers to the conjugate acid and CB refers to the conjugate base.

The problem gives you the pK_a, so convert that to a K_a:

$$K_a = 10^{-pK_a} = 10^{-3.46} = 3.5 \times 10^{-4}$$

In this problem, the equilibrium equation and the change in the concentrations are

$$HOCN \rightleftharpoons H^+ + OCN^-$$

$$0.015 - x \quad +x \quad +x$$

Entering this information in the K_a expression gives you the following:

$$K_a = \frac{[H^+][OCN^-]}{[HOCN]}$$

$$3.5 \times 10^{-4} = \frac{[x][x]}{[0.015 - x]}$$

This is a quadratic equation, and you can solve it as such. But before doing the math, think about the problem logically. If the change in the denominator is insignificant, you can drop the $-x$ and make the problem easier to solve.

A simple check is to compare the exponent on the K to the exponent on the concentration (in scientific notation). In this case, the exponent on the K is –4, and the exponent on the concentration is –2. If the exponent on the K is at least 3 less than the exponent on the concentration, you can assume that $-x$ in the denominator is insignificant. That doesn't occur in this case, so you have to solve this equation as a quadratic.

Write the quadratic equation in standard form:

$$(3.5 \times 10^{-4})(0.015 - x) = x^2$$

$$(5.25 \times 10^{-6}) - (3.5 \times 10^{-4})x = x^2$$

$$x^2 + (3.5 \times 10^{-4})x - (5.25 \times 10^{-6}) = 0$$

Then solve for x using the quadratic formula. Here, $a = 1$, $b = 3.5 \times 10^{-4}$, and $c = -5.25 \times 10^{-6}$:

$$x = \frac{-b \pm \sqrt{b^2 - 4ac}}{2a}$$

$$= \frac{-(3.5 \times 10^{-4}) \pm \sqrt{(3.5 \times 10^{-4})^2 - 4(1)(-5.25 \times 10^{-6})}}{2(1)}$$

$$= 2.1 \times 10^{-3} = [H^+]$$

You can neglect the negative root because negative concentrations aren't possible.

Finally, find the pH using the pH definition:

$$pH = -\log[H^+] = -\log[2.1 \times 10^{-3}] = 2.68$$

945. 1.64

Methylamine isn't one of the strong bases (NaOH, KOH, LiOH, RbOH, CsOH, $Ca(OH)_2$, $Sr(OH)_2$, and $Ba(OH)_2$), so it must be a weak base. Weak base calculations need a base equilibrium constant, K_b. The generic form of every K_b problem is

$$CB \rightleftharpoons OH^- + CA$$

$$K_b = \frac{[OH^-][CA]}{[CB]}$$

where CA refers to the conjugate acid and CB refers to the conjugate base. For balancing purposes, water may be present in the equilibrium chemical equation; however, water is also the solvent, so it shouldn't be in the equilibrium expression.

The problem gives you the pK_b, so convert that to a K_b:

$$K_b = 10^{-pK_b} = 10^{-3.28} = 5.2 \times 10^{-4}$$

In this problem, the equilibrium equation and the change in the concentrations are

$$CH_3NH_2 + H_2O \rightleftharpoons OH^- + CH_3NH_3^+$$
$$1.0 - x \qquad\qquad +x \qquad +x$$

Entering this information in the K_b expression gives you the following:

$$K_b = \frac{[OH^-][CH_3NH_3^+]}{[CH_3NH_2]}$$

$$5.2 \times 10^{-4} = \frac{[x][x]}{[1.0 - x]}$$

This is a quadratic equation, and you can solve it as such. But before doing the math, think about the problem logically. If the change in the denominator is insignificant, you can drop the $-x$ and make the problem easier to solve.

A simple check is to compare the exponent on the K to the exponent on the concentration (in scientific notation). In this case, the exponent on the K is -4, and the exponent on the concentration is 0. If the exponent on the K is at least 3 less than the exponent on the concentration (as it is in this case), you can assume that $-x$ in the denominator is insignificant. That means you can rewrite the K equation as

$$5.2 \times 10^{-4} = \frac{[x][x]}{[1.0]}$$

$$x = 2.3 \times 10^{-2} = [OH^-]$$

The x value is sufficiently small that $1.0 - x \approx 1.0$, which validates the assumption.

Finally, find the pOH using the pOH definition:

$$pOH = -\log[OH^-] = -\log[2.3 \times 10^{-2}] = 1.64$$

946. 1.33

$NaHSO_4$ is a strong electrolyte that completely separates into Na^+ and HSO_4^-. The hydrogen sulfate ion, due to the presence of an acidic hydrogen, isn't one of the strong acids (HNO_3, HCl, HBr, HI, $HClO_3$, $HClO_4$, and H_2SO_4), so it must be a weak acid. Weak acid calculations need an acid equilibrium constant, K_a. The generic form of every K_a problem is

$$CA \rightleftharpoons H^+ + CB$$

$$K_a = \frac{[H^+][CB]}{[CA]}$$

where CA refers to the conjugate acid and CB refers to the conjugate base. In this case, the equilibrium equation and the change in the concentrations are

$$HSO_4^- \rightleftharpoons H^+ + SO_4^{2-}$$
$$0.25 - x \quad +x \quad +x$$

Entering this information in the K_a expression gives you the following:

$$K_a = \frac{[H^+][SO_4^{2-}]}{[HSO_4^-]}$$

$$1.1 \times 10^{-2} = \frac{[x][x]}{[0.25 - x]}$$

This is a quadratic equation, and you can solve it as such. But before doing the math, think about the problem logically. If the change in the denominator is insignificant, you can drop the $-x$ and make the problem easier to solve.

A simple check is to compare the exponent on the K to the exponent on the concentration (in scientific notation). In this case, the exponent on the K is −2, and the exponent on the concentration is −1. If the exponent on the K is at least 3 less than the exponent on the concentration, you can assume that $-x$ in the denominator is insignificant. That doesn't occur in this case, so you have to solve this equation as a quadratic.

Write the quadratic equation in standard form:

$$(1.1 \times 10^{-2})(0.25 - x) = x^2$$

$$(2.75 \times 10^{-3}) - (1.1 \times 10^{-2})x = x^2$$

$$x^2 + (1.1 \times 10^{-2})x - (2.75 \times 10^{-3}) = 0$$

Then solve for x using the quadratic formula. Here, $a = 1$, $b = 1.1 \times 10^{-2}$, and $c = -2.75 \times 10^{-3}$:

$$x = \frac{-b \pm \sqrt{b^2 - 4ac}}{2a}$$

$$= \frac{-(1.1 \times 10^{-2}) \pm \sqrt{(1.1 \times 10^{-2})^2 - 4(1)(-2.75 \times 10^{-3})}}{2(1)}$$

$$= 4.7 \times 10^{-2} = [H^+]$$

You can neglect the negative root because negative concentrations aren't possible.

Finally, use the pH definition to find the pH:

$$pH = -\log[H^+] = -\log[4.7 \times 10^{-2}] = 1.33$$

947. 9.38

Calcium acetate is a strong electrolyte, which completely separates into Ca^{2+} and $C_2H_3O_2^-$. The acetate ion is a conjugate base of a weak acid. The acetate ion isn't one of the strong bases (NaOH, KOH, LiOH, RbOH, CsOH, $Ca(OH)_2$, $Sr(OH)_2$, and $Ba(OH)_2$), so it must be a weak base. Weak base calculations need a base equilibrium constant, K_b. The generic form of every K_b problem is

$$CB \rightleftharpoons OH^- + CA$$

$$K_b = \frac{[OH^-][CA]}{[CB]}$$

where CA refers to the conjugate acid and CB refers to the conjugate base. For balancing purposes, water may be present in the equilibrium chemical equation; however, water is also the solvent, so it shouldn't be in the equilibrium expression.

The problem gives you the K_a, so convert the K_a to K_b using the following relationship, where $K_w = 1.00 \times 10^{-14}$:

$$K_w = K_a K_b$$

$$K_b = \frac{K_w}{K_a} = \frac{1.00 \times 10^{-14}}{1.7 \times 10^{-5}} = 5.9 \times 10^{-10}$$

The formula of calcium acetate, $Ca(C_2H_3O_2)_2$, contains two acetate ions; therefore, the concentration of the acetate ion is twice that of the calcium acetate, or 1.0 M. In this problem, the equilibrium equation and the change in the concentrations are

$$C_2H_3O_2^- + H_2O \rightleftharpoons OH^- + HC_2H_3O_2$$
$$1.0 - x +x +x$$

Entering this information in the K_b expression gives you the following:

$$K_b = \frac{[OH^-][HC_2H_3O_2]}{[C_2H_3O_2^-]}$$

$$5.9 \times 10^{-10} = \frac{[x][x]}{[1.0 - x]}$$

This is a quadratic equation, and you can solve it as such. But before doing the math, think about the problem logically. If the change in the denominator is insignificant, you can drop the $-x$ and make the problem easier to solve.

A simple check is to compare the exponent on the K to the exponent on the concentration (in scientific notation). In this case, the exponent on the K is -10, and the exponent on the concentration is 0. If the exponent on the K is at least 3 less than the exponent on the concentration (as it is in this case), you can assume that $-x$ in the denominator is insignificant. That means you can rewrite the K equation as

$$5.9 \times 10^{-10} = \frac{[x][x]}{[1.0]}$$

$$x = 2.4 \times 10^{-5} = [OH^-]$$

The x value is sufficiently small that $1.0 - x \approx 1.0$, which validates the assumption.

Next, find the pOH using the pOH definition:

$$pOH = -\log[OH^-] = -\log[2.4 \times 10^{-5}] = 4.62$$

Finally, find the pH using the relationship $pK_w = pH + pOH$, where $pK_w = 14.000$:

$$pH = pK_w - pOH = 14.000 - 4.62 = 9.38$$

948. 2.500×10^{-3} mol HCl

First, add the information from the problem to the balanced chemical equation:

$$HCl(aq) + NaOH(aq) \rightarrow NaCl(aq) + H_2O(aq)$$

    ? mol   0.1000 M

          25.00 mL

Next, change the molarity to moles by multiplying the molarity by the volume. The conversion is easier to see if you write the molarity unit M in terms of its definition (mol/L). The given volume is in milliliters, so remember that a liter is 1,000 mL:

$$mol = MV$$

$$= \left(\frac{mol}{L}\right) L$$

$$= \left(\frac{mol}{1,000\ mL}\right) mL$$

$$= \left(\frac{0.1000\ mol\ NaOH}{1,000\ mL}\right)(25.00\ mL)$$

$$= 2.500 \times 10^{-3}\ mol\ NaOH$$

Now convert from moles of NaOH to moles of HCl using the mole ratio, which is based on the coefficients from the balanced chemical equation. Here are all the calculations for this problem:

$$mol\ NaOH/mL \rightarrow mol\ NaOH \rightarrow mol\ HCl$$

$$\left(\frac{0.1000\ mol\ NaOH}{1,000\ mL}\right)\left(\frac{25.00\ mL}{1}\right)\left(\frac{1\ mol\ HCl}{1\ mol\ NaOH}\right) = 2.500 \times 10^{-3}\ mol\ HCl$$

949. 0.1771 M HClO

First, add the information from the question to the balanced chemical equation:

$$HClO(aq) + KOH(aq) \rightarrow KClO(aq) + H_2O(aq)$$

    ? M   0.1250 M

   25.00 mL   35.42 mL

Next, change the molarity to moles by multiplying the molarity by the volume. The conversion is easier to see if you write the molarity unit M in terms of its definition (mol/L). The given volume is in milliliters, so remember that a liter is 1,000 mL:

$$\text{mol} = MV$$

$$= \left(\frac{\text{mol}}{\cancel{L}} \right) \cancel{L}$$

$$= \left(\frac{\text{mol}}{1{,}000 \ \cancel{mL}} \right) \cancel{mL}$$

$$= \left(\frac{0.1250 \ \text{mol KOH}}{1{,}000 \ \cancel{mL}} \right) (35.42 \ \cancel{mL})$$

$$= 4.4275 \times 10^{-3} \ \text{mol KOH}$$

Now convert from moles of KOH to moles of HClO using the mole ratio, which is based on the coefficients from the balanced chemical equation:

$$\left(\frac{0.1250 \ \cancel{\text{mol KOH}}}{1{,}000 \ \cancel{mL}} \right) \left(\frac{35.42 \ \cancel{mL}}{1} \right) \left(\frac{1 \ \text{mol HClO}}{1 \ \cancel{\text{mol KOH}}} \right) = 4.4275 \times 10^{-3} \ \text{mol HClO}$$

Dividing by the volume of the HClO solution (in liters) yields the molarity. Here are all the calculations for this problem:

$$\text{mol KOH/mL} \rightarrow \text{mol KOH} \rightarrow \text{mol HClO} \rightarrow \text{mol HClO/L}$$

$$\left(\frac{0.1250 \ \cancel{\text{mol KOH}}}{1{,}000 \ \cancel{mL}} \right) \left(\frac{35.42 \ \cancel{mL}}{1} \right) \left(\frac{1 \ \text{mol HClO}}{1 \ \cancel{\text{mol KOH}}} \right) \left(\frac{1}{0.02500 \ \text{L}} \right) = 0.1771 \ \text{M HClO}$$

---

950.   **0.2123 M HClO$_2$**

First, add the information from the problem to the balanced chemical equation:

$$\text{HClO}_2(\text{aq}) + \text{NaOH}(\text{aq}) \rightarrow \text{NaOH}(\text{aq}) + \text{H}_2\text{O}(\text{aq})$$

   ? M   0.1350 M

25.00 mL   39.32 mL

Next, change the molarity to moles by multiplying the molarity by the volume. The conversion is easier to see if you write the molarity unit M in terms of its definition (mol/L). The given volume is in milliliters, so remember that a liter is 1,000 mL:

$$\text{mol} = MV$$

$$= \left(\frac{\text{mol}}{\cancel{L}} \right) \cancel{L}$$

$$= \left(\frac{\text{mol}}{1{,}000 \ \cancel{mL}} \right) \cancel{mL}$$

$$= \left(\frac{0.1350 \ \text{mol NaOH}}{1{,}000 \ \cancel{mL}} \right) (39.32 \ \cancel{mL})$$

$$= 5.3083 \times 10^{-3} \ \text{mol NaOH}$$

Now convert from moles of NaOH to moles of HClO$_2$ using the mole ratio, which is based on the coefficients from the balanced chemical equation:

$$\left(\frac{0.1350 \ \cancel{\text{mol NaOH}}}{1{,}000 \ \cancel{mL}} \right) \left(\frac{39.32 \ \cancel{mL}}{1} \right) \left(\frac{1 \ \text{mol HClO}_2}{1 \ \cancel{\text{mol NaOH}}} \right) = 5.3083 \times 10^{-3} \ \text{mol HClO}_2$$

Dividing by the volume of the $HClO_2$ solution (in liters) yields the molarity. Here are all the calculations for this problem:

$$\text{mol NaOH/mL} \rightarrow \text{mol NaOH} \rightarrow \text{mol } HClO_2 \rightarrow \text{mol } HClO_2/\text{L}$$

$$\left(\frac{0.1350 \text{ mol NaOH}}{1,000 \text{ mL}} \right)\left(\frac{39.32 \text{ mL}}{1} \right)\left(\frac{1 \text{ mol } HClO_2}{1 \text{ mol NaOH}} \right)\left(\frac{1}{0.02500 \text{ L}} \right) = 0.2123 \text{ M } HClO_2$$

---

951.   **1.250×10^{-2} mol LiOH**

First, add the information from the problem to the balanced chemical equation:

$$H_2SO_4(aq) + 2LiOH(aq) \rightarrow Li_2SO_4(aq) + 2H_2O(aq)$$

0.2500 M   ? mol

25.00 mL

Next, change the molarity to moles by multiplying the molarity by the volume. The conversion is easier to see if you write the molarity unit M in terms of its definition (mol/L). The given volume is in milliliters, so remember that a liter is 1,000 mL:

$$\text{mol} = MV$$

$$= \left(\frac{\text{mol}}{L} \right) L$$

$$= \left(\frac{\text{mol}}{1,000 \text{ mL}} \right) \text{mL}$$

$$= \left(\frac{0.2500 \text{ mol } H_2SO_4}{1,000 \text{ mL}} \right)(25.00 \text{ mL})$$

$$= 6.250 \times 10^{-3} \text{ mol } H_2SO_4$$

Now convert from moles of H_2SO_4 to moles of LiOH using the mole ratio, which is based on the coefficients from the balanced chemical equation. Here are all the calculations for this problem:

$$\text{mol } H_2SO_4/\text{mL} \rightarrow \text{mol } H_2SO_4 \rightarrow \text{mol LiOH}$$

$$\left(\frac{0.2500 \text{ mol } H_2SO_4}{1,000 \text{ mL}} \right)\left(\frac{25.00 \text{ mL}}{1} \right)\left(\frac{2 \text{ mol LiOH}}{1 \text{ mol } H_2SO_4} \right) = 1.250 \times 10^{-2} \text{ mol LiOH}$$

---

952.   **0.09496 M $Ca(OH)_2$**

First, add the information from the problem to the balanced chemical equation:

$$2HNO_3(aq) + Ca(OH)_2(aq) \rightarrow Ca(NO_3)_2(aq) + 2H_2O(aq)$$

0.1275 M   25.00 mL

37.24 mL   ? M

Next, change the molarity to moles by multiplying the molarity by the volume. The conversion is easier to see if you write the molarity unit M in terms of its definition (mol/L). The given volume is in milliliters, so remember that a liter is 1,000 mL:

$$\text{mol} = MV$$

$$= \left(\frac{\text{mol}}{L} \right) L$$

$$= \left(\frac{\text{mol}}{1,000 \text{ mL}} \right) \text{mL}$$

$$= \left(\frac{0.1275 \text{ mol HNO}_3}{1,000 \text{ mL}} \right) (37.24 \text{ mL})$$

$$= 2.7374 \times 10^{-3} \text{ mol HNO}_3$$

Now convert from moles of HNO_3 to moles of $Ca(OH)_2$ using the mole ratio, which is based on the coefficients from the balanced chemical equation:

$$\left(\frac{0.1275 \text{ mol HNO}_3}{1,000 \text{ mL}} \right) \left(\frac{37.24 \text{ mL}}{1} \right) \left(\frac{1 \text{ mol Ca(OH)}_2}{2 \text{ mol HNO}_3} \right) = 4.7481 \times 10^{-3} \text{ mol Ca(OH)}_2$$

Dividing by the volume of the $Ca(OH)_2$ solution (in liters) yields the molarity. Here are all the calculations for this problem:

$$\text{mol HNO}_3/\text{mL} \rightarrow \text{mol HNO}_3 \rightarrow \text{mol Ca(OH)}_2 \rightarrow \text{mol Ca(OH)}_2/\text{L}$$

$$\left(\frac{0.1275 \text{ mol HNO}_3}{1,000 \text{ mL}} \right) \left(\frac{37.24 \text{ mL}}{1} \right) \left(\frac{1 \text{ mol Ca(OH)}_2}{2 \text{ mol HNO}_3} \right) \left(\frac{1}{0.02500 \text{ L}} \right) = 0.09496 \text{ M Ca(OH)}_2$$

---

953. **0.5889 M Sr(OH)$_2$**

First, add the information from the problem to the balanced chemical equation:

$$2H_3PO_4(aq) + 3Sr(OH)_2(aq) \rightarrow Sr_3(PO_4)_2(aq) + 6H_2O(aq)$$

39.26 mL   25.00 mL

0.2500 M   ? M

Next, change the molarity to moles by multiplying the molarity by the volume. The conversion is easier to see if you write the molarity unit M in terms of its definition (mol/L). The given volume is in milliliters, so remember that a liter is 1,000 mL:

$$\text{mol} = MV$$

$$= \left(\frac{\text{mol}}{L} \right) L$$

$$= \left(\frac{\text{mol}}{1,000 \text{ mL}} \right) \text{mL}$$

$$= \left(\frac{0.2500 \text{ mol H}_3\text{PO}_4}{1,000 \text{ mL}} \right) (39.26 \text{ mL})$$

$$= 9.815 \times 10^{-3} \text{ mol H}_3\text{PO}_4$$

Now convert from moles of H_3PO_4 to moles of $Sr(OH)_2$ using the mole ratio, which is based on the coefficients from the balanced chemical equation:

$$\left(\frac{0.2500 \text{ mol H}_3\text{PO}_4}{1,000 \text{ mL}} \right) \left(\frac{39.26 \text{ mL}}{1} \right) \left(\frac{3 \text{ mol Sr(OH)}_2}{2 \text{ mol H}_3\text{PO}_4} \right) = 1.47225 \times 10^{-2} \text{ mol Sr(OH)}_2$$

Dividing by the volume of the $Sr(OH)_2$ solution (in liters) yields the molarity. Here are all the calculations for this problem:

$$\text{mol } H_3PO_4/mL \rightarrow \text{mol } H_3PO_4 \rightarrow \text{mol } Sr(OH)_2 \rightarrow \text{mol } Sr(OH)_2/L$$

$$\left(\frac{0.2500 \text{ mol } H_3PO_4}{1,000 \text{ mL}}\right)\left(\frac{39.26 \text{ mL}}{1}\right)\left(\frac{3 \text{ mol } Sr(OH)_2}{2 \text{ mol } H_3PO_4}\right)\left(\frac{1}{0.02500 \text{ L}}\right) = 0.5889 \text{ M } Sr(OH)_2$$

954.

3.500×10^{-3} mol HNO_2

The problem doesn't give you an equation, so the first step is to write a balanced chemical equation:

$$HNO_2(aq) + KOH(aq) \rightarrow KNO_2(aq) + H_2O(l)$$

Add the information from the problem to the balanced chemical equation:

$$HNO_2(aq) + KOH(aq) \rightarrow KNO_2(aq) + H_2O(l)$$

$$\text{? mol} \quad 0.1000 \text{ M}$$
$$35.00 \text{ mL}$$

Next, change the molarity to moles by multiplying the molarity by the volume. The conversion is easier to see if you write the molarity unit M in terms of its definition (mol/L). The given volume is in milliliters, so remember that a liter is 1,000 mL:

$$\text{mol} = MV$$

$$= \left(\frac{\text{mol}}{L}\right)L$$

$$= \left(\frac{\text{mol}}{1,000 \text{ mL}}\right)\text{mL}$$

$$= \left(\frac{0.1000 \text{ mol } KOH}{1,000 \text{ mL}}\right)(35.00 \text{ mL})$$

$$= 3.500 \times 10^{-3} \text{ mol } KOH$$

Now convert from moles of KOH to moles of HNO_2 using the mole ratio, which is based on the coefficients from the balanced chemical equation. Here are all the calculations for this problem:

$$\text{mol } KOH/mL \rightarrow \text{mol } KOH \rightarrow \text{mol } HNO_2$$

$$\left(\frac{0.1000 \text{ mol } KOH}{1,000 \text{ mL}}\right)\left(\frac{35.00 \text{ mL}}{1}\right)\left(\frac{1 \text{ mol } HNO_2}{1 \text{ mol } KOH}\right) = 3.500 \times 10^{-3} \text{ mol } HNO_2$$

955.

0.4277 M RbOH

The problem doesn't give you an equation, so the first step is to write a balanced chemical equation:

$$H_2SO_4(aq) + 2RbOH(aq) \rightarrow Rb_2SO_4(aq) + 2H_2O(l)$$

Add the information from the problem to the balanced chemical equation:

$$H_2SO_4(aq)+2RbOH(aq)\rightarrow Rb_2SO_4(aq)+2H_2O(l)$$

43.29 mL   25.00 mL

0.1235 M   ? M

Next, change the molarity to moles by multiplying the molarity by the volume. The conversion is easier to see if you write the molarity unit M in terms of its definition (mol/L). The given volume is in milliliters, so remember that a liter is 1,000 mL:

$$mol = MV$$

$$= \left(\frac{mol}{L}\right)L$$

$$= \left(\frac{mol}{1,000\ mL}\right)mL$$

$$= \left(\frac{0.1235\ mol\ H_2SO_4}{1,000\ mL}\right)(43.29\ mL)$$

$$= 1.06926\times10^{-3}\ mol\ H_2SO_4$$

Now convert from moles of H_2SO_4 to moles of RbOH using the mole ratio, which is based on the coefficients from the balanced chemical equation:

$$\left(\frac{0.1235\ mol\ H_2SO_4}{1,000\ mL}\right)\left(\frac{43.29\ mL}{1}\right)\left(\frac{2\ mol\ RbOH}{1\ mol\ H_2SO_4}\right)=5.346315\times10^{-3}\ mol\ RbOH$$

Dividing by the volume of the RbOH solution (in liters) yields the molarity. Here are all the calculations for this problem:

$$mol\ H_2SO_4/mL \rightarrow mol\ H_2SO_4 \rightarrow mol\ RbOH \rightarrow mol\ RbOH/L$$

$$\left(\frac{0.1235\ mol\ H_2SO_4}{1,000\ mL}\right)\left(\frac{43.29\ mL}{1}\right)\left(\frac{2\ mol\ RbOH}{1\ mol\ H_2SO_4}\right)\left(\frac{1}{0.02500\ L}\right)=0.4277\ M\ RbOH$$

956.   **1.865×10^{-2} M H_3PO_4**

The problem doesn't give you an equation, so the first step is to write a balanced chemical equation:

$$2H_3PO_4(aq)+3Ca(OH)_2(aq)\rightarrow Ca_3(PO_4)_2(s)+6H_2O(l)$$

Add the information from the problem to the balanced chemical equation:

$$2H_3PO_4(aq)+3Ca(OH)_2(aq)\rightarrow Ca_3(PO_4)_2(s)+6H_2O(l)$$

? M   0.1000 M

100.0 mL   27.98 mL

Next, change the molarity to moles by multiplying the molarity by the volume. The conversion is easier to see if you write the molarity unit M in terms of its definition (mol/L). The given volume is in milliliters, so remember that a liter is 1,000 mL:

$$\text{mol} = MV$$

$$= \left(\frac{\text{mol}}{\cancel{L}}\right)\cancel{L}$$

$$= \left(\frac{\text{mol}}{1,000\ \cancel{mL}}\right)\cancel{mL}$$

$$= \left(\frac{0.1000\ \text{mol Ca(OH)}_2}{1,000\ \cancel{mL}}\right)\left(27.98\ \cancel{mL}\right)$$

$$= 2.798\times10^{-3}\ \text{mol Ca(OH)}_2$$

Now convert from moles of $Ca(OH)_2$ to moles of H_3PO_4 using the mole ratio, which is based on the coefficients from the balanced chemical equation:

$$\left(\frac{0.1000\ \text{mol }\cancel{Ca(OH)_2}}{1,000\ \cancel{mL}}\right)\left(\frac{27.98\ \cancel{mL}}{1}\right)\left(\frac{2\ \text{mol H}_3\text{PO}_4}{3\ \text{mol }\cancel{Ca(OH)_2}}\right) = 1.865333\times10^{-3}\ \text{mol H}_3\text{PO}_4$$

Dividing by the volume of the H_3PO_4 solution (in liters) yields the molarity. Here are all the calculations for this problem:

$$\text{mol Ca(OH)}_2/\text{mL} \rightarrow \text{mol Ca(OH)}_2 \rightarrow \text{mol H}_3\text{PO}_4 \rightarrow \text{mol H}_3\text{PO}_4/\text{L}$$

$$\left(\frac{0.1000\ \text{mol }\cancel{Ca(OH)_2}}{1,000\ \cancel{mL}}\right)\left(\frac{27.98\ \cancel{mL}}{1}\right)\left(\frac{2\ \text{mol H}_3\text{PO}_4}{3\ \text{mol }\cancel{Ca(OH)_2}}\right)\left(\frac{1}{0.1000\ \text{L}}\right) = 1.865\times10^{-2}\ \text{M H}_3\text{PO}_4$$

---

957.   0.6170 M NH_3

The problem doesn't give you an equation, so the first step is to write a balanced chemical equation:

$$H_2SO_4(aq) + 2NH_3(aq) \rightarrow (NH_4)_2SO_4(aq)$$

Add the information from the problem to the balanced chemical equation:

$$H_2SO_4(aq) + 2NH_3(aq) \rightarrow (NH_4)_2SO_4(aq)$$

0.3215 M   ? M

47.98 mL   50.00 mL

Next, change the molarity to moles by multiplying the molarity by the volume. The conversion is easier to see if you write the molarity unit M in terms of its definition (mol/L). The given volume is in milliliters, so remember that a liter is 1,000 mL:

$$\text{mol} = MV$$

$$= \left(\frac{\text{mol}}{\cancel{L}}\right)\cancel{L}$$

$$= \left(\frac{\text{mol}}{1,000\ \cancel{mL}}\right)\cancel{mL}$$

$$= \left(\frac{0.3215\ \text{mol H}_2\text{SO}_4}{1,000\ \cancel{mL}}\right)\left(47.98\ \cancel{mL}\right)$$

$$= 1.542557\times10^{-2}\ \text{mol H}_2\text{SO}_4$$

Now convert from moles of H_2SO_4 to moles of NH_3 using the mole ratio, which is based on the coefficients from the balanced chemical equation:

$$\left(\frac{0.3215 \text{ mol } H_2SO_4}{1{,}000 \text{ mL}}\right)\left(\frac{47.98 \text{ mL}}{1}\right)\left(\frac{2 \text{ mol } NH_3}{1 \text{ mol } H_2SO_4}\right) = 3.085114 \times 10^{-2} \text{ mol } NH_3$$

Dividing by the volume of the NH_3 solution (in liters) yields the molarity. Here are all the calculations for this problem:

$$\text{mol } H_2SO_4/\text{mL} \rightarrow \text{mol } H_2SO_4 \rightarrow \text{mol } NH_3 \rightarrow \text{mol } NH_3/L$$

$$\left(\frac{0.3215 \text{ mol } H_2SO_4}{1{,}000 \text{ mL}}\right)\left(\frac{47.98 \text{ mL}}{1}\right)\left(\frac{2 \text{ mol } NH_3}{1 \text{ mol } H_2SO_4}\right)\left(\frac{1}{0.05000 \text{ L}}\right) = 0.6170 \text{ M } NH_3$$

---

958. **0.2303 g $HC_2H_3O_2$**

The problem doesn't give you an equation, so the first step is to write a balanced chemical equation:

$$HC_2H_3O_2(aq) + NaOH(aq) \rightarrow NaC_2H_3O_2(aq) + H_2O(l)$$

Add the information from the problem to the balanced chemical equation:

$$HC_2H_3O_2(aq) + NaOH(aq) \rightarrow NaC_2H_3O_2(aq) + H_2O(l)$$
$$\quad ? \text{ g} \qquad\quad 0.09527 \text{ M}$$
$$\qquad\qquad\quad 39.95 \text{ mL}$$

Next, change the molarity to moles by multiplying the molarity by the volume. The conversion is easier to see if you write the molarity unit M in terms of its definition (mol/L). The given volume is in milliliters, so remember that a liter is 1,000 mL:

$$\text{mol} = MV$$

$$= \left(\frac{\text{mol}}{L}\right)L$$

$$= \left(\frac{\text{mol}}{1{,}000 \text{ mL}}\right)\text{mL}$$

$$= \left(\frac{0.09527 \text{ mol } NaOH}{1{,}000 \text{ mL}}\right)(39.95 \text{ mL})$$

$$= 3.8060365 \times 10^{-3} \text{ mol } NaOH$$

Now convert from moles of $NaOH$ to moles of $HC_2H_3O_2$ using the mole ratio, which is based on the coefficients from the balanced chemical equation:

$$\left(\frac{0.09527 \text{ mol } NaOH}{1{,}000 \text{ mL}}\right)\left(\frac{39.95 \text{ mL}}{1}\right)\left(\frac{1 \text{ mol } HC_2H_3O_2}{1 \text{ mol } NaOH}\right) = 3.8060365 \times 10^{-3} \text{ mol } HC_2H_3O_2$$

To finish the problem, multiply the moles by the molar mass. Here are all the calculations for this problem:

$$\text{mol } NaOH/\text{mL} \rightarrow \text{mol } NaOH \rightarrow \text{mol } HC_2H_3O_2 \rightarrow \text{mol } HC_2H_3O_2/L$$

$$\left(\frac{0.09527 \text{ mol } NaOH}{1{,}000 \text{ mL}}\right)\left(\frac{39.95 \text{ mL}}{1}\right)\left(\frac{1 \text{ mol } HC_2H_3O_2}{1 \text{ mol } NaOH}\right)\left(\frac{60.52 \text{ g } HC_2H_3O_2}{1 \text{ mol } HC_2H_3O_2}\right)$$

$$= 0.2303 \text{ g } HC_2H_3O_2$$

959. **0.05333 g NH_3**

The problem doesn't give you an equation, so the first step is to write a balanced chemical equation:

$$HCl(aq) + NH_3(aq) \rightarrow NH_4Cl(aq)$$

Add the information from the problem to the balanced chemical equation:

$$HCl(aq) + NH_3(aq) \rightarrow NH_4Cl(aq)$$

0.08271 M   ? g

37.86 mL

Next, change the molarity to moles by multiplying the molarity by the volume. The conversion is easier to see if you write the molarity unit M in terms of its definition (mol/L). The given volume is in milliliters, so remember that a liter is 1,000 mL:

$$mol = MV$$

$$= \left(\frac{mol}{L} \right) L$$

$$= \left(\frac{mol}{1,000 \; mL} \right) mL$$

$$= \left(\frac{0.08271 \; mol \; HCl}{1,000 \; mL} \right) (37.86 \; mL)$$

$$= 3.1314006 \times 10^{-3} \; mol \; HCl$$

Now convert from moles of HCl to moles of NH_3 using the mole ratio, which is based on the coefficients from the balanced chemical equation:

$$\left(\frac{0.08271 \; mol \; HCl}{1,000 \; mL} \right) \left(\frac{37.86 \; mL}{1} \right) \left(\frac{1 \; mol \; NH_3}{1 \; mol \; HCl} \right) = 3.1314006 \times 10^{-3} \; mol \; NH_3$$

To finish the problem, multiply the moles by the molar mass. Here are all the calculations for this problem:

$$mol \; HCl/mL \rightarrow mol \; HCl \rightarrow mol \; NH3 \rightarrow g \; NH3$$

$$\left(\frac{0.08271 \; mol \; HCl}{1,000 \; mL} \right) \left(\frac{37.86 \; mL}{1} \right) \left(\frac{1 \; mol \; NH_3}{1 \; mol \; HCl} \right) \left(\frac{17.031 \; g \; NH_3}{1 \; mol \; NH_3} \right) = 0.05333 \; g \; NH_3$$

960. **HNO_2**

A buffer solution must contain a conjugate acid-base pair of a weak acid or base. Sodium nitrite ($NaNO_2$) contains the nitrite ion, which is the conjugate base of nitrous acid (HNO_2), a weak acid.

961. **$(NH_4)_2SO_4$**

A buffer solution must contain a conjugate acid-base pair of a weak acid or base. Ammonia (NH_3) is a weak base, so to create a buffer, you need to add the conjugate acid. The conjugate acid of ammonia acid is the ammonium ion $\left(NH_4^+ \right)$.

962. KHCO$_3$

A buffer solution must contain a conjugate acid-base pair of a weak acid or base. Sodium carbonate (Na$_2$CO$_3$) contains the carbonate ion, which is the conjugate base of the hydrogen carbonate ion $\left(HCO_3^-\right)$, a weak acid.

963. 4.58

This is a buffer solution problem, which is easiest to solve with the Henderson-Hasselbalch equation:

$$pH = pK_a + \log \frac{CB}{CA}$$

The conjugate acid, CA, is acetic acid (0.75 M), and the conjugate base, CB, is the acetate ion (0.50 M). The acetate ion comes from sodium acetate, which is a strong electrolyte.

Entering the given information in the Henderson-Hasselbalch equation gives you the answer:

$$pH = 4.76 + \log \frac{0.50}{0.75} = 4.58$$

You can also solve any buffer problem as a K_a or K_b problem. This normally takes more time than using the Henderson-Hasselbalch equation and results in the same answer.

Acetic acid isn't of the strong acids (HNO$_3$, HCl, HBr, HI, HClO$_3$, HClO$_4$, and H$_2$SO$_4$), so it must be a weak acid. Weak acid calculations need an acid equilibrium constant, K_a. The generic form of every K_a problem is

$$CA \rightleftharpoons H^+ + CB$$

$$K_a = \frac{\left[H^+\right]\left[CB\right]}{\left[CA\right]}$$

The problem gives you the pK_a, so convert the pK_a to a K_a:

$$K_a = 10^{-pK_a} = 10^{-4.76} = 1.7 \times 10^{-5}$$

In this case, the equilibrium equation and the change in the concentrations (calculated previously) are

$$HC_2H_3O_2 \rightleftharpoons H^+ + C_2H_3O_2^-$$
$$0.75 - x \quad +x \quad 0.50 + x$$

Entering this information in the K_a expression gives you the following:

$$K_a = \frac{\left[H^+\right]\left[C_2H_3O_2^-\right]}{\left[HC_2H_3O_2\right]}$$

$$1.7 \times 10^{-5} = \frac{\left[x\right]\left[0.50 + x\right]}{\left[0.75 - x\right]}$$

This is a quadratic equation, and you can solve it as such. But before doing the math, think about the problem logically. If the changes in the numerator and denominator are insignificant, as they are here, you can drop the $+x$ and $-x$, simplifying the problem:

$$1.7 \times 10^{-5} = \frac{[x][0.50]}{[0.75]}$$

$$x = 2.6 \times 10^{-5} = [H^+]$$

The x value is sufficiently small that $1.0 - x \approx 1.0$, which validates the assumption.

Finally, find the pH using the pH definition:

$$pH = -\log[H^+] = -\log[2.6 \times 10^{-5}] = 4.58$$

964.   **4.90**

This is a buffer solution problem, which is easiest to solve with the Henderson-Hasselbalch equation:

$$pH = pK_a + \log\frac{CB}{CA}$$

The conjugate acid, CA, is the ammonium ion (0.35 M), and the conjugate base, CB, is ammonia (0.25 M). The ammonium ion comes from ammonium chloride, which is a strong electrolyte.

Entering the given information in the Henderson-Hasselbalch equation gives you the answer:

$$pOH = 4.75 + \log\frac{0.35}{0.25} = 4.90$$

You can also solve any buffer problem as a K_a or K_b problem (see Question 963 for an example of this approach). This normally takes more time than using the Henderson-Hasselbalch equation and results in the same answer.

965.   **3.32**

This is a buffer solution problem, which is easiest to solve with the Henderson-Hasselbalch equation:

$$pH = pK_a + \log\frac{CB}{CA}$$

The conjugate acid, CA, is hydrofluoric acid (0.25 M), and the conjugate base, CB, is the fluoride ion (0.35 M). The fluoride ion comes from sodium fluoride, which is a strong electrolyte.

Entering the given information in the Henderson-Hasselbalch equation gives you the answer:

$$pH = 3.17 + \log\frac{0.35}{0.25} = 3.32$$

You can also solve any buffer problem as a K_a or K_b problem (see Question 963 for an example of this approach).

966. 3.10

This is a buffer solution problem, which is easiest to solve with the Henderson-Hasselbalch equation:

$$pOH = pK_b + \log\frac{CA}{CB}$$

The conjugate acid, CA, is the methylammonium ion (0.50 M), and the conjugate base, CB, is methylamine (0.75 M). The methylammonium ion comes from methylammonium chloride, which is a strong electrolyte.

Entering the given information in the Henderson-Hasselbalch equation gives you the following:

$$pOH = 3.28 + \log\frac{0.50}{0.75} = 3.10$$

You can also solve any buffer problem as a K_a or K_b problem (see Question 963 for an example of this approach).

967. 9.40

This is a buffer solution problem, which is easiest to solve with the Henderson-Hasselbalch equation:

$$pOH = pK_b + \log\frac{CA}{CB}$$

The conjugate acid, CA, is the ammonium ion (0.25 M), and the conjugate base, CB, is ammonia (0.35 M). The ammonium ion comes from ammonium chloride, which is a strong electrolyte.

Entering the given information in the Henderson-Hasselbalch equation gives you the following:

$$pOH = 4.75 + \log\frac{0.25}{0.35} = 4.60$$

To convert the pOH to pH, use the relationship $pH + pOH = pK_w$, where $pK_w = 14.000$:

$$pH = pK_w - pOH = 14.000 - 4.60 = 9.40$$

You can also solve any buffer problem as a K_a or K_b problem (see Question 963 for an example of this approach).

968. 11.02

This is a buffer solution problem, which is easiest to solve with the Henderson-Hasselbalch equation:

$$pOH = pK_b + \log\frac{CA}{CB}$$

The conjugate acid, CA, is the methylammonium ion (0.25 M), and the conjugate base, CB, is methylamine (0.50 M). The methylammonium ion comes from methylammonium chloride, which is a strong electrolyte.

$$pOH = 3.28 + \log \frac{0.25}{0.50} = 2.98$$

To finish the problem, convert the pOH to pH using the relationship pH + pOH = pK_w, where pK_w = 14.000:

$$pH = pK_w - pOH = 14.000 - 2.98 = 11.02$$

You can also solve any buffer problem as a K_a or K_b problem (see Question 963 for an example of this approach).

969.   **2.69**

The problem doesn't give you an equation, so the first step to write a balanced chemical equation:

$$NaOH(aq) + HF(aq) \rightarrow Na^+(aq) + F^-(aq) + H_2O(l)$$

Write the sodium fluoride as the separated ions, because the fluoride ion will be necessary for later calculations.

Adding the given amounts of material to the balanced chemical equation gives you the following:

$$NaOH(aq) + HF(aq) \rightarrow Na^+(aq) + F^-(aq) + H_2O(l)$$

$$0.25 \text{ mol} \quad 1.00 \text{ mol}$$

Quantities of two reactants are present, so this is a limiting reactant problem. Because the stoichiometry is 1:1, the substance in the lesser amount is the limiting reactant.

The limiting reactant will go to zero and take an equal amount of the other reactant with it. This means 0.75 mol of HF will remain. The reactant will produce an equal number of moles of product. The product of interest is the fluoride ion, and 0.25 mol of this ion will form. Dividing each of the mole values by the volume of solution gives the molarities: The solution is 0.75 M HF and 0.25 M F$^-$.

You now have a buffer solution problem, which is easiest to solve with the Henderson-Hasselbalch equation:

$$pH = pK_a + \log \frac{CB}{CA}$$

The conjugate acid, CA, is hydrofluoric acid (0.75 M), and the conjugate base, CB, is the fluoride ion (0.25 M). The fluoride ion comes from sodium fluoride, which is a strong electrolyte.

Entering the given information in the Henderson-Hasselbalch equation gives you the answer:

$$pH = 3.17 + \log \frac{0.25}{0.75} = 2.69$$

You can also solve any buffer problem as a K_a or K_b problem (see Question 963 for an example of this approach). This normally takes more time than using the Henderson-Hasselbalch equation and results in the same answer.

970. 9.55

The problem doesn't give you an equation, so the first step in this problem is to write a balanced chemical equation:

$$NH_3(aq) + HCl(aq) \rightarrow NH_4^+(aq) + Cl^-(aq)$$

Write the ammonium as separated ions, because the fluoride ion will be necessary for later calculations.

Adding the given amounts of material to the balanced chemical equation gives you the following:

$$NH_3(aq) + HCl(aq) \rightarrow NH_4^+(aq) + Cl^-(aq)$$
$$\qquad 0.75 \text{ mol} \quad 0.25 \text{ mol}$$

Quantities of two reactants are present, so this is a limiting reactant problem. Because the stoichiometry is 1:1, the substance in the lesser amount is the limiting reactant.

The limiting reactant will go to zero and take an equal amount of the other reactant with it. This means 0.50 mol of NH_3 will remain. The reactant will produce an equal number of moles of product. The product of interest is the ammonium ion, and 0.25 mol of this ion will form. Dividing each of the mole values by the volume of solution gives the molarities: The solution is 0.50 M NH_3 and 0.25 M NH_4^+.

You now have a buffer solution problem, which is easiest to solve with the Henderson-Hasselbalch equation:

$$pOH = pK_b + \log\frac{CA}{CB}$$

The conjugate acid, CA, is the ammonium ion (0.25 M), and the conjugate base, CB, is ammonia (0.50 M). The ammonium ion comes from ammonium chloride, which is a strong electrolyte.

Entering the given information in the Henderson-Hasselbalch equation gives you the following:

$$pOH = 4.75 + \log\frac{0.25}{0.50} = 4.45$$

To convert the pOH to pH and finish the problem, use the relationship pH + pOH = pK_w, where pK_w = 14.000:

$$pH = pK_w - pOH = 14.000 - 4.45 = 9.55$$

You can also solve any buffer problem as a K_a or K_b problem (see Question 963 for an example of this approach).

971. 2.91

The initial pH is the pH before the addition of any base; therefore, only the acetic acid will affect the pH of the solution.

Acetic acid isn't one of the strong acids (HNO_3, HCl, HBr, HI, $HClO_3$, $HClO_4$, and H_2SO_4), so it must be a weak acid. Weak acid calculations need an acid equilibrium constant, K_a. The generic form of every K_a problem is

$$CA \rightleftharpoons H^+ + CB$$

$$K_a = \frac{[H^+][CB]}{[CA]}$$

where CA refers to the conjugate acid and CB refers to the conjugate base. In this case, the equilibrium equation and the change in the concentrations are

$$HC_2H_3O_2 \rightleftharpoons H^+ + C_2H_3O_2^-$$
$$0.08750 - x \quad +x \quad +x$$

Entering this information in the K_a expression gives you the following:

$$K_a = \frac{[H^+][C_2H_3O_2^-]}{[HC_2H_3O_2]}$$

$$1.7 \times 10^{-5} = \frac{[x][x]}{[0.08750 - x]}$$

This is a quadratic equation, and you can solve it as such. But before doing the math, think about the problem logically. If the change in the denominator is insignificant, you can drop the $-x$ and make the problem easier to solve.

A simple check is to compare the exponent on the K to that of the concentration (in scientific notation). In this case, the exponent on the K is –5, and the exponent on the concentration is –2. If the exponent on the K is at least 3 less than the exponent on the concentration (as it is in this case), you can assume that $-x$ in the denominator is insignificant. Therefore, you can rewrite the K equation as

$$1.7 \times 10^{-5} = \frac{[x][x]}{[0.08750]}$$

$$x = 1.22 \times 10^{-3} = [H^+]$$

The x value is sufficiently small that $1.0 - x \approx 1.0$, which validates the assumption.

Finally, find the pH using the pH definition:

$$pH = -\log[H^+] = -\log[1.22 \times 10^{-3}] = 2.91$$

972. 11.11

The initial pH is the pH before the addition of any HCl; therefore, the only substance present that will affect the pH is the ammonia.

Ammonia isn't one of the strong bases (NaOH, KOH, LiOH, RbOH, CsOH, Ca(OH)$_2$, Sr(OH)$_2$, and Ba(OH)$_2$), so it must be a weak base. Weak base calculations need a base equilibrium constant, K_b. The generic form of every K_b problem is

$$CB \rightleftharpoons OH^- + CA$$

$$K_b = \frac{[OH^-][CA]}{[CB]}$$

where CA refers to the conjugate acid and CB refers to the conjugate base. For balancing purposes, water may be present in the equilibrium chemical equation; however, water is also the solvent, so it shouldn't be in the equilibrium expression.

The problem gives you the pK_b, so convert the pK_a to the K_b:

$$K_b = 10^{-4.75} = 1.8 \times 10^{-5}$$

In this case, the equilibrium equation and the change in the concentrations are

$$NH_3 + H_2O \rightleftharpoons OH^- + NH_4^+$$
$$0.08750 - x \qquad +x \quad +x$$

Entering this information in the K_b expression gives you the following:

$$K_b = \frac{\left[OH^-\right]\left[NH_4^+\right]}{\left[NH_3\right]}$$

$$1.8 \times 10^{-5} = \frac{[x][x]}{[0.08750 - x]}$$

This is a quadratic equation, and you can solve it as such. But before doing the math, think about the problem logically. If the change in the denominator is insignificant, you can drop the $-x$ and make the problem easier to solve.

A simple check is to compare the exponent on the K to the exponent on the concentration (in scientific notation). In this case, the exponent on the K is -5, and the exponent on the concentration is -2. If the exponent on the K is at least 3 less than the exponent on the concentration (as it is in this case), you can assume that $-x$ in the denominator is insignificant. That means you can rewrite the K equation as

$$1.8 \times 10^{-5} = \frac{[x][x]}{[0.08750]}$$

$$x = 1.3 \times 10^{-3} = \left[OH^-\right]$$

The x value is sufficiently small that $1.0 - x \approx 1.0$, which validates the assumption.

Find the pOH using the pOH definition:

$$pOH = -\log\left[OH^-\right] = -\log\left[1.3 \times 10^{-3}\right] = 2.89$$

Finally, find the pH using the relationship $pK_w = pH + pOH$, where $pK_w = 14.000$:

$$pH = pK_w - pOH = 14.000 - 2.89 = 11.11$$

---

973. 7.00

Sodium hydroxide is one of the strong bases (NaOH, KOH, LiOH, RbOH, CsOH, $Ca(OH)_2$, $Sr(OH)_2$, and $Ba(OH)_2$), and hydrochloric acid is one of the strong acids (HNO_3, HCl, HBr, HI, $HClO_3$, $HClO_4$, and H_2SO_4). No calculations are necessary for strong acid–strong base titrations, because the equivalence point is always at pH = 7.00.

---

974. 4.76

First, add the information from the problem to the balanced chemical equation:

$$Ba(OH)_2(aq) + 2HC_2H_3O_2(aq) \rightarrow Ba^{2+}(aq) + 2C_2H_3O_2^-(aq) + 2H_2O(l)$$
$$\quad 0.05000\ M \qquad 0.08800\ M$$
$$\quad 11.00\ mL \qquad\ 25.00\ mL$$

The problem gives you the quantities of two reactants, so you need to determine which reactant is the limiting reactant.

First, find the number of moles of each reactant. Change the molarity to moles by multiplying the molarity by the volume. The conversion is easier to see if you write the molarity unit M in terms of its definition (mol/L). The given volume is in milliliters, so remember that a liter is 1,000 mL:

$$\text{mol} = MV = \left(\frac{\text{mol}}{\text{L}}\right)\text{L} = \left(\frac{\text{mol}}{1,000 \text{ mL}}\right)\text{mL}$$

$$\left(\frac{0.08800 \text{ mol HC}_2\text{H}_3\text{O}_2}{1,000 \text{ mL}}\right)(25.00 \text{ mL}) = 2.200 \times 10^{-3} \text{ mol HC}_2\text{H}_3\text{O}_2$$

$$\left(\frac{0.05000 \text{ mol Ba(OH)}_2}{1,000 \text{ mL}}\right)(11.00 \text{ mL}) = 5.500 \times 10^{-4} \text{ mol Ba(OH)}_2$$

Due to the stoichiometry of the reaction, 5.500×10^{-4} mol of $Ba(OH)_2$ will react completely with twice as many moles (1.100×10^{-3} mol) of $HC_2H_3O_2$; because more $HC_2H_3O_2$ is present, the $Ba(OH)_2$ is the limiting reactant.

The amount of limiting reactant will decrease to zero, taking some of the excess reactant away and producing some of the product. The only product of importance is the conjugate base of the weak acid ($HC_2H_3O_2$).

The moles of the nonlimiting reactant (conjugate acid) remaining after the reaction is the original number of moles minus the moles reacted:

$$\left(2.200 \times 10^{-3} \text{ mol HC}_2\text{H}_3\text{O}_2\right) - \left[\left(\frac{5.500 \times 10^{-4} \text{ mol Ba(OH)}_2}{1}\right)\left(\frac{2 \text{ mol HC}_2\text{H}_3\text{O}_2}{1 \text{ mol Ba(OH)}_2}\right)\right]$$

$$= 1.100 \times 10^{-3} \text{ mol HC}_2\text{H}_3\text{O}_2$$

Next, use the moles of the limiting reactant to determine the moles of acetate ion (conjugate base) formed:

$$\left(\frac{5.500 \times 10^{-4} \text{ mol Ba(OH)}_2}{1}\right)\left(\frac{2 \text{ mol C}_2\text{H}_3\text{O}_2^{-}}{1 \text{ mol Ba(OH)}_2}\right) = 1.100 \times 10^{-3} \text{ mol C}_2\text{H}_3\text{O}_2^{-}$$

At this point, the volume of the solution is 25.00 mL + 11.00 mL = 36.00 mL = 0.03600 L.

Next, determine the concentrations of the conjugate acid and the conjugate base by dividing the moles of each substance by the volume of the solution in liters:

$$\left(\frac{1.100 \times 10^{-3} \text{ mol HC}_2\text{H}_3\text{O}_2}{0.03600 \text{ L}}\right) = 3.0556 \times 10^{-2} \text{ M HC}_2\text{H}_3\text{O}_2$$

$$\left(\frac{1.100 \times 10^{-3} \text{ mol C}_2\text{H}_3\text{O}_2^{-}}{0.03600 \text{ L}}\right) = 3.0556 \times 10^{-2} \text{ M C}_2\text{H}_3\text{O}_2^{-}$$

This is now a buffer solution problem, which is easiest to solve with the Henderson-Hasselbalch equation:

$$pH = pK_a + \log\frac{CB}{CA}$$

The conjugate acid, CA, is acetic acid (3.0556×10^{-2} M), and the conjugate base, CB, is the acetate ion (3.0556×10^{-2} M). The acetate ion comes from sodium acetate, which is a strong electrolyte.

Entering the given information in the Henderson-Hasselbalch equation gives you the final answer:

$$\text{pH} = 4.76 + \log \frac{3.0556 \times 10^{-2} \text{ M}}{3.0556 \times 10^{-2} \text{ M}} = 4.76$$

Note this is a special situation. Whenever the concentrations of the conjugate base and conjugate acid are the same, pH = pK_a

You can also solve any buffer problem as a K_a or K_b problem. This normally takes more time than using the Henderson-Hasselbalch equation and results in the same answer.

Acetic acid isn't one of the strong acids (HNO_3, HCl, HBr, HI, $HClO_3$, $HClO_4$, and H_2SO_4), so it must be a weak acid. Weak acid calculations need an acid equilibrium constant, K_a. The generic form of every K_a problem is

$$CA \rightleftharpoons H^+ + CB$$

$$K_a = \frac{\left[H^+\right]\left[CB\right]}{\left[CA\right]}$$

The problem gives you the pK_a, so convert the pK_a to a K_a:

$$K_a = 10^{-pK_a} = 10^{-4.76} = 1.7 \times 10^{-5}$$

In this case, the equilibrium equation and the change in the concentrations (calculated previously) are

$$
\begin{array}{ccccc}
HC_2H_3O_2 & \rightleftharpoons & H^+ & + & C_2H_3O_2{}^- \\
3.0556 \times 10^{-2} \text{ M} - x & & +x & & 3.0556 \times 10^{-2} \text{ M} + x
\end{array}
$$

Entering this information in the K_a expression gives you the following:

$$K_a = \frac{\left[H^+\right]\left[C_2H_3O_2{}^-\right]}{\left[HC_2H_3O_2\right]}$$

$$1.7 \times 10^{-5} = \frac{\left[x\right]\left[3.0556 \times 10^{-2} \text{ M} + x\right]}{\left[3.0556 \times 10^{-2} \text{ M} - x\right]}$$

This is a quadratic equation, and you can solve it as such. But before doing the math, think about the problem logically. If the changes in the numerator and denominator are insignificant, as they are here, you can drop the $+x$ and $-x$, simplifying the problem:

$$1.7 \times 10^{-5} = \frac{\left[x\right]\left[3.0556 \times 10^{-2} \text{ M}\right]}{\left[3.0556 \times 10^{-2} \text{ M}\right]}$$

$$x = 1.7 \times 10^{-5} = \left[H^+\right]$$

The x value is sufficiently small that $1.0 - x \approx 1.0$, which validates the assumption.

Finally, find the pH using the pH definition:

$$\text{pH} = -\log\left[H^+\right] = -\log\left[1.7 \times 10^{-5}\right] = 4.76$$

975. 9.25

First, add the information from the problem to the balanced chemical equation:

$$H_2SO_4(aq) + 2NH_3(aq) \rightarrow (NH_4)_2SO_4(aq)$$

0.05000 M   0.08800 M

11.00 mL   25.00 mL

The problem gives you the quantities of two reactants, so you need to determine which reactant is the limiting reactant.

First, find the number of moles of each reactant. Change the molarities to moles by multiplying the molarity by the volume. The conversion is easier to see if you write the molarity unit M in terms of its definition (mol/L). The given volume is in milliliters, so remember that a liter is 1,000 mL:

$$mol = MV = \left(\frac{mol}{\cancel{L}}\right)\cancel{L} = \left(\frac{mol}{1,000\,\cancel{mL}}\right)\cancel{mL}$$

$$\left(\frac{0.05000 \text{ mol } H_2SO_4}{1,000 \text{ mL}}\right)(11.00 \text{ mL}) = 5.500 \times 10^{-4} \text{ mol } H_2SO_4$$

$$\left(\frac{0.08800 \text{ mol } NH_3}{1,000 \text{ mL}}\right)(25.00 \text{ mL}) = 2.200 \times 10^{-3} \text{ mol } NH_3$$

Due to the stoichiometry of the reaction, 5.500×10^{-4} mol of H_2SO_4 will react completely with twice as many moles (1.100×10^{-3} mol) of NH_3; because more NH_3 is present, H_2SO_4 is the limiting reactant.

The amount of limiting reactant will decrease to zero, taking some of the excess reactant away and producing some of the product. The only product of importance is the conjugate acid of the weak base (NH_3).

The moles of nonlimiting reactant remaining (conjugate acid) after the reaction is the original number of moles minus the moles reacted:

$$\left(2.200 \times 10^{-3} \text{ mol } NH_3\right) - \left[\left(\frac{5.500 \times 10^{-4} \text{ mol } H_2SO_4}{1}\right)\left(\frac{2 \text{ mol } NH_3}{1 \text{ mol } H_2SO_4}\right)\right]$$

$$= 1.100 \times 10^{-3} \text{ mol } NH_3$$

Next, use the moles of the limiting reactant to determine the moles of ammonium ion (conjugate acid) formed:

$$\left(\frac{5.500 \times 10^{-4} \text{ mol } H_2SO_4}{1}\right)\left(\frac{2 \text{ mol } NH_4^+}{1 \text{ mol } H_2SO_4}\right) = 1.100 \times 10^{-3} \text{ mol } NH_4^+$$

At this point, the volume of the solution is 25.00 mL + 11.00 mL = 36.00 mL = 0.03600 L.

Next, determine the concentrations of the conjugate acid and the conjugate base by dividing the moles of the substance by the volume of the solution in liters:

$$\left(\frac{1.100 \times 10^{-3} \text{ mol } NH_3}{0.03600 \text{ L}}\right) = 3.0556 \times 10^{-2} \text{ M } NH_3$$

$$\left(\frac{1.100 \times 10^{-3} \text{ mol } NH_4^+}{0.03600 \text{ L}}\right) = 3.0556 \times 10^{-2} \text{ M } NH_4^+$$

This is now a buffer solution problem, which is easiest to solve by using the Henderson-Hasselbalch equation:

$$pOH = pK_b + \log \frac{CA}{CB}$$

The conjugate acid, CA, is the ammonium ion (3.0556×10^{-2} M), and the conjugate base, CB, is ammonia (3.0556×10^{-2} M).

Entering the given information in the Henderson-Hasselbalch equation gives you the pOH:

$$pOH = 4.75 + \log \frac{3.0556 \times 10^{-2} \text{ M}}{3.0556 \times 10^{-2} \text{ M}} = 4.75$$

Finally, find the pH using the relationship, $pK_w = pH + pOH$, where $pK_w = 14.000$:

$$pH = pK_w - pOH = 14.000 - 4.75 = 9.24$$

You can also solve any buffer problem as a K_a or K_b problem (see Question 974 for an example of this approach). This normally takes more time than using the Henderson-Hasselbalch equation and results in the same answer.

---

976.   5.76

First, add the information from the problem to the balanced chemical equation:

$$Ba(OH)_2(aq) + 2HC_2H_3O_2(aq) \rightarrow Ba^{2+}(aq) + 2C_2H_3O_2^-(aq) + 2H_2O(l)$$

| 0.05000 M | 0.08800 M |
| 20.00 mL | 25.00 mL |

The problem gives you the quantities of two reactants, so you need to determine which one is the limiting reactant.

First, find the number of moles of each reactant. Change the molarities to moles by multiplying the molarity by the volume. The conversion is easier to see if you write the molarity unit M in terms of its definition (mol/L). The given volume is in milliliters, so remember that a liter is 1,000 mL:

$$mol = MV = \left(\frac{mol}{L}\right)L = \left(\frac{mol}{1,000 \text{ mL}}\right)mL$$

$$\left(\frac{0.08800 \text{ mol } HC_2H_3O_2}{1,000 \text{ mL}}\right)(25.00 \text{ mL}) = 2.200 \times 10^{-3} \text{ mol } HC_2H_3O_2$$

$$\left(\frac{0.05000 \text{ mol } Ba(OH)_2}{1,000 \text{ mL}}\right)(20.00 \text{ mL}) = 1.000 \times 10^{-3} \text{ mol } Ba(OH)_2$$

Due to the stoichiometry of the reaction, 1.000×10^{-3} mol $Ba(OH)_2$ will react completely with twice as many moles (2.000×10^{-3} mol) of $HC_2H_3O_2$; because more $HC_2H_3O_2$ is present, the $Ba(OH)_2$ is the limiting reactant.

The amount of limiting reactant will decrease to zero, taking some of the excess reactant away and producing some of the product. The only product of importance is the conjugate base of the weak acid ($HC_2H_3O_2$).

The moles of nonlimiting reactant remaining (conjugate acid) after the reaction is the original number of moles minus the moles reacted:

$$\left(2.200\times10^{-3} \text{ mol HC}_2\text{H}_3\text{O}_2\right)-\left[\left(\frac{1.000\times10^{-3} \text{ mol Ba(OH)}_2}{1}\right)\left(\frac{2 \text{ mol HC}_2\text{H}_3\text{O}_2}{1 \text{ mol Ba(OH)}_2}\right)\right]$$

$$= 2.000\times10^{-4} \text{ mol HC}_2\text{H}_3\text{O}_2$$

Next, use the moles of the limiting reactant to determine the moles of acetate ion (conjugate base) formed:

$$\left(\frac{1.000\times10^{-3} \text{ mol Ba(OH)}_2}{1}\right)\left(\frac{2 \text{ mol C}_2\text{H}_3\text{O}_2^-}{1 \text{ mol Ba(OH)}_2}\right)=2.000\times10^{-3} \text{ mol C}_2\text{H}_3\text{O}_2^-$$

At this point, the volume of the solution is 25.00 mL + 20.00 mL = 45.00 mL = 0.04500 L.

Next, determine the concentrations of the conjugate acid and the conjugate base by dividing the moles of the substance by the volume of the solution in liters:

$$\left(\frac{2.000\times10^{-4} \text{ mol HC}_2\text{H}_3\text{O}_2}{0.04500 \text{ L}}\right)=4.444\times10^{-3} \text{ M HC}_2\text{H}_3\text{O}_2$$

$$\left(\frac{2.000\times10^{-3} \text{ mol C}_2\text{H}_3\text{O}_2^-}{0.04500 \text{ L}}\right)=4.444\times10^{-2} \text{ M C}_2\text{H}_3\text{O}_2^-$$

This is now a buffer solution problem, which is easiest to solve with the Henderson-Hasselbalch equation:

$$\text{pH} = \text{p}K_a + \log\frac{\text{CB}}{\text{CA}}$$

The conjugate acid, CA, is acetic acid (4.444×10^{-3} M), and the conjugate base, CB, is the acetate ion (4.444×10^{-3} M). The acetate ion comes from sodium acetate, which is a strong electrolyte.

Entering the given information in the Henderson-Hasselbalch equation gives you the final answer:

$$\text{pH} = 4.76 + \log\frac{4.444\times10^{-2} \text{ M}}{4.444\times10^{-3} \text{ M}} = 5.76$$

You can also solve any buffer problem as a K_a or K_b problem (see Question 974 for an example of this approach). This normally takes more time than using the Henderson-Hasselbalch equation and results in the same answer.

977. 8.72

First, add the information from the problem to the balanced chemical equation:

$$\text{H}_2\text{SO}_4(\text{aq})+2\text{NH}_3(\text{aq})\rightarrow(\text{NH}_4)_2\text{SO}_4(\text{aq})$$

| 0.05000 M | 0.08800 M |
| 17.00 mL | 25.00 mL |

The problem gives you the quantities of two reactants, so you need to determine which one is the limiting reactant.

First, find the number of moles of each reactant. Change the molarities to moles by multiplying the molarity by the volume. The conversion is easier to see if you write the molarity unit M in terms of its definition (mol/L). The given volume is in milliliters, so remember that a liter is 1,000 mL:

$$\text{mol} = MV = \left(\frac{\text{mol}}{\cancel{L}}\right)\cancel{L} = \left(\frac{\text{mol}}{1,000\,\cancel{\text{mL}}}\right)\cancel{\text{mL}}$$

$$\left(\frac{0.05000 \text{ mol } H_2SO_4}{1,000\,\cancel{\text{mL}}}\right)(17.00\,\cancel{\text{mL}}) = 8.500 \times 10^{-4} \text{ mol } H_2SO_4$$

$$\left(\frac{0.08800 \text{ mol } NH_3}{1,000\,\cancel{\text{mL}}}\right)(25.00\,\cancel{\text{mL}}) = 2.200 \times 10^{-3} \text{ mol } NH_3$$

Due to the stoichiometry of the reaction, 8.500×10^{-4} mol of H_2SO_4 will react completely with twice as many moles (1.700×10^{-3} mol) of NH_3; because more NH_3 is present, the H_2SO_4 is the limiting reactant.

The amount of limiting reactant will decrease to zero, taking some of the excess reactant away and producing some of the product. The only product of importance is the conjugate acid of the weak base (NH_3).

The moles of nonlimiting reactant remaining (conjugate acid) after the reaction is the original number of moles minus the moles reacted:

$$\left(2.200 \times 10^{-3} \text{ mol } NH_3\right) - \left[\left(\frac{8.500 \times 10^{-4}\,\cancel{\text{mol } H_2SO_4}}{1}\right)\left(\frac{2 \text{ mol } NH_3}{1\,\cancel{\text{mol } H_2SO_4}}\right)\right]$$

$$= 5.000 \times 10^{-4} \text{ mol } NH_3$$

Next, use the moles of the limiting reactant to determine the moles of ammonium ion (conjugate acid) formed:

$$\left(\frac{8.500 \times 10^{-4}\,\cancel{\text{mol } H_2SO_4}}{1}\right)\left(\frac{2 \text{ mol } NH_4^+}{1\,\cancel{\text{mol } H_2SO_4}}\right) = 1.700 \times 10^{-3} \text{ mol } NH_4^+$$

At this point, the volume of the solution is 25.00 mL + 17.00 mL = 42.00 mL = 0.04200 L.

Now determine the concentrations of the conjugate acid and the conjugate base by dividing the moles of the substance by the volume of the solution in liters:

$$\left(\frac{5.000 \times 10^{-4} \text{ mol } NH_3}{0.04200 \text{ L}}\right) = 1.190 \times 10^{-2} \text{ M } NH_3$$

$$\left(\frac{1.700 \times 10^{-3} \text{ mol } NH_4^+}{0.04200 \text{ L}}\right) = 4.048 \times 10^{-2} \text{ M } NH_4^+$$

This is now a buffer solution problem, which is easiest to solve with the Henderson-Hasselbalch equation:

$$\text{pOH} = pK_b + \log\frac{CA}{CB}$$

The conjugate acid, CA, is the ammonium ion (4.048×10^{-2} M), and the conjugate base, CB, is ammonia (1.190×10^{-2} M).

Entering the given information in the Henderson-Hasselbalch equation gives you the following:

$$\text{pOH} = 4.75 + \log\frac{4.048 \times 10^{-2} \text{ M}}{1.190 \times 10^{-2} \text{ M}} = 5.28$$

Finally, get the pH by using the relationship $pK_w = pH + pOH$, where $pK_w = 14.000$:

$$pH = pK_w - pOH = 14.000 - 5.28 = 8.72$$

You can also solve any buffer problem as a K_a or K_b problem (see Question 974 for an example of this approach). This normally takes more time than using the Henderson-Hasselbalch equation and results in the same answer.

---

978.   **8.72**

The problem doesn't give you an equation, so the first step is to write a balanced chemical equation:

$$Ba(OH)_2(aq) + 2HC_2H_3O_2(aq) \rightarrow Ba^{2+}(aq) + 2C_2H_3O_2^-(aq) + 2H_2O(l)$$

Then add the information from the problem to the balanced chemical equation:

$$Ba(OH)_2(aq) + 2HC_2H_3O_2(aq) \rightarrow Ba^{2+}(aq) + 2C_2H_3O_2^-(aq) + 2H_2O(l)$$

   0.05000 M   0.08800 M

                  25.00 mL

This reaction is at the equivalence point, so none of the reactants will remain. Therefore, the only thing present that might affect the pH is the conjugate base of the weak acid, which is the acetate ion.

You need to calculate the molarity of the acetate ion in the solution at the equivalence point, and to do this, you need to know the moles present and the total volume of the solution. But you have more information about the acetic acid, so begin calculations with the acetic acid.

First, change the molarity of the acetic acid to moles by multiplying the molarity by the volume. The conversion is easier to see if you write the molarity unit M in terms of its definition (mol/L). The given volume is in milliliters, so remember that a liter is 1,000 mL.

$$mol = MV$$

$$= \left(\frac{mol}{L} \right) L$$

$$= \left(\frac{mol}{1,000 \, mL} \right) mL$$

$$= \left(\frac{0.08800 \, mol \, HC_2H_3O_2}{1,000 \, mL} \right) (25.00 \, mL)$$

$$= 2.200 \times 10^{-3} \, mol \, HC_2H_3O_2$$

From the moles of acetic acid, you can determine both the number of moles of acetate ion formed and the volume of barium hydroxide solution necessary to do this.

The moles of acetate ion (the conjugate base) formed comes from multiplying the moles of acetic acid by the mole ratio from the balanced chemical equation:

$$\left(\frac{2.200 \times 10^{-3} \, mol \, HC_2H_3O_2}{1} \right) \left(\frac{2 \, mol \, C_2H_3O_2^-}{2 \, mol \, HC_2H_3O_2} \right) = 2.200 \times 10^{-3} \, mol \, C_2H_3O_2^-$$

Find the volume of barium hydroxide solution added using the moles of acetic acid, the mole ratio from the balanced chemical equation, and the molarity of the barium hydroxide solution:

$$\left(\frac{2.200\times10^{-3}\ \text{mol HC}_2\text{H}_3\text{O}_2}{1}\right)\left(\frac{1\ \text{mol Ba(OH)}_2}{2\ \text{mol HC}_2\text{H}_3\text{O}_2}\right)\left(\frac{1{,}000\ \text{mL}}{0.05000\ \text{mol Ba(OH)}_2}\right)=22.00\ \text{mL}$$

At this point, the volume of the solution is 25.00 mL + 22.00 mL = 47.00 mL = 0.04700 L.

Next, determine the concentration of the conjugate base by dividing the moles of the substance by the volume of the solution in liters:

$$\left(\frac{2.200\times10^{-3}\ \text{mol C}_2\text{H}_3\text{O}_2^-}{0.04700\ \text{L}}\right)=4.6809\times10^{-2}\ \text{M C}_2\text{H}_3\text{O}_2^-$$

Next is the equilibrium part of the problem. The acetate ion is the conjugate base of a weak acid, so it will be part of a K_b equilibrium. The generic form of every K_b problem is

$$CB \rightleftharpoons OH^- + CA$$

$$K_b=\frac{\left[OH^-\right]\left[CA\right]}{\left[CB\right]}$$

where CA refers to the conjugate acid and CB refers to the conjugate base. For balancing purposes, water may be present in the equilibrium chemical equation; however, water is also the solvent, so it shouldn't be in the equilibrium expression.

You need to convert the pK_a given in the problem to a K_b. This is a two-part conversion. First find pK_b using the relationship $pK_w = pK_a + pK_b$, where $pK_w = 14.000$:

$$pK_b = pK_w - pK_a = 14.000 - 4.76 = 9.24$$

Then change pK_b to K_b:

$$K_b=10^{-pK_b}=10^{-9.24}=5.75\times10^{-10}$$

In this problem, the equilibrium equation and the change in the concentrations are

$$C_2H_3O_2^- + H_2O \rightleftharpoons OH^- + HC_2H_3O_2$$
$$4.6809\times10^{-2}-x \qquad +x \qquad +x$$

Entering this information in the K_b expression gives you the following:

$$K_b=\frac{\left[OH^-\right]\left[HC_2H_3O_2\right]}{\left[C_2H_3O_2^-\right]}$$

$$5.75\times10^{-10}=\frac{[x][x]}{\left[4.6809\times10^{-2}-x\right]}$$

This is a quadratic equation, and you can solve it as such. But before doing the math, think about the problem logically. If the change in the denominator is insignificant, you can drop the $-x$ and make the problem easier to solve.

A simple check is to compare the exponent on the K to the exponent on the concentration (in scientific notation). In this case, the exponent on the K is –10, and the exponent on the concentration is –2. If the exponent on the K is at least 3 less than the exponent on the concentration (as it is in this case), you can assume that $-x$ in the denominator is insignificant. That means you can rewrite the K equation as

$$5.75\times10^{-10}=\frac{[x][x]}{\left[4.6809\times10^{-2}\right]}$$

$$x=5.19\times10^{-6}=\left[OH^-\right]$$

The x value is sufficiently small that $1.0 - x \approx 1.0$, which validates the assumption.

Find the pOH using the pOH definition:

$$pOH = -\log\left[OH^-\right] = -\log\left[5.19\times10^{-6}\right] = 5.28$$

Finally, find the pH using the relationship $pK_w = pH + pOH$:

$$pH = pK_w - pOH = 14.000 - 5.28 = 8.72$$

---

979.   9.19

The problem doesn't give you an equation, so the first step is to write a balanced chemical equation:

$$Ba(OH)_2(aq) + 2HC_2H_3O_2(aq) \rightarrow Ba^{2+}(aq) + 2C_2H_3O_2^-(aq) + 2H_2O(l)$$

Then add the information from the problem to the balanced chemical equation:

$$Ba(OH)_2(aq) + 2HC_2H_3O_2(aq) \rightarrow Ba^{2+}(aq) + 2C_2H_3O_2^-(aq) + 2H_2O(l)$$

    0.4800 M   0.7600 M

            25.00 mL

This reaction is at the equivalence point, so none of the reactants will remain. Therefore, the only thing present that might affect the pH is the acetate ion, the conjugate base of the weak acid. You need to calculate the molarity of the acetate ion in the solution at the equivalence point, which requires knowing the moles present and the total volume of the solution. The problem gives you more information about the acetic acid, so begin your calculations with acetic acid.

Change the molarity of the acetic acid to moles by multiplying the molarity by the volume. The conversion is easier to see if you write the molarity unit M in terms of its definition (mol/L). The given volume is in milliliters, so remember that a liter is 1,000 mL.

$$mol = MV$$
$$= \left(\frac{mol}{L}\right)L$$
$$= \left(\frac{mol}{1,000\,mL}\right)mL$$
$$= \left(\frac{0.7600\,mol\,HC_2H_3O_2}{1,000\,mL}\right)(25.00\,mL)$$
$$= 1.900\times10^{-2}\,mol\,HC_2H_3O_2$$

From the moles of acetic acid, you can determine both the number of moles of acetate ions formed and the volume of barium hydroxide solution.

The moles of acetate ion (conjugate base) formed comes from multiplying the moles of acetic acid by the mole ratio from the balanced chemical equation:

$$\left(\frac{1.900\times10^{-2}\,mol\,HC_2H_3O_2}{1}\right)\left(\frac{2\,mol\,C_2H_3O_2^-}{2\,mol\,HC_2H_3O_2}\right) = 1.900\times10^{-2}\,mol\,C_2H_3O_2^-$$

Find the volume of barium hydroxide solution added using the moles of acetic acid, the mole ratio from the balanced chemical equation, and the molarity of the barium hydroxide solution:

$$\left(\frac{1.900\times10^{-2} \text{ mol } \cancel{HC_2H_3O_2}}{1}\right)\left(\frac{1 \text{ mol } \cancel{Ba(OH)_2}}{2 \text{ mol } \cancel{HC_2H_3O_2}}\right)\left(\frac{1{,}000 \text{ mL}}{0.4800 \text{ mol } \cancel{Ba(OH)_2}}\right) = 19.79 \text{ mL}$$

At this point, the volume of the solution is 25.00 mL + 19.79 mL = 44.79 mL = 0.04479 L.

Now determine the concentration of the conjugate base by dividing the moles of the substance by the volume of the solution in liters:

$$\left(\frac{1.900\times10^{-2} \text{ mol } C_2H_3O_2^{-}}{0.04479 \text{ L}}\right) = 0.4242 \text{ M } C_2H_3O_2^{-}$$

Next is the equilibrium part of the problem. The acetate ion is the conjugate base of a weak acid, so it will be part of a K_b equilibrium. The generic form of every K_b problem is

$$CB \rightleftharpoons OH^{-} + CA$$

$$K_b = \frac{\left[OH^{-}\right]\left[CA\right]}{\left[CB\right]}$$

where CA refers to the conjugate acid and CB refers to the conjugate base. For balancing purposes, water may be present in the equilibrium chemical equation; however, water is also the solvent, so it shouldn't be in the equilibrium expression.

You need to convert the pK_a given in the problem to a K_b. This is a two-part conversion. First find pK_b using the relationship $pK_w = pK_a + pK_b$, where $pK_w = 14.000$:

$$pK_b = pK_w - pK_a = 14.000 - 4.76 = 9.24$$

Then change pK_b to K_b:

$$K_b = 10^{-pK_b} = 10^{-9.24} = 5.75\times10^{-10}$$

In this problem, the equilibrium equation and the change in the concentrations are

$$C_2H_3O_2^{-} + H_2O \rightleftharpoons OH^{-} + HC_2H_3O_2$$
$$0.4242 \text{ M} \qquad\quad +x \qquad +x$$

Entering this information in the K_a expression gives you the following:

$$K_b = \frac{\left[OH^{-}\right]\left[HC_2H_3O_2\right]}{\left[C_2H_3O_2^{-}\right]}$$

$$5.75\times10^{-10} = \frac{[x][x]}{[0.4242 - x]}$$

This is a quadratic equation, and you can solve it as such. But before doing the math, think about the problem logically. If the change in the denominator is insignificant, you can drop the –x and make the problem easier to solve.

A simple check is to compare the exponent on the K to that of the concentration (in scientific notation). In this case, the exponent on the K is –10, and the exponent on the concentration is –1. If the exponent on the K is at least 3 less than the exponent on the concentration (as it is in this case), you can assume that –x in the denominator is insignificant. That means you can rewrite the K equation as

$$5.75\times10^{-10} = \frac{[x][x]}{[0.4242]}$$

$$x = 1.56\times10^{-5} = \left[OH^{-}\right]$$

The x value is sufficiently small that $1.0 - x \approx 1.0$, which validates the assumption.

Find the pOH using the pOH definition:

$$pOH = -\log\left[OH^-\right] = -\log\left[1.56 \times 10^{-5}\right] = 4.81$$

Finally, find the pH using the relationship $pK_w = pH + pOH$:

$$pH = pK_w - pOH = 14.000 - 4.81 = 9.19$$

980.   5.26

The problem doesn't give you an equation, so the first step is to write a balanced chemical equation:

$$2NH_3(aq) + H_2SO_4(aq) \rightarrow 2NH_4^+(aq) + SO_4^{2-}(aq)$$

Then add the information from the problem to the balanced chemical equation:

$$2NH_3(aq) + H_2SO_4(aq) \rightarrow 2NH_4^+(aq) + SO_4^{2-}(aq)$$

0.09800 M   0.06000 M

25.00 mL

This reaction is at the equivalence point, so none of the reactants will remain. Therefore, the only thing present that might affect the pH is the ammonium ion, the conjugate acid of the weak base.

You need to calculate the molarity of the ammonium ion in the solution at the equivalence point, which requires knowing the moles present and the total volume of the solution. The problem gives you more information about the ammonia, so begin calculations with ammonia.

Change the molarity of the ammonia to moles by multiplying the molarity by the volume. The conversion is easier to see if you write the molarity unit in terms of its definition (mol/L). The given volume is in milliliters, so remember that a liter is 1,000 mL.

$$mol = MV$$

$$= \left(\frac{mol}{L}\right)L$$

$$= \left(\frac{mol}{1,000\ mL}\right)mL$$

$$= \left(\frac{0.09800\ mol\ NH_3}{1,000\ mL}\right)(25.00\ mL)$$

$$= 2.450 \times 10^{-3}\ mol\ NH_3$$

From the moles of ammonia, you can determine both the number of moles of ammonium ion formed and the volume of the sulfuric acid solution.

The moles of ammonium ion (conjugate acid) formed comes from multiplying the moles of ammonia by the mole ratio from the balanced chemical equation:

$$\left(\frac{2.450 \times 10^{-3}\ mol\ NH_3}{1}\right)\left(\frac{2\ mol\ NH_4^+}{2\ mol\ NH_3}\right) = 2.450 \times 10^{-3}\ mol\ NH_4^+$$

Find the volume of barium hydroxide solution added using the moles of ammonia, the mole ratio from the balanced chemical equation, and the molarity of the sulfuric acid solution:

$$\left(\frac{2.450\times10^{-3}\ \text{mol NH}_3}{1}\right)\left(\frac{1\ \text{mol H}_2\text{SO}_4}{2\ \text{mol NH}_3}\right)\left(\frac{1,000\ \text{mL}}{0.06000\ \text{mol H}_2\text{SO}_4}\right)=20.42\ \text{mL}$$

At this point, the volume of the solution is 25.00 mL + 20.42 mL = 45.42 mL = 0.04542 L.

Now determine the concentration of the conjugate base by dividing the moles of the substance by the volume of the solution in liters:

$$\left(\frac{2.450\times10^{-3}\ \text{mol NH}_4^+}{0.04542\ \text{L}}\right)=5.394\times10^{-2}\ \text{M NH}_4^+$$

Next is the equilibrium part of the problem. The ammonium ion is the conjugate acid of a weak base, so it will be part of a K_a equilibrium. The generic form of every K_a problem is

$$CA \rightleftharpoons H^+ + CB$$

$$K_a = \frac{[H^+][CB]}{[CA]}$$

where CA refers to the conjugate acid and CB refers to the conjugate base.

You need to convert the pK_a given in the problem to a K_b. This is a two-part conversion. First find pK_b using the relationship $pK_w = pK_a + pK_b$, where $pK_w = 14.000$:

$$pK_b = pK_w - pK_a = 14.000 - 4.75 = 9.25$$

Then change pK_b to K_b:

$$K_b = 10^{-pK_b} = 10^{-9.25} = 5.62\times10^{-10}$$

In this case, the equilibrium equation and the change in the concentrations are

$$NH_4^+ \qquad \rightleftharpoons H^+ + NH_3$$
$$5.394\times10^{-2}-x \quad +x \quad +x$$

Entering this information in the K_a expression gives you the following:

$$K_a = \frac{[H^+][NH_3]}{[NH_4^+]}$$

$$5.62\times10^{-10} = \frac{[x][x]}{[5.394\times10^{-2}-x]}$$

This is a quadratic equation, and you can solve it as such. But before doing the math, think about the problem logically. If the change in the denominator is insignificant, you can drop the $-x$ and make the problem easier to solve.

A simple check is to compare the exponent on the K to that of the concentration (in scientific notation). In this case, the exponent on the K is –10, and the exponent on the concentration is –2. If the exponent on the K is at least 3 less than the exponent on the concentration (as it is in this case), you can assume that $-x$ in the denominator is insignificant, simplifying the K equation to

$$5.62 \times 10^{-10} = \frac{[x][x]}{\left[5.394 \times 10^{-2}\right]}$$

$$x = 5.51 \times 10^{-6} = \left[H^+\right]$$

The x value is sufficiently small that $1.0 - x \approx 1.0$, which validates the assumption.

Finally, find the pH using the pH definition:

$$pH = -\log\left[H^+\right] = -\log\left[5.51 \times 10^{-6}\right] = 5.26$$

---

981. origin

The coordinates of the *origin* are (0, 0).

---

982. interpolation

Finding a point that falls between the minimum and maximum values plotted on a line or curve is *interpolation;* you're finding a value inside your data.

Interpolation is an important method for estimating values based on data trends from the experiment. You can find an approximate value for a point that falls between the tested data points without doing an additional experiment, which can save a significant amount of time.

---

983. extrapolation

Estimating a value that falls below the minimum values or above the maximum values but follows the trend shown in the graph is called *extrapolation.*

Extrapolation allows you to make an educated guess about a result of a future experiment. The method assumes that the trend observed in the given values will continue beyond this range, which isn't always a valid assumption. One important aspect of science is that a theory should be able to predict future events (extrapolation).

---

984. the origin

The line crosses the x-axis when the volume equals zero. This graph should pass through the origin because zero volume corresponds to zero mass.

---

985. absolute zero

This graph is a plot of Charles's gas law. The x-intercept represents the point where the volume of an ideal gas would be zero, which occurs when the temperature is absolute zero.

---

986. speed

The slope is *rise over run:* the change in y-values (distance) divided by the change in x-values (time). Therefore, the line has units of distance/time, such as miles per hour, which correspond to units of speed.

987. density

The slope is *rise over run:* the change in *y*-values (mass) divided by the change in *x*-values (volume). Therefore, the line has units of mass/volume, such as grams per cubic centimeter, which correspond to units of density. *Density* is the mass of something divided by its volume.

988. heat capacity

The slope is *rise over run:* the change in *y*-values (heat energy) divided by the change in *x*-values (temperature). Therefore, the line has units of energy/temperature, such as joules per degree Celsius. This unit corresponds to *heat capacity,* which is the amount of energy necessary to increase the temperature of a substance by 1°C.

989. straight line with a positive slope

A *directly proportional* relationship produces a straight line with a positive slope. You can write the relationship from Charles's law as $V \propto T$, where $\propto$ means *proportional.* With this relationship, as the gas's temperature increases, the volume increases by the same factor. For example, if you double the temperature, the volume doubles as well, and the volume triples if you triple the temperature.

This law assumes that the pressure and the amount of gas are constant.

990. a line curving downward

A graph showing an *inversely proportional* relationship is a smooth curve that slopes downward. You can write the relationship from Boyle's law as $P \propto \frac{1}{V}$, or PV = constant.

With this relationship, if the pressure doubles, the volume is halved, and if you triple the pressure, you get a third of the volume.

This law assumes that the temperature and the amount of gas are constant.

991. a line curving upward

The volume is related to the edge of a cube *(s)*. The relationship is $V = s^3$. Cubic relationships give curved lines. The graph curves upward because a larger cube has more volume.

992. 27 pennies

After plotting the data, draw the line of best fit. For a linear set of data, the line of best fit is a straight line that splits the difference between the data points. Points will be equally distributed on each side of this line. In science, graphs very rarely connect the dots.

The following graph shows the plotted data points and the best-fit line. Your graph should have individual points above and below this line.

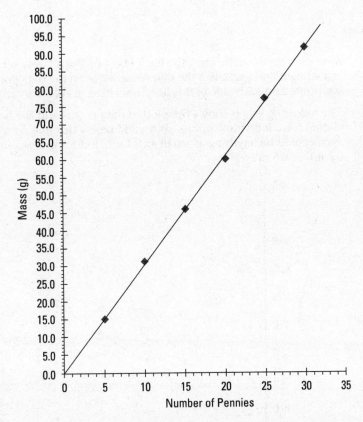

To answer this question, go up to about 84.1 on the *y*-axis and then go right until you hit the line of best fit. Next, go down to the *x*-axis and read the value.

Note: The smallest division on your *x*-axis could be as small as 0.1 penny, but for practicality, it will probably be 1 penny or larger to avoid cutting a penny into pieces.

993. the average mass of one penny

To find the slope of a line, you choose two points on the line and divide the difference in the *y*-coordinates by the difference in the *x*-coordinates. The *y*-values represent mass, and the *x*-values represent the number of pennies:

$$m = \frac{\Delta y}{\Delta x} = \frac{y_2 - y_1}{x_2 - x_1}$$

$$= \frac{\text{mass}}{\text{number of pennies}}$$

$$= \frac{\text{grams}}{\text{penny}}$$

The units (grams/penny) indicate that the slope represents the average mass of one penny.

994. 4.6 mL

After plotting the data, draw the line of best fit. For a linear set of data, the line of best fit is a straight line that splits the difference between the data points. Points will be equally distributed on each side of this line. In science, graphs very rarely connect the dots.

The following graph shows the plotted data points and the best-fit line. Your graph should have individual points above and below this line. The smallest division on your x-axis could be anything as small as 0.1 mL, but for practicality, it's more likely to be 0.2 mL or 0.5 mL.

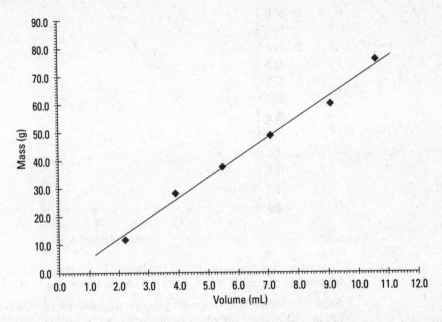

To answer this question, go up to about 31.0 g on the y-axis and then go right until you hit the line of best fit. Next, go down to the x-axis and read the value.

995. 70.1 g

After plotting the data, draw the line of best fit. For a linear set of data, the line of best fit is a straight line that splits the difference between the data points. Points will be equally distributed on each side of this line. In science, graphs very rarely connect the dots.

The following graph shows the plotted data points and the best-fit line. Your graph should have individual points above and below this line. The smallest division on your y-axis could be anything as small as 0.01 g, but for practicality, it's more likely to be 0.1 g, 0.5 g, or 1 g.

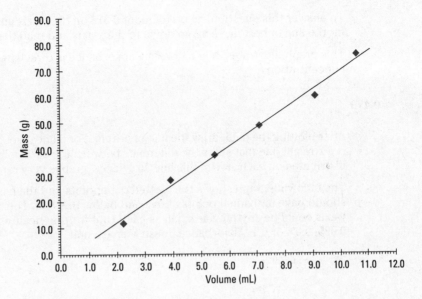

To answer this question, go over to about 10.0 mL on the *x*-axis and then go up until you hit the line of best fit. Next, go over to the *y*-axis and read the value.

---

996.   **0.083 M**

After plotting the data, draw the line of best fit. For a linear set of data, the line of best fit is a straight line that splits the difference between the data points. Points will be equally distributed on each side of this line. In science, graphs very rarely connect the dots.

The following graph shows the plotted data points and the best-fit line. Your graph should have individual points above and below this line. The smallest division on your *x*-axis could be anything as small as 0.001 M, but for practicality, it's more likely to be 0.0025 M or 0.0050 M.

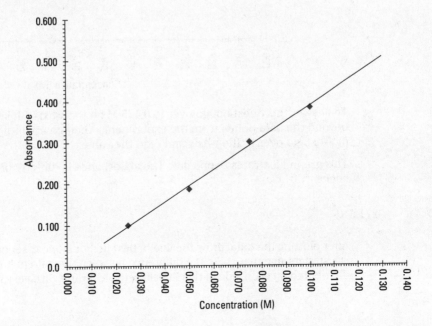

To answer this question, go up to about 0.318 on the *y*-axis and then go right until you hit the line of best fit. Next, go down to the *x*-axis and read the value.

This graph illustrates Beer's law. The absorbance is directly proportional to the concentration.

997.   **0.479**

After plotting the data, draw the line of best fit. For a linear set of data, the line of best fit is a straight line that splits the difference between the data points. Points will be equally distributed on each side of this line. In science, graphs very rarely connect the dots.

The following graph shows the plotted data points and the best-fit line. Your graph should have individual points above and below this line. The smallest division on your *y*-axis could be anything as small as 0.001, but for practicality, it's more likely to be 0.025, 0.05, or 0.1. Absorbance doesn't have a unit.

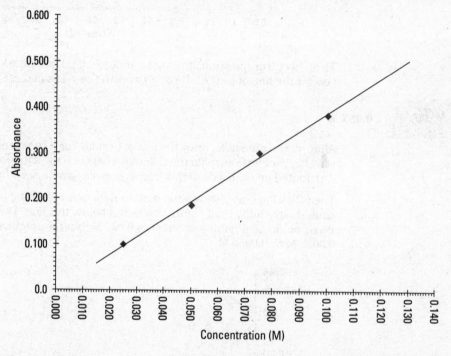

To answer this question, go over to 0.125 M on the *x*-axis (this is an extrapolated value, beyond the data points from the experiment); then go up until you hit the line of best fit. Next, go over to the *y*-axis and read the value.

This graph illustrates Beer's law. The absorbance is directly proportional to the concentration.

998.   **−271.5°C**

After plotting the data, draw the line of best fit. For a linear set of data, the line of best fit is a straight line that splits the difference between the data points. Points will be equally distributed on each side of this line. In science, graphs are very rarely connect the dots.

The following graph shows the best-fit line. Your graph should have individual points above and below this line. The smallest division on your *x*-axis could be anything as small as 0.1°C, but for practicality, it's more likely to be 10°C or 25°C (although numbers this big will make it difficult to estimate the value).

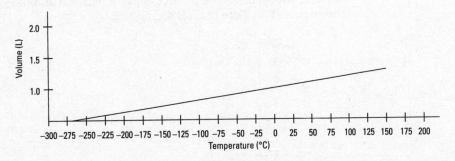

To answer this question, go over on the line of best fit until it intersects with the *x*-axis and read the value.

This graph illustrates Charles's law and the method for determining *absolute zero*, the lowest possible temperature. Data from other experiments have shown that the actual value of absolute zero, which is 0 kelvins on the Kelvin scale, is –273.15°C.

999. **the line has a gradual positive slope, shoots up, and then levels out**

This graph is shaped like an *s*-curve that has had the top and bottom pulled so that the middle of the *s* is almost a vertical line:

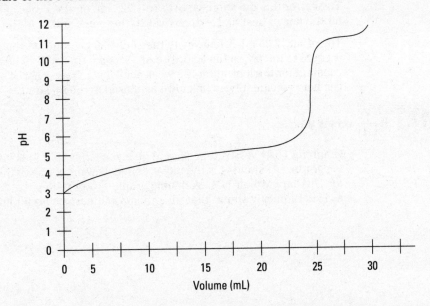

The graph shows data from an acid-base titration in which the base is the titrant. In an acid-base titration, the pH of the solution is plotted versus the volume of a substance added. The center of the sharp rise is the equivalence point of the titration.

1,000. 4.64

Plot the data and draw the line of best fit, which is a smooth curve. The smallest division on your y-axis could be anything as small as 0.01, but for practicality, it's more likely to be 1.00. Note that pH has no unit.

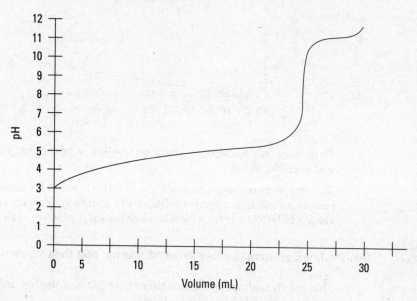

To answer this question, go over to 12.25 mL on the x-axis and then go up until you hit the line of best fit. Next, go over to the y-axis and read the value.

The pH at 12.25 mL is halfway to the equivalence point. At the *equivalence point*, the pH is equal to the pK_a of the acid. The pK_a is equal to $-\log K_a$ where K_a is the equilibrium constant for the ionization of a weak acid. If you have a list of possible acids, the K_a value can help you identify an unknown acid used in the titration.

1,001. $\dfrac{1}{[A]}$ **versus time**

Graphing $1/[A]$ versus time should give you a line that looks like the one in the following graph. $[A]$ stands for the molar concentration of one of the reactants, so the units for $1/[A]$ are M^{-1}, or $1/M$. Sketching graphs (or graphing the data with a program such as Excel) quickly shows that the other combinations aren't linear.

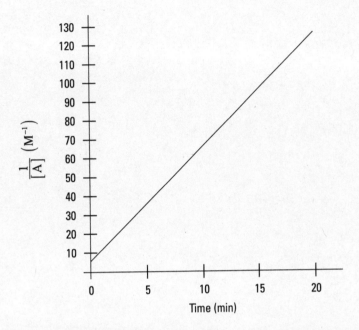

A second-order kinetics reaction always gives you a positive linear plot of $1/[A]$ versus time. Identifying this relationship is a useful way to show that a reaction follows second-order kinetics. The slope of the line is related to the rate constant for the reaction. A first-order reaction would give you a negative linear plot for $\ln[A]$ versus time, and a zero-order reaction would give you a negative linear plot for $[A]$ versus time.

Appendix

The Periodic Table of Elements

*U*se this periodic table as a reference for various problems throughout the book.

Index

Workspace

Workspace

Workspace

About the Authors

Heather Hattori became interested in chemistry her junior year in high school when a coach told her that drafting and architecture weren't for girls. Luckily, her chemistry and physics teachers had no such misconceptions. While teaching high school, she completed her master's degree in chemistry and biology and rediscovered her love for the outdoors. When not dreaming about working on a PhD or writing fiction, Heather crafts and volunteers in her community. Her parting words to you are "Practice, practice, practice so that you can follow your dreams — some things in life are necessary to learn even if you don't think you'll ever need them."

Richard H. Langley grew up in southwestern Ohio. He attended Miami University in Oxford, Ohio, where he earned bachelor's degrees in chemistry and mineralogy and a master's degree in chemistry. He next went to the University of Nebraska in Lincoln, where he received his doctorate in chemistry. He took a postdoctoral position at Arizona State University in Tempe, Arizona, and then became a visiting assistant professor at the University of Wisconsin–River Falls. He has taught at Stephen F. Austin State University in Nacogdoches, Texas, since 1982.

He is the author of *500 Physical Chemistry Questions* and coauthor of numerous other books, including *Chemistry for the Utterly Confused, Organic Chemistry II For Dummies,* and *Biochemistry For Dummies.*

Dedication

From Rich: I would like to thank Adrienne Soliz for her friendship and support.

Authors' Acknowledgments

From Heather: I would like to thank my parents and all the students I've ever taught before. Without life and the ability to test problems on students during 20 years of teaching, coming up with the fodder for this book would have been very unlikely. I would especially like to thank my very patient editors Chrissy Guthrie, Lindsey Lefevere, and Danielle Voirol at Wiley, who held it all together through a very long series of unfortunate incidents in my life that happened during the writing of this book. Thanks also to technical reviewers Lacey Moss and Jason Dunham for their careful review of the questions and solutions and to my agent Grace Freedson, who manages all the other important details in all of my "special projects." And lastly, I want to thank my college general chemistry professor "Doc" Richard H. Langley and teaching mentor Dr. John T. Moore, without whom this book would not have been possible.

From Rich: I would like to thank Lindsay Lefevere, Danielle Voirol, and especially Chrissy Guthrie for their patience and help. In addition, I would like to thank Lacey Moss and Jason Dunham for their suggestions.

Publisher's Acknowledgments

Executive Editor: Lindsay Sandman Lefevere

Senior Project Editor: Christina Guthrie

Senior Copy Editor: Danielle Voirol

Technical Editors: Jason Dunham, Lacey Moss

Project Coordinators: Katie Crocker, Sheree Montgomery

Project Managers: Jay Kern, Laura Moss-Hollister

Cover Image: ©iStockphoto.com/webking

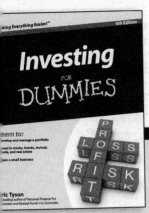

Math & Science

Algebra I For Dummies,
2nd Edition
978-0-470-55964-2

Anatomy and Physiology
For Dummies,
2nd Edition
978-0-470-92326-9

Astronomy For Dummies,
3rd Edition
978-1-118-37697-3

Biology For Dummies,
2nd Edition
978-0-470-59875-7

Chemistry For Dummies,
2nd Edition
978-1-1180-0730-3

Pre-Algebra Essentials
For Dummies
978-0-470-61838-7

Microsoft Office

Excel 2013 For Dummies
978-1-118-51012-4

Office 2013 All-in-One
For Dummies
978-1-118-51636-2

PowerPoint 2013
For Dummies
978-1-118-50253-2

Word 2013 For Dummies
978-1-118-49123-2

Music

Blues Harmonica
For Dummies
978-1-118-25269-7

Guitar For Dummies,
3rd Edition
978-1-118-11554-1

iPod & iTunes
For Dummies,
10th Edition
978-1-118-50864-0

Programming

Android Application
Development For Dummies,
2nd Edition
978-1-118-38710-8

iOS 6 Application
Development For Dummies
978-1-118-50880-0

Java For Dummies,
5th Edition
978-0-470-37173-2

Religion & Inspiration

The Bible For Dummies
978-0-7645-5296-0

Buddhism For Dummies,
2nd Edition
978-1-118-02379-2

Catholicism For Dummies,
2nd Edition
978-1-118-07778-8

Self-Help & Relationships

Bipolar Disorder
For Dummies,
2nd Edition
978-1-118-33882-7

Meditation For Dummies,
3rd Edition
978-1-118-29144-3

Seniors

Computers For Seniors
For Dummies,
3rd Edition
978-1-118-11553-4

iPad For Seniors
For Dummies,
5th Edition
978-1-118-49708-1

Social Security
For Dummies
978-1-118-20573-0

Smartphones & Tablets

Android Phones
For Dummies
978-1-118-16952-0

Kindle Fire HD
For Dummies
978-1-118-42223-6

NOOK HD For Dummies,
Portable Edition
978-1-118-39498-4

Surface For Dummies
978-1-118-49634-3

Test Prep

ACT For Dummies,
5th Edition
978-1-118-01259-8

ASVAB For Dummies,
3rd Edition
978-0-470-63760-9

GRE For Dummies,
7th Edition
978-0-470-88921-3

Officer Candidate Tests
For Dummies
978-0-470-59876-4

Physician's Assistant E
For Dummies
978-1-118-11556-5

Series 7 Exam
For Dummies
978-0-470-09932-2

Windows 8

Windows 8 For Dummie
978-1-118-13461-0

Windows 8 For Dummie
Book + DVD Bundle
978-1-118-27167-4

Windows 8 All-in-One
For Dummies
978-1-118-11920-4

Available in print and e-book formats.

Take Dummies with you everywhere you go!

Whether you're excited about e-books, want more from the web, must have your mobile apps, or swept up in social media, Dummies makes everything easier .

Visit Us

Like Us

Follow Us

Watch Us

Join Us

Pin Us

Circle Us

Shop Us

For Du.

Dummies products make life easier

- DIY
- Consumer Electronics
- Crafts

- Software
- Cookware
- Hobbies

- Videos
- Music
- Games
- and More!

For more information, go to **Dummies.com**® and search the store by category

FOR
DUMMIES
A Wiley Brand